Communications
in Computer and Information Science

Series Editors

Gang Li, *School of Information Technology, Deakin University, Burwood, VIC, Australia*

Joaquim Filipe, *Polytechnic Institute of Setúbal, Setúbal, Portugal*

Ashish Ghosh, *Indian Statistical Institute, Kolkata, West Bengal, India*

Zhiwei Xu, *Chinese Academy of Sciences, Beijing, China*

Rationale
The CCIS series is devoted to the publication of proceedings of computer science conferences. Its aim is to efficiently disseminate original research results in informatics in printed and electronic form. While the focus is on publication of peer-reviewed full papers presenting mature work, inclusion of reviewed short papers reporting on work in progress is welcome, too. Besides globally relevant meetings with internationally representative program committees guaranteeing a strict peer-reviewing and paper selection process, conferences run by societies or of high regional or national relevance are also considered for publication.

Topics
The topical scope of CCIS spans the entire spectrum of informatics ranging from foundational topics in the theory of computing to information and communications science and technology and a broad variety of interdisciplinary application fields.

Information for Volume Editors and Authors
Publication in CCIS is free of charge. No royalties are paid, however, we offer registered conference participants temporary free access to the online version of the conference proceedings on SpringerLink (http://link.springer.com) by means of an http referrer from the conference website and/or a number of complimentary printed copies, as specified in the official acceptance email of the event.

CCIS proceedings can be published in time for distribution at conferences or as post-proceedings, and delivered in the form of printed books and/or electronically as USBs and/or e-content licenses for accessing proceedings at SpringerLink. Furthermore, CCIS proceedings are included in the CCIS electronic book series hosted in the SpringerLink digital library at http://link.springer.com/bookseries/7899. Conferences publishing in CCIS are allowed to use Online Conference Service (OCS) for managing the whole proceedings lifecycle (from submission and reviewing to preparing for publication) free of charge.

Publication process
The language of publication is exclusively English. Authors publishing in CCIS have to sign the Springer CCIS copyright transfer form, however, they are free to use their material published in CCIS for substantially changed, more elaborate subsequent publications elsewhere. For the preparation of the camera-ready papers/files, authors have to strictly adhere to the Springer CCIS Authors' Instructions and are strongly encouraged to use the CCIS LaTeX style files or templates.

Abstracting/Indexing
CCIS is abstracted/indexed in DBLP, Google Scholar, EI-Compendex, Mathematical Reviews, SCImago, Scopus. CCIS volumes are also submitted for the inclusion in ISI Proceedings.

How to start
To start the evaluation of your proposal for inclusion in the CCIS series, please send an e-mail to ccis@springer.com.

Tzu-wei Tsai · Kuohsiang Chen ·
Toshimasa Yamanaka · Shinichi Koyama ·
Simon Schütte · Anitawati Mohd Lokman
Editors

Kansei Engineering and Emotion Research

10th International Conference, KEER 2024
Taichung, Taiwan, November 20–23, 2024
Proceedings, Part I

Editors
Tzu-wei Tsai
National Taichung University of Science and Technology
Taichung, Taiwan

Toshimasa Yamanaka
University of Tsukuba
Tsukuba, Japan

Simon Schütte
Linköping University
Linköping, Sweden

Kuohsiang Chen
National Cheng Kung University
Tainan, Taiwan

Shinichi Koyama
University of Tsukuba
Tsukuba, Japan

Anitawati Mohd Lokman
Universiti Teknologi MARA
Shah Alam, Selangor, Malaysia

ISSN 1865-0929 ISSN 1865-0937 (electronic)
Communications in Computer and Information Science
ISBN 978-981-97-9889-6 ISBN 978-981-97-9890-2 (eBook)
https://doi.org/10.1007/978-981-97-9890-2

© The Editor(s) (if applicable) and The Author(s), under exclusive license
to Springer Nature Singapore Pte Ltd. 2024

This work is subject to copyright. All rights are solely and exclusively licensed by the Publisher, whether the whole or part of the material is concerned, specifically the rights of translation, reprinting, reuse of illustrations, recitation, broadcasting, reproduction on microfilms or in any other physical way, and transmission or information storage and retrieval, electronic adaptation, computer software, or by similar or dissimilar methodology now known or hereafter developed.
The use of general descriptive names, registered names, trademarks, service marks, etc. in this publication does not imply, even in the absence of a specific statement, that such names are exempt from the relevant protective laws and regulations and therefore free for general use.
The publisher, the authors and the editors are safe to assume that the advice and information in this book are believed to be true and accurate at the date of publication. Neither the publisher nor the authors or the editors give a warranty, expressed or implied, with respect to the material contained herein or for any errors or omissions that may have been made. The publisher remains neutral with regard to jurisdictional claims in published maps and institutional affiliations.

This Springer imprint is published by the registered company Springer Nature Singapore Pte Ltd.
The registered company address is: 152 Beach Road, #21-01/04 Gateway East, Singapore 189721, Singapore

If disposing of this product, please recycle the paper.

Preface

The 10th International Conference on Kansei Engineering and Emotion Research (KEER 2024) was organised by the Japan Society of Kansei Engineering (JSKE), the Taiwan Institute of Kansei (TIK), the European Kansei Group (EKG), and the Malaysia Association of Kansei Engineering (MAKE). This book contains the collection of Special Session papers that were accepted at the conference.

KEER 2024 was conducted in Taichung, Taiwan. This conference has evolved into a premier hub for the exchange of information among researchers, scientists, engineers, and practitioners regarding the implementation of the Kansei and Emotion approach. KEER 2024 featured nine concurrent Special Session tracks that addressed both traditional and emerging topics in the field of Kansei and Emotion. These tracks included Kansei Issues in Cross-Cultural Design, Kansei in Senses and Interaction, Kansei Approach to Sustainability Society, Innovative Design for Cultural Sustainability, Image and Media in Kansei Design, and more. All tracks described research conducted in a variety of fields, emphasising the potential benefits it could have on the quality of life in a wide range of real-world scenarios. KEER 2024 offered a platform for scholars, researchers, and practitioners to exchange their research, innovation, and expertise in related fields.

The KEER 2024 conference received 154 paper submissions from 7 countries in 3 regions, namely the Asia Pacific, USA, and Europe. Other accepted papers were published in the regular conference proceedings, while 57 papers were published and presented as Special Session papers. The ratio of acceptance of Special Session papers was 37%, and submissions were reviewed by 88 experts from a variety of countries, including Japan, Taiwan, Malaysia, Vietnam, China, the USA, and Sweden.

We are grateful to the KEER International Board, the sponsors and supporters, chapter contributors, and the local and international committee for their hard work and support in ensuring the success of KEER 2024. We trust that the papers included in this book will be beneficial to the audience and serve as a helpful reference for all individuals who are interested in addressing any of the research areas that have been covered in this book.

Tzu-wei Tsai
Kuohsiang Chen
Toshimasa Yamanaka
Shinichi Koyama
Simon Schütte
Anitawati Mohd Lokman

The Organizing Committee

Honorary Chairs

Tung-Shou Chen — President of National Taichung University of Science and Technology, Taiwan

Kuohsiang Chen — National Cheng Kung University, Taiwan

Conference Chair

Tzu-Wei Tsai — Dean of Design College, National Taichung University of Science and Technology, Taiwan
President of Taiwan Institute of Kansei (TIK)

Conference Co-chairs

Chih-Wei Wen — Head of the Department of Commercial Design, Taiwan

Chung-Hua Chu — Head of the Department of Multimedia Design, Taiwan

Hsien-Jung Wu — Head of the Department of Creative Product Design, Taiwan

Pei-Yu Chen — Head of the Department of Interior Design, Taiwan

Secretariat

Secretary-General

Wei-Lun Lee — National Taichung University of Science and Technology, Taiwan

Secretary

Yu-Ting Feng — National Yunlin University of Science and Technology, Taiwan

Executive Committee

Cheng-Chung Chen	Tunghai University, Taiwan
Chia-An Lin	National Taichung University of Science and Technology, Taiwan
Chien-Hua Cheng	National Taichung University of Science and Technology, Taiwan
Chi-Sen Hung	National Taichung University of Science and Technology, Taiwan
Chuan-Po Wang	National Taichung University of Science and Technology, Taiwan
Chun-Chun Hou	National Taichung University of Science and Technology, Taiwan
Chung-Hua Chu	National Taichung University of Science and Technology, Taiwan
Chung-Yi Chang	National Taichung University of Science and Technology, Taiwan
Hsiao-Wei Huang	National Taichung University of Science and Technology, Taiwan
Hsien-Jung Wu	National Taichung University of Science and Technology, Taiwan
Hsu-Lien, Chiu	National Taichung University of Science and Technology, Taiwan
Hsiao-Chen You	National Taichung University of Science and Technology, Taiwan
Li-Hui Lee	Tunghai University, Taiwan
I-Chia Tsai	National Taichung University of Science and Technology, Taiwan
Jeff C. Chen	National Kaohsiung Normal University, Taiwan
Meng-Chieh Jeffrey Lee	National Taichung University of Science and Technology, Taiwan
Min-Chi Chiu	National Taichung University of Science and Technology, Taiwan
Kuang-Chih Lo	National Taichung University of Science and Technology, Taiwan
Shu-Jen Chao	National Taichung University of Science and Technology, Taiwan
Tzu-Yun Hsu	National Taichung University of Science and Technology, Taiwan
Yen-Liang Wu	National Taichung University of Science and Technology, Taiwan
Ying-Hsiu Huang	National Kaohsiung Normal University, Taiwan

Yu-Han Wang — National Taichung University of Science and Technology, Taiwan
Yung-Ting Chen — National Kaohsiung Normal University, Taiwan
Wang-Chin Tsai — National Yunlin University of Science and Technology, Taiwan

KEER International Steering Committee

Anitawati Mohd Lokman — Universiti Teknologi MARA, Shah Alam (MAKE)
Hideyoshi Yanagisawa — University of Tokyo, Japan (JSKE)
Jue Zhang — Kogakuin University, Japan (JSKE)
Kamijo Masayoshi — Shinshu University, Japan (JSKE)
Kiersten Muenchinger — University of Oregon, USA (EKG)
Kuo-Hsiang Chen — National Cheng Kung University, Taiwan (TIK)
Lluís Marco-Almagro — Universitat Politècnica de Catalunya|BarcelonaTech, Spain (EKG)
Jeff C. Chen — National Kaohsiung Normal University, Taiwan (TIK)
Nazlina Shaari — Universiti Putra Malaysia (MAKE)
Noor Afiza Mat Razali — National Defence University Malaysia (MAKE)
Pierre Levy — Eindhoven University of Technology, Netherlands (EKG)
Shinichi Koyama — University of Tsukuba, Japan (JSKE)
Simon Schutte — Linköping University, Sweden (EKG)
Toshimasa Yamanaka — University of Tsukuba, Japan (JSKE)
Tzu-Wei Tsai — National Taichung University of Science and Technology, (TIK)
Suomiya Bao — University of Tsukuba (JSKE)
Shigekazu Ishihara — Hiroshima International University (JSKE)

Special Session Chairs

Anitawati Mohd Lokman — Universiti Teknologi MARA, Malaysia
Ansel Tsai — National Taiwan University of Arts, Taiwan
Kuo-Hsiang Chen — National Cheng Kung University, Taiwan
Meng-Chieh Jeffrey Lee — National Taichung University of Science and Technology, Taiwan
Po Hsien Lin — National Taiwan University of Arts, Taiwan
Teng-Wen Chang — National Yunlin University of Science and Technology, Taiwan

Tzu-Wei Tsai	National Taichung University of Science and Technology, Taiwan
Yong Long Chen	National Taichung University of Science and Technology, Taiwan
Wang-Chin Tsai	National Yunlin University of Science and Technology, Taiwan

Scientific Committee

Abdur Rahman	Universiti Teknologi MARA, Malaysia
Ana Hadiana	Indonesian Institute of Sciences, Indonesia
Anitawati Mohd Lokman	Universiti Teknologi MARA, Malaysia
Ansel Tsai	National Taiwan University of Arts, Taiwan
Azhar Abdul Aziz	Universiti Teknologi MARA, Malaysia
Chao-Ming Wang	National Yunlin University of Science and Technology, Taiwan
Cheng-Chung Chen	Tunghai University, Taiwan
Cheng-Min Tsai	National Taiwan University of Arts, Taiwan
Cheng-Yuan Chang	National United University, Taiwan
Chia-Han Yang	National Cheng Kung University, Taiwan
Chia-Ling Chang	National Taitung University, Taiwan
Chia-Chen Lu	Tunghai University, Taiwan
Chi-Hua Wu	Southern Taiwan University of Science and Technology, Taiwan
Chi-Sen Hung	National Taichung University of Science and Technology, Taiwan
Chih-Long Lin	National Taiwan University of Arts, Taiwan
Chih-Wei Shiu	National Taitung University, Taiwan
Chien-Chih Ni	Lunghwa University of Science and Technology, Taiwan
Chien-Hua Cheng	National Taichung University of Science and Technology, Taiwan
Chun-Chao Lin	Shu-Te University, Taiwan
Chun-Cheng Hsu	National Yang Ming Chiao Tung University, Taiwan
Chun-Chih Chen	National Kaohsiung Normal University, Taiwan
Chun-Juei Chou	National Cheng Kung University, Taiwan
Chun-Ming Lien	National Taipei University of Business, Taiwan
Chyun-Chau Lin	Shu-Te University, Taiwan
Chris K. W. Su	National Kaohsiung University of Science and Technology, Taiwan

The Organizing Committee

Dian-Li Chen	National Taiwan Craft Research and Development, Taiwan
Fang-Wu Tung	National Taiwan University of Science and Technology, Taiwan
Feng-Jung Liu	Cheng Shiu University, Taiwan
Fu Jen Wang	National Chin-Yi University of Technology, Taiwan
Guan-Ze Liao	National Tsing Hua University, Taiwan
Han-Yu Lin	National Kaohsiung Normal University, Taiwan
Hsi-Jen Chen	National Cheng Kung University, Taiwan
Hsiao-Wei Huang	National Taichung University of Science and Technology, Taiwan
Hsiao-Chen You	National Taichung University of Science and Technology, Taiwan
Hui-Jiun Hu	National Chiayi University, Taiwan
Hui-Yun Yen	Chinese Culture University, Taiwan
Huang Cheng-Yong	National Dong Hwa University, Taiwan
Huang-Yin Chen	National Yunlin University of Science and Technology, Taiwan
I-Ying Chiang	National Tsing Hua University, Taiwan
Indra Griha Tofik Isa	Politeknik Negeri Sriwijaya, Indonesia
Indri Ariyanti	Politeknik Negeri Sriwijaya, Indonesia
Jeff C. Chen	National Kaohsiung Normal University, Taiwan
Jiun-Chih Her	National Taitung University, Taiwan
Jiun-Jhy Her	National Taitung University, Taiwan
Jo-Yu Kuo	National Taipei University of Technology, Taiwan
Kai-Shuan Shen	Fo Guang University, Taiwan
Kim C. K. Lee	National Cheng Kung University, Taiwan
Kiyomi Yoshioka	Meisei University, Japan
Kuo-Hsiang Chen	National Cheng Kung University, Taiwan
Kuo-Liang Huang	Sichuan Fine Arts Institute, China
Kuo-Ping Lin	Tunghai University, Taiwan
Kuo-Pin Chang	National Taichung University of Science and Technology, Taiwan
Lluís Marco-Almagro	Universitat Politècnica de Catalunya, Spain
Li-Hui Lee	Tunghai University, Taiwan
Li-Shu Lu	National Yunlin University of Science and Technology, Taiwan
Min-Chi Chiu	National Taichung University of Science and Technology, Taiwan
Min-Yuan Ma	National Cheng Kung University, Taiwan
Ming-Feng Wang	National Pingtung University of Science and Technology, Taiwan

Ming-Shih Chen	Tunghai University, Taiwan
M. Otsuka	Kanto Gakuin University, Japan
Nazlina Shaari	Universiti Putra Malaysia, Malaysia
Novi Purnama Sari	Politeknik Negeri Jakarta, Indonesia
Ping-Hsuan Han	National Taipei University of Technology, Taiwan
Pei-Jung Cheng	National Chengchi University, Taiwan
Rain Chen	Southern Taiwan University of Science and Technology, Taiwan
Roshaliza Rosli	Universiti Teknologi MARA, Malaysia
Saidatul Rahah Hamidi	Universiti Teknologi MARA, Malaysia
Sheng-Fen Chien	National Cheng Kung University, Taiwan
Shih-Ting Tsai	Nanhua University, Taiwan
Shing-Ru-Yang	Tainan University of Technology, Taiwan
Shinichi Koyama	University of Tsukuba, Japan
Suomiya Bao	University of Tsukuba, Japan
Surya Sumarni Hussein	Universiti Teknologi MARA, Malaysia
Tai-Jui Wang	Chinese Culture University, Taiwan
Takashi Sakamoto	National Institute of Advanced Industrial Science and Technology, Japan
Teng-Wen Chang	National Yunlin University of Science and Technology, Taiwan
Tien-Li Chen	National Taiwan Craft Research and Development Institute, Taiwan
Tham Kwok Wai	National University of Singapore, Singapore
Tse-Wei Hsu	Southern Taiwan University of Science and Technology, Taiwan
Tzu-Wei Tsai	National Taichung University of Science and Technology, Taiwan
Wan-Ling Chang	National Cheng Kung University, Taiwan
Wang-Chin Tsai	National Yunlin University of Science and Technology, Taiwan
Wei Liu	Beijing Normal University, China
Wei-Lun Lee	National Taichung University of Science and Technology, Taiwan
Wei-Sheng Tai	National Taichung University of Science and Technology, Taiwan
Wen-Tsong Huang	Chung Yuan Christian University, Taiwan
Wen-Yen Lin	National Taichung University of Science and Technology, Taiwan
Wen-Zhong Su	National Taiwan University of Arts, Taiwan
Yi-Chen Chen	Shih Chien University, Taiwan
Yi-Kang Sun	National Taiwan University of Arts, Taiwan
Yi-Ting Huang	National Taipei University of Technology, Taiwan

Yi-Chia Tsai — National Taichung University of Science and Technology, Taiwan
Yuan-Yen Chang — National Taichung University of Science and Technology, Taiwan
Yu-Han Wang — National Taichung University of Science and Technology, Taiwan
Yu-Hsu Lee — National Yunlin University of Science and Technology, Taiwan
Yu-Ju Lin — National Taipei University of Business, Taiwan
Yung-Chuan Ma — National Yunlin University of Science and Technology, Taiwan
Yung-Ting Chen — National Kaohsiung Normal University, Taiwan
Yao-Xun Chang — Ming Chuan University, Taiwan
Yuichiro Kinoshita — University of Yamanashi, Japan
Yusuke Nagamori — Tokyo City University, Japan
Ying-Hsiu Huang — National Kaohsiung Normal University, Taiwan
Yen-Liang Wu — National Taichung University of Science and Technology, Taiwan

International Reviewers

Akihiro Ogino — Kyoto Sangyo University, Japan
Ana Hadiana — Institute of Sciences, Indonesia
Anitawati Mohd Lokman — Universiti Teknologi MARA, Malaysia
Azhar Abd Aziz — Universiti Teknologi MARA, Malaysia
Carole Favart — Toyota Motor Europe, Belgium
Erina Kakehashi — Keio University, Japan
Hideyoshi Yanagisawa — University of Tokyo, Japan
Hiroaki Yoshida — Shinshu University, Japan
Hiroko Shoji — Chuo University, Japan
Hiroshi Nunokawa — Iwate Prefectural University, Japan
Hiroyuki Yamada — University of Tsukuba, Japan
Hisao Shiizuka — Kogakuin University, Japan
Indra Griha Tofik Isa — Politeknik Negeri Sriwijaya, Indonesia
Indri Aryanti — Politeknik Negeri Sriwijaya, Indonesia
Jue Zhang — Kogakuin University, Japan
Kazuhisa Takemura — Waseda University, Japan
Kazutaka Ueda — University of Tokyo, Japan
Kazuya Oizumi — University of Tokyo, Japan
Keiichi Muramatsu — Saitama University, Japan
Kiyomi Yoshioka — Meisei University, Japan

Kumiko Miura	Waseda University, Japan
Kuniko Otomo	University of Tsukuba, Japan
Kuohsiang Chen	National Cheng Kung University, Taiwan
Kuo-Liang Huang	Sichuan Fine Arts Institute, China
KyoungOk Kim	Shinshu University, Japan
Lluis Marco-Almagro	Universitat Politècnica de Catalunya|BarcelonaTech, Spain
Masayoshi Kamijo	Shinshu University, Japan
Masayuki Takatera	Shinshu University, Japan
Masayuki Otsuka	Kanto Gakuin University, Japan
Michiko Ohkura	Shibaura Institute of Technology, Japan
Mizuki Nakajima	Tokyo Denki University, Japan
Naoki Takahashi	Chuo University, Japan
Nazlina Shaari	Universiti Putra Malaysia, Malaysia
Noorafiza Mat Razali	National Defense University, Malaysia
Noriko Shingaki Seijo	University, Japan
Novi Purnama Sari	Politeknik Negeri Jakarta, Indonesia
Nur'aina Daud	Universiti Teknologi MARA, Malaysia
Pierre Levy Conservatoire	National des Arts et Métiers, France
Rodrigo Queiroz Kuhni Fernandes	University of Tsukuba, Japan
Roshaliza Rosli	Malaysia Digital Economy Corporation, Malaysia
Ryoichi Tamura	Kyushu University, Japan
Saidatul Rahah Hamidi	Universiti Teknologi MARA, Malaysia
Shamsiah Abdul Kadir	Universiti Kebangsaan Malaysia, Malaysia
Shigekazu Ishihara	Hiroshima International University, Japan
Shinichi Koyama	University of Tsukuba, Japan
Shogo Okamoto	Nagoya University, Japan
Simon Schütte	Linköping University, Sweden
So Masuko	Rakuten Research Institute, Japan
Surya Sumarni Hussein	Universiti Teknologi MARA, Malaysia
Takahiro Yokoi Tsukuba	University of Technology, Japan
Takashi Sakamoto	National Institute of Advanced Industrial Science and Technology, Japan
Takeo Kato	Keio University, Japan
Tomoharu Ishikawa	Utsunomiya University, Japan
Toshimasa Yamanaka	University of Tsukuba, Japan
Wonseok Yang	Shibaura Institute of Technology, Japan
Yosuke Horiba	Shinshu University, Japan
Yuichiro Kinoshita	University of Yamanashi, Japan
Yuri Hamada	Chuo University, Japan
Yusuke Ashizawa	Shibaura Institute of Technology, Japan
Yusuke Nagamori	Tokyo City University, Japan

Local Reviewers

Chang-Wei Lee	National Taiwan University of Art, Taiwan
Cheng-Chung Chen	Tunghai University, Taiwan
Cheng-Yuan Chang	National United University, Taiwan
Chien-Hua Cheng	National Taichung University of Science and Technology, Taiwan
Chien-Jou Yang	National Taiwan University of Science and Technology, Taiwan
Chien-Yu Lin	National University of Tainan, Taiwan
Ching-Chien	Liang Southern Taiwan University of Science and Technology, Taiwan
Chi-Sen Hung	National Taichung University of Science and Technology, Taiwan
Chuan-Po Wang	National Taichung University of Science and Technology, Taiwan
Chun-Cheng Hsu	National Yang Ming Chiao Tung University, Taiwan
Chun-Chih Chen	National Kaohsiung Normal University, Taiwan
Chun-Chun Hou	National Taichung University of Science and Technology, Taiwan
Chun-Chun Wei	National Taipei University of Business, Taiwan
Chung-Shing Wang	Tunghai University, Taiwan
Chung-Yi Chang	National Taichung University of Science and Technology, Taiwan
Chun-Heng Ho	National Cheng Kung University, Taiwan
Chyun-Chau Lin	Shu-Te University, Taiwan
Hsiao-Wei Huang	National Taichung University of Science and Technology, Taiwan
Hsien-Chih Chuang	Chinese Culture University, Taiwan
Hsi-Jen Chen	National Cheng Kung University, Taiwan
Hsuan Lin	Chaoyang University of Technology, Taiwan
I-Chia Tsai	National Taichung University of Science and Technology, Taiwan
I-Jue Lee	National Taipei University of Science and Technology, Taiwan
Jeff C. Chen	National Kaohsiung Normal University, Taiwan
Jia-An Lin	National Taichung University of Science and Technology, Taiwan
Jiun-Jhy Her	National Taitung University, Taiwan
Kai-Shuan Shen	Fo Guang University, Taiwan
Kevin C. Tseng	National Taipei University of Technology, Taiwan
Kuo-Ping Lin	Tunghai University, Taiwan

Kuo-Wei Su	National Kaohsiung University of Science and Technology, Taiwan
Li-Hui Lee	Tunghai University, Taiwan
Li-Shu Lu	National Yunlin University of Science and Technology, Taiwan
Meng-Chieh Lee	National Taichung University of Science and Technology, Taiwan
Meng-Hsueh Hsieh	National Cheng Kung University, Taiwan
Ming-Feng Wang	National Pingtung University of Science and Technology, Taiwan
Ming-Shih Chen	Tunghai University, Taiwan
Min-Yuan Ma	National Cheng Kung University, Taiwan
Pei-Jung Cheng	National Chengchi University, Taiwan
Sheng-Fen Chien	National Cheng Kung University, Taiwan
Sheng-Jung Ou	Chaoyang University of Technology, Taiwan
Shyue-Ran Li	National Pingtung University, Taiwan
Teen-Hang Meen	National Formosa University, Taiwan
Tien-Li Chen	National Taiwan Craft Research and Development Institute, Taiwan
Tzu-Wei Tsai	National Taichung University of Science and Technology, Taiwan
Wang-Chin Tsai	National Yunlin University of Science and Technology, Taiwan
Wei Lin Feng	Chia University, Taiwan
Wei-Lun Lee	National Taichung University of Science and Technology, Taiwan
Wen-Hwa Cheng	National Formosa University, Taiwan
Yang-Cheng Lin	National Cheng Kung University, Taiwan
Yen-Liang Wu	National Taichung University of Science and Technology, Taiwan
Yeong-Long Chen	National Taichung University of Science and Technology, Taiwan
Yi-Chen Chen	Shih Chien University, Taiwan
Ying-Hsiu Huang	National Kaohsiung Normal University, Taiwan
Yu-Han Wang	National Taichung University of Science and Technology, Taiwan
Yung-Ting Chen	National Kaohsiung Normal University, Taiwan

Exhibition Curator

Chi-Sen Hung — National Taichung University of Science and Technology, Taiwan
Chun-Chun Hou — National Taichung University of Science and Technology, Taiwan

Organizer

National Taichung University of Science and Technology

Implementers

College of Design, National Taichung University of Science and Technology
Taiwan Institute of Kansei
Kansei Engineering and Emotion Research (KEER)

Co-organizers

Japanese Society of Kansei Engineering (JSKE)
Malaysia Association of Kansei Engineering (MAKE)
European Kansei Group (EKG)
Taiwan Association of Digital Media Design
Design Crossover Association
Ergonomics Society of Taiwan
The Graphic Association of the Republic China

Advisory Organizations

Taichung City Government
National Science and Technology Council, Taiwan
Ministry of Education, Taiwan

Contents – Part I

Emotion Research in Southeast Asia (SEA): Bridging Cultures, Advancing Design

Emotional Preferences in Metaverse Library Interface: A Kansei Analysis 3
   Nik Azlina Nik Ahmad, Anitawati Mohd Lokman,
   Ahmad Iqbal Hakim Suhaimi, and Munaisyah Abdullah

Developing the Concept of Emotion for Rendang Packaging Design Using
Kansei Engineering 15
   Novi Purnama Sari, Wiwi Prastiwinarti, Anitawati Mohd Lokman,
   Iqbal Yamin, Ade Isna, Lytta Yennia, and Raditya Naufal

Exploring Emotion Classification for Children with Autism in Response
to Robot Movement: A Preliminary Case Study in Malaysia 28
   Fatin Nadhirah Zabani, Jo Anne Saw, Nur'aina Daud, Azhar Abd Aziz,
   and Anitawati Mohd Lokman

Implementation of the K-Means Genetic Algorithm to Determine
the Design Concept for Toast Bread Packaging by Kansei Engineering 41
   Novi Purnama Sari, Wiwi Prastiwinarti, Rachmadita Dwi Pramesti,
   Lytta Yennia Putri, and Rafi Ramdan Permana

Design Guide for Emotional Evocative Student's Leadership Program 53
   Zaiha Ahmad, Zuraeda Ibrahim, Nordiana Ibrahim,
   Nur Nafishah Azmi, Mohd Sazili Shahibi, and Anitawati Mohd Lokman

Extraction of Toast Packaging Design Elements Using Long Short Term
Memory-Neural Network with Kansei Engineering Approach 67
   Wiwi Prastiwinarti, Novi Purnama Sari, Rafi Ramdan Permana,
   and Lytta Yennia

Assistive Tool for Emotion and Importance Quadrant LEIQ™ 78
   Nur Batrisyia Damia Mustafa, Saidatul Rahah Hamidi,
   Surya Sumarni Hussein, and Shuhaida Mohamed Shuhidan

Key Determinants of Value-Based Management for ICT Project Delivery:
Insights from Kansei Engineering and Emotion Research 93
   Surya Sumarni Hussein, Nur Hanis Solehah Mohd Rosli,
   and Azran Ahmad

Exploring Research Direction for Human Metaverse Interaction 104
    I Made Bambang Ariawan, Shamsiah Abd Kadir, Afiza Ismail, and Anitawati Mohd Lokman

Enhancement of Kansei Model for Political Security Threat Prediction Using Bi-LSTM 116
    Liyana Safra Zaabar, Khairul Khalil Ishak, and Noor Afiza Mat Razali

The Integration of Kansei Engineering and Artificial Intelligence Based on Methodology and Application Perspective: A Review 129
    Wen-Tsai Sung and Indra Griha Tofik Isa

Competencies Required by Different Positions for Innovation

An Initial Exploration of Virtual Reality for Emotional Expression: The Case of "inescapable Grief" 143
    Yao-Xun Chang, Ki-Yan Ma, and Hei-Man Hung

Combined Bidirectional Long Short-Term Memory and Mel-Frequency Cepstral Coefficients with Convolution Neural Network Using Triplet Loss for Speaker Recognition 155
    Young-Long Chen, Jing-Fong Ciou, Chih-Han Lin, and Shih-Sheng Lien

An Improved YOLOv5 Model with FRB Method for Product Surface Defect Detection 166
    Young-Long Chen, Chuan-Cheng Chung, and Li-Hong Qin

Application of Digital Gamification Systems in Intelligent Automated Learning 177
    Chein-Hui Lee, Evelyn Saputri, and Min-Chi Chiu

Exploring the Intersection of Kansei Engineering and Affective Computing in Digital Media Design Research

Constructing a Curatorial Awareness-Based Sustainable Development Model—A Kansei Engineering Perspective 191
    Huang-Yin Chen and Teng-Wen Chang

Design For/From Pray-The Building Process of Virtual Pilgrimage Site 204
    Chia Hui Nico Lo

A Research of Kansei Engineering and Affective Computing for Generating Game Controller 216
    Yinghsiu Huang and Syue-Ting Lung

Applying Design Puzzle and Kansei Analysis to Analyze the Emotional
Factors in the Bionic Modeling Wooden Toy Design Process by Designers 232
    Chen-Syuan Lin and Teng-Wen Chang

Ontology Construction and Sentiment Computation Analysis in Animated
Series: A Case Study of Mobile Suit Gundam 247
    Tse-Wei Hsu

Image and Media in Kansei Design

A Comparative Study Assessing the Effectiveness of Machine Learning
Technology Versus the Questionnaire Method in Product Aesthetics
Surveys 263
    Chun-Wei Chen

Cognitive Research on AI-Assisted Generation of Animations with Diverse
Artistic Styles 276
    Chia-Ling Chang

A Study About Visual Movement of Applying Proximity of Gestalt
Psychology on Straight Line 289
    Tsu-Min Hsiang and Wei-Ming Liao

Application of the Immersive Virtual Training: The First Immersive
Simulation Cave System to the First Responder Training in Taiwan 304
    Hao-Yang Chen, Pey-Yune Hu, and Lien-Shang Wu

Applying the Story Context Analysis Method in Teaching Immersive
Media Development 323
    Tzu-Wei Tsai, Su-Ting Tasi, and Shao-Han Liao

Innovative Design for Cultural Sustainability

A Study on Qualia's Experiential Model from the Concept of Service
Innovation Design: A Case of 7-Eleven Service Experience in Taiwan 337
    Yi-Hang Lin and Po-Hsien Lin

The Impact of Reusing Old Houses on Community Sustainable
Development: A Case Study of Longquan Street in Taipei City 348
    Chang-Wei Chang, Ying-Shueh Shih, and Yi-Fu Hsu

Generative Artificial Intelligence to Enhance the Sustainability
of Traditional Crafts: The Case of Ceramic Teapots 362
    Yi-Fu Hsu, Chang-Wei Chang, and Chih-Long Lin

Innovative Frontiers in Visual Arts: AI's Role in Interdisciplinary
Collaboration 374
   Jen-Feng Chen, Yun-Song Chu, and Po-Hsien Lin

Impact of the Audience's Aesthetic Perceptions on the Traditional
Dance-Drama: Eternal Love Across the Magpie Bridge 389
   Tze-Fei Huang

Research on Perceptual Differences in Corporate Identity Systems 403
   Jhih-Ling Jiang and Rungtai Lin

Exploration of Mural Creation in Eastern Zhejiang Art Mural Villages 415
   Bai-Hui Du

Author Index 431

Contents – Part II

Kansei Approach to Sustainability Society

SmartTM: An IoT-Based Smart Trash Management System 3
   Khuat Duc Anh, Vu Kim Anh, Nguyen Duy Minh,
   Luong Nguyen Viet Khoa, Ngo Truong Minh, and Phan Duy Hung

A Preliminary Study on Generative AI Drawing Systems in Design
Courses: A Case Study of Introduction to Design 14
   Shih-Cheng Fann

The Investigation of How Brand Design Language Impacts Young
Motorcycle Enthusiasts: Yamaha and Honda as Example 28
   Hsin-Wei Huang and Yu-Hsu Lee

Using the ALESSI Evaluation Model to Explore the Shape Structure
of Parametric Design Products 39
   Yen-Ting Chen and Yu-Hsu Lee

Preliminary Research on Artificial Intelligence-Assisted Furniture
Design-Take the Torii Entrance Cabinet as an Example 51
   YiLing Lo

Kansei in Senses and Interaction

Exploring Emotional Expression Design in LINE Stickers: Utilizing
Kansei Engineering and Emotional Design for Market and Brand
Enhancement 67
   Shih-Yun Lu, Yi-Cheng Lin, and Wei-Her Hsieh

Elderly Acceptance of Virtual Character Exercise Videos 80
   Cheih Ying Chen and Ping Chia Hsiang

The Impact of Voice Design on User Perception: A Case Study of Campus
Counseling Voice Services 92
   Ting-Cheng Chang and Hsiao-Chen You

Kansei Evaluation of Car Interiors in 360° Photo Viewing: Differences
Between Flat Display and Virtual Reality 103
   Yong-Yi Cheng, Jo-Yu Kuo, and Han Chen

Exploring the Conceptual Design of Soul Communication Devices
from a Speculative Design Perspective 115
   Jing Lun Chua and Chien-Kuo Teng

The Needs and Design of Home Mobility Assistance for the Elderly
from the Perspective of Lifestyle 129
   I-Chia Tsai and Yu-Tien Chang

Designing Effective Career Websites: A Study on User Experience and Job
Seeker Engagement 144
   Wen-Tsong Huang

Kansei Issues in Cross-Cultural Design

Soft Movement Space Research and Product Creation Under the Aging
Experience of Emotional Design 159
   Yu Chao and Chien-Kuo Teng

Exploring Product Innovation Design Based on the Systematic Design
Method and TRIZ Theory 171
   Cunfeng Pei and Xin Cao

Exploring the Appeal of Pokémon Ga-Olé from the Perspective
of Virtual-Real Interaction 183
   Chih-Kuan Lin, Chi-Cheng Cheng, and Kai-Shuan Shen

Exploring Cultural Heritage and Innovation: Portuguese Azulejos Tile Art
in Macao's Creative Design 194
   Cailin Huang, Ke Song, and Hok Kun Wan

A Kansei Engineering Approach to Virtual Personality of Embodied Voice
Assistants 206
   Hsiao-Chen You, Ding-Xiang Luo, and Ling-Yu Ho

Cross-Cultural Bamboo Craft Design Collaboration: An Interview Review
of Taiwanese Design Assistants and Nicaraguan Craftsmen 220
   Hung Yu Chang

A Study on Visual Schema for the History of Automotive Style - Apply
Arts History Method to Design History 232
   Chang-Yu Pan

Wellbeing/Experience Quality of Life/Healthcare

Motivation Assisting System for Physical Exercise Based on Classifying
Users' Attitudes 243
    Taiyo Sunny Kojima, Shugo Ono, Atsunari Suzuki, Etsuko Ogasawara,
    Naoki Takahashi, Toru Nakata, Takashi Sakamoto, Fumitake Sakaori,
    and Toshikazu Kato

How Advanced Technology Design Meets Senior Adults' Emotional
Needs: A Case Study of Japanese Senior Adults' Experience 254
    Yu-Han Wang, Satoshi Muraki, Jeewon Choi, and Yuk-Wa Fan

Optimizing Learning Experience: Innovative Design of Distance Learning
Auxiliary Products for Elementary School Students 266
    Rui Zhu, Tien-Li Chen, Chun-Yueh Hou, and Yu Chuang

A Study on the Relationship Between Indoor Thermal Comfort
and the Physical and Psychological Perception of the Elderly 277
    Chung-Yi Chang, Jenn-Ann Tan, Meng-Chieh Jeffrey Lee,
    and Tain-Junn Cheng

An Initial Investigation of Heart Rate Variability (HRV) Related to Foot
Bathing Water Temperature and Vibration Relaxation for Elderly People 294
    Wei-Lun Lee, Yi-Tung Lin, Meng-Chieh, Jeffrey Lee, and Tain-Junn Cheng

Study of the Correlation Between Indoor Air Quality and Brain Waves
of Elderly People in a Nursing Center 306
    Yi Chun Kuo, Shih Chiao Sun, Meng Chieh Lee, and Tain Junn Cheng

Author Index 319

Emotion Research in Southeast Asia (SEA): Bridging Cultures, Advancing Design

Emotional Preferences in Metaverse Library Interface: A Kansei Analysis

Nik Azlina Nik Ahmad[1], Anitawati Mohd Lokman[2]([✉]), Ahmad Iqbal Hakim Suhaimi[2], and Munaisyah Abdullah[1]

[1] Software Engineering Section, Universiti Kuala Lumpur, Malaysian Institute of Information Technology, 50250 Kuala Lumpur, Malaysia
`nikazlina@unikl.edu.my`
[2] College of Computing, Informatics and Mathematics, Universiti Teknologi MARA, 40450 Shah Alam, Malaysia
`anitawati@uitm.edu.my`

Abstract. While metaverse technology continues to advance, catching user attention ultimately relies on the appealing qualities of its user interface (UI), making it essential to include emotional perception. A well-designed UI aligned with user preferences fosters user engagement and contributes to product success. However, creating a UI that prioritizes emotions remains an ongoing challenge. For that reason, this study attempts to explore the relationship between UI design factors and the user perception of the demand experience. This research utilizes *Kansei* Engineering (KE) method to investigate user preferences on the metaverse library UI and examine the correlation between emotional responses. A three-step methodology consisting of instrument, evaluation, and analysis was applied in this study. The evaluation utilized a total of 28 specimens and 60 *Kansei* Words (KWs). The data were analyzed using multivariate statistical analysis; Cronbach's Alpha, Principal Component Analysis (PCA) and Factor Analysis (FA) to identify the significant UI design for the metaverse library. The results of this research take the form of recommendations for UI design concepts based on the most dominant emotional factors. The data analysis results revealed that "radiance", "orderliness", and "conciseness" were the most significant emotional factors for the metaverse library application. Through the application of KE analysis, the emotionally demanding UI concepts within the metaverse library can be determined. This analysis provides valuable guidance for designing or redesigning the library interface, which in turn, contributes to an enhanced metaverse experience.

Keywords: Emotional Design · Emotive Keywords · *Kansei* Engineering · Library Application · Metaverse Experience

1 Introduction

The emergence of the metaverse has sparked interest in applications that enhance virtual experiences, with the concept of a metaverse library emerges as particularly promising. This application will give users access to a variety of digital content in an immersive virtual environment [1]. However, designing a metaverse library comes with its own set of challenge, one of which is UI design concerns [2]. In order to tackle this challenge, it is essential to use a methodical approach such as the *Kansei* Engineering (KE) method. KE offers a methodical approach that holds promise in addressing the challenges inherent in designing the UI [3]. By leveraging KE methodology, designers can meticulously map out user emotional preferences, thus laying a strong foundation for UI design. This methodical approach ensures that the UI is able to consistently adjust to subjective user perceptions, leading to improved user engagement [4–6].

Despite the growing interest in emotion research and the widespread attention given to the metaverse, the exploration at the intersection of these two areas is surprisingly scarce. Although previous research has explored the methods that can be applied for emotional design in the metaverse [7], the study did not provide any design recommendations for the metaverse. In contrast, [8] evaluated the user experience (UX) of new metaverse-based learning system for undergraduates, assessing emotions to gauge satisfaction. Similarly, [9] examined how emotional experiences affect users' likelihood to continue using commercial metaverse apps. An additional area of investigation can be found in the work of [10], which examined the emotional keywords applicable to digital and pervasive applications design, focusing on *Kansei* words (KWs). However, the scope of the past research did not include the identification of design factors that correlate with user emotions in metaverse contexts, resulting in a substantial gap in UX and design efficacy research. To bridge the gap, this study delved into the realm of metaverse library applications in an Augmented Reality (AR) environment, as the subject of investigation. This study aimed to enhance UX in the metaverse library by integrating emotional factors into the evaluation process and providing optimal design recommendations for deeper engagement and satisfaction. It investigates how user emotional factors influence the metaverse library interface to identify key factors crucial for UI design or redesign.

2 Background of Study

2.1 Metaverse Library

Metaverse applications that incorporate augmented reality (AR) have emerged as revolutionary innovations, introducing a paradigm shift in the way digital interactions are conducted. Metaverse is a persistent multiuser environment emerging physical reality with digital virtuality that is based on the convergence of technologies such as augmented reality [11, 12]. Many sectors, including education and libraries, are now realizing the potential benefits and opportunities it brings [13]. With the growing popularity of the metaverse, a transformative shift in education is anticipated, necessitating libraries to modernize their services to align with technological advancements and participate in this mixed reality revolution [14, 15].

Nevertheless, achieving a balance between the necessity for technical developments and UX requirements is important when shaping the concept for advanced library technology. Therefore, providing a solid foundation for a virtual environment requires the establishment of a platform that facilitates efficient user interaction via a well-designed user interface [16]. This perspective is further corroborated by [17], indicating the prevailing consensus among scholars regarding metaverse interface. This highlights the critical need for a meticulous development of UI in metaverse environments to improve user engagement and satisfaction. Thus, [17] suggested for continual refinement of UI specifications in the design of the metaverse, emphasizing the emotional aspect [18].

2.2 Emotion Measurement

As technology continues to evolve alongside human behavior, researchers have turned their attention to explore the field of emotion measurement, recognizing its versatile applications across multiple fields [19], including the metaverse [20]. One of the popular method to measure users' emotional responses towards a product is through *Kansei* Engineering [21–23]. KE is recognized for its capability to shape users' emotions, referred to as "*kansei*," into the design of products [24]. Prior research employing KE in digital contexts has primarily focused on refining the UI design of mobile applications [25], web applications [26, 27], and pervasive applications [10]. This research [25] conducted a *Kansei* evaluation to identify the most suitable design features for a mobile commerce application, in which they have identified 42 potential design elements and selected 8 final elements for inclusion. Alternatively, [26] focused on enhancing web interface UX using color concepts, resulting in guidelines for incorporating 'Excited' and 'Stylish' *Kansei* emotive keywords. In another context, [27] conducted a multivariate analysis to identify the most effective design elements for online sales product pages. However, apart from other studies that focused on UI design, [10] argues that *Kansei* research should go beyond UI design and also focus on building the domain-specific KWs database. Collaborating with industry experts, they identified and validated 175 emotive keywords for diverse applications, including the metaverse. Encouraged by the accomplishments of previous research in this field and the absence of research in the junction of emotional metaverse interface design, this study employed the KE methodology to determine ideal design perceptions of the metaverse library, drawing on insights from users' emotional responses.

3 Methodology

This section elaborates the methodology and procedures used to investigate users' emotional preferences regarding the metaverse library interface. Adapting the *Kansei* Engineering method, the study utilizes KE analysis to conduct multivariate statistical analysis. Structured into three distinct phases, this study aims to provide a comprehensive view of users' emotional responses in the context of metaverse libraries. Each step of the procedure, encompassing instrument, evaluation, and analysis, is illustrated in Fig. 1 and elaborated upon in the following subsections.

3.1 Instrument

Specimens

When practicing KE, it is important to choose specimens from the field domain that

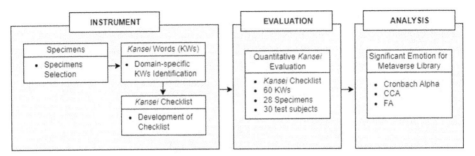

Fig. 1. The research method

effectively capture the many features and characteristics pertinent to the intended design goals [28]. The scope of this research is specifically directed towards the AR perspective, with the objective to investigate users' emotional preferences within the metaverse library context. Therefore, this study was provided with 28 AR-based mobile library applications that are relevant to the study.

Kansei Words (KWs)
The identification of relevant *Kansei* Words (KWs) for use in the research context is another important aspect of KE investigations. The KWs in this study were developed by synthesizing information from various sources [29] such as journals, conference proceedings, periodicals, and magazines, followed by a series of validation process by the language and technical experts, as detailed in [10]. These sources include adjectives that convey users' perceptions and emotions towards metaverse library. In this stage, a list of contextually relevant keywords is identified, laying the foundation for the next phase that focuses on the development of a *Kansei* checklist or questionnaire.

Kansei Checklist
The subsequent step involves converting the KWs into the semantic differential (SD) scale structure known as *Kansei* checklist. The KWs were organized into a five-level scale pattern that spans from 1 to 5. The SD Scale was then transformed into a Google Form to facilitate the process of gathering data online.

3.2 Evaluation

Prior to full-scale implementation, it is necessary to conduct a pilot evaluation in order to determine the reliability and validity of the research methods and data collection procedures [30–32]. In this pilot evaluation, 30 undergraduate students were selected to be involved as test subjects. The instruments used for the evaluation consisted of 28 specimens and 60 KWs in SD scale form. The evaluation took place in a computer lab of a local university, where participants were briefed and provided with a video demonstration of each specimen. The participants were also given access to an online *Kansei* checklist via Google Form, enabling them to input their responses according to the specimens provided to them.

3.3 Analysis

Following the collection of the evaluation data, a multivariate statistical analysis was performed on the averaged data. The analysis was carried out using Cronbach's Alpha, Principal Correlation Analysis (PCA) and Factor Analysis (FA) in order to identify which specimens evoke specific emotions and to determine the significant emotional factors for metaverse library interface design.

4 Results and Discussion

4.1 Instrument

Specimens

Determining the specimens for metaverse library involve a step of collecting initial pool of 32 specimens, followed by a screening procedure to identify the most suitable specimens for the study context. As a results of this procedure, there were a total of 28 final specimens of the AR-based mobile library applications. The selection of specimens was made based on design variations. Different application designs were selected to capture a range of user emotions and user responses; the basis of KE which relies on diverse specimens to understand how different visual elements elicit specific emotional responses from users. The specimen samples, referred to as SMP1, SMP2, and SMP3, along with their snapshot images and design criteria, are presented in Table 1.

Table 1. Selected Specimen Samples.

ID	SMP1	SMP2	SMP3
Snapshot			
Criteria	• Concise information presentation • Tooltips • Search bar • Horizontal layout	• Bright label of visual guides • Icons • Clean design • Horizontal layout	• Bold navigation cues • Simple direction • High contrast background • Vertical layout

Kansei Words (KWs)

In order to obtain the most pertinent KWs for this research, the KWs were formulated in consultation with UI/UX designers, AR professionals, librarian as well as language experts. As a result, 60 domain-specific emotional keywords were finalized, each designated with a sequential code (such as KW.1, KW.2, continuing up to KW.60) to facilitate easy identification and organization, as presented in Table 2. These keywords represent the range of emotions that users may encounter when interacting with the metaverse library, including both positive and negative experiences. These KWs will undergo data analysis to identify the most significant keywords that are strongly related with the specimens under investigation. Table 2 list all the emotional keywords associated with the metaverse library.

Table 2. *Kansei* Words for metaverse library.

Code	*Kansei* Word	Code	*Kansei* Word	Code	*Kansei* Words
KW.1	Appealing	KW.2	Attractive	KW.3	Balanced
KW.4	Boring	KW.5	Bright	KW.6	Cheerful
KW.7	Clean	KW.8	Clear	KW.9	Cluttered
KW.10	Colorful	KW.11	Complicated	KW.12	Comprehensible
KW.13	Concise	KW.14	Confusing	KW.15	Consistent
KW.16	Cool	KW.17	Creative	KW.18	Crowded
KW.19	Dull	KW.20	Easy	KW.21	Elegant
KW.22	Engaged	KW.23	Enjoyable	KW.24	Exciting
KW.25	Fanciful	KW.26	Fascinating	KW.27	Fresh
KW.28	Friendly	KW.29	Fun	KW.30	Guided
KW.31	Harmonious	KW.32	Helpful	KW.33	Inconsistent
KW.34	Informative	KW.35	Interactive	KW.36	Interesting
KW.37	Joyful	KW.38	Latest	KW.39	Messy
KW.40	Minimalist	KW.41	Modern	KW.42	Neat
KW.43	New	KW.44	Nice	KW.45	Organized
KW.46	Outdated	KW.47	Precise	KW.48	Prestigious
KW.49	Professional	KW.50	Realistic	KW.51	Satisfactory
KW.52	Simple	KW.53	Soft	KW.54	Sophisticated
KW.55	Straightforward	KW.56	Trendy	KW.57	Unclear
KW.58	Understandable	KW.59	Up-to-date	KW.60	Vibrant

Kansei Checklist

This study employs all 60 KWs obtained in the prior step. The 60 KWs were organized into a checklist, employing a Likert scale ranging from 1 to 5 to streamline the data collection process. The checklist included keywords such as "not appealing - appealing" and "not attractive - attractive" to represent a scale of 1–5.

4.2 Evaluation

This phase encompassed the evaluation of participants' emotional responses towards the specimens, aiming to determine their ability to elicit user emotions and identify the specific emotional responses evoked towards the metaverse library specimens. The evaluation was conducted using an online Google Form, supplemented by instructional guidance delivered through a demonstration of the specimens in a controlled computer lab environment. The data obtained from this procedure will serve as the dataset for analysis, as detailed in Sect. 4.3.

4.3 Analysis

In order to perform the statistical analysis, the participants' evaluation responses are recapitulated into an average value and used as the basis for the subsequent step in the process. The subsequent step involves data processing, which involves the application of multivariate statistics to the data obtained from the questionnaires. A Cronbach's Alpha, PCA, and FA are employed in this process.

Cronbach's Alpha

Cronbach's Alpha provides a valuable means to assess the reliability of data. Data is considered reliable for use in subsequent multivariate analysis calculations when Cronbach's Alpha yields a value greater than 0.7 [33]. Therefore, the first analysis involves using Cronbach's Alpha to assess the reliability of the collected data. In this study, the calculation of Cronbach's Alpha yielded a value of 0.829, exceeding the threshold of 0.700. This indicates that the data is considered reliable and suitable for use in subsequent analysis procedures.

Principal Component Analysis (PCA)

PCA is a method used to visually represent the direction and strength of emotions within the *Kansei* structure. In this study, PCA was used to identify the connection between emotions or KWs and specimens [34]. The outcomes of the PCA are illustrated through graphical representations in Figs. 2 and 3. In Fig. 2, the distribution of emotions is observable in the PCA loading plot. In general, participants perceive specimens along the positive x-axis as having a more favorable design, characterized by qualities like 'neat' and 'comprehensible', whereas those along the negative x-axis are evaluated less favorably, such as 'crowded' and 'boring'.

The following biplot, shown in Fig. 3 is used to visualize the correlations between specimens and KWs. The findings indicate that specimens L, K, and A are positioned along the positive axis of Component This suggests that these specimens have a suitable UI based on users' emotional preferences. It also implies that the UIs of specimens A, K, and L could be recommended as guidelines for designing metaverse libraries.

Factor Analysis (FA)

FA was conducted to determine which emotion had a significant value. According to the results of the analysis, the first three factors listed in Table 3 are regarded as significant. It is evident from the table that the first factor accounts for 37.457% of the variables, the

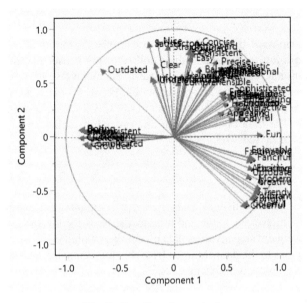

Fig. 2. Loading plot analysis.

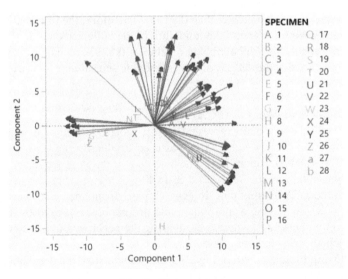

Fig. 3. Biplot analysis.

second factor accounts for 20.239%, and the third factors accounts for 14.993%. These three factors account for the majority of factor contributions, comprising 72.689%. This indicates that Factor 1, Factor 2, and Factor 3 have a significant impact on the emotions of the respondents.

Table 3. Factor variability.

Factor	Variance	Percentage %	Cumulative %
1	7.200	37.457	37.457
2	3.890	20.239	57.696
3	2.882	14.993	72.689

To further comprehend the data, factor analysis is then calculated to provide insights into how strongly each variable contributes to the factor. Table 4 presents the results of factor analysis, providing insight into the significant emotion that was discovered. Variables with high scores are perceived as significant factors in metaverse library interface. A reference score of 0.85 was established for the investigation, in accordance with the recommendation of [35].

Table 4. Factor analysis.

Factor 1		Factor 2		Factor 3	
Vibrant	0.932	Neat	0.980	Comprehensible	0.970
Colorful	0.932	Minimalist	0.977	Helpful	0.954
Modern	0.927	Organized	0.973	Understandable	0.934
Trendy	0.925	Balanced	0.966	Informative	0.901
Cheerful	0.923	Clean	0.964	Clear	0.866
Up-to-date	0.919	Professional	0.945		
Bright	0.901	Fresh	0.936		
Creative	0.898	Simple	0.934		
Attractive	0.878				

Table 4 shows that the *Kansei* concept is structured by the three identified factors. The first factor consists of Vibrant, Colorful, Modern, Trendy, Cheerful, Up-to-date, Bright, Creative, and Attractive. The research labels this factor as the concept of 'Radiance', which implies that the UI of the metaverse library should be designed with high color quality. The second factor consists of Neat, Minimalist, Organized, Balanced, Clean, Professional, Fresh, and Simple. The research labels this factor as the concept of 'Orderliness', which denotes the need for the UI to be well-structured. While the third factor consists of Comprehensible, Helpful, Understandable, Informative, and Clear. The research labels this factor as the concept of 'Conciseness', which indicates the need for UI to present information in a succinct manner. These three factors explain 72.689% of the total data, and considered to be indicators to the design of metaverse library that embeds a targeted emotional concept. Therefore, it is recommended to incorporate these three emotional design factors into the design of the metaverse library, with reference to the variations of design elements observed across the identified specimens A, K, and L. Adopting this design strategy will effectively elicit the intended emotional reaction from users, hence increasing their satisfaction with the metaverse experience.

5 Conclusion

By employing *Kansei* analysis, this study investigated the underlying factors influencing user emotional preferences within the metaverse library interface. The study reveals a significant correlation between UI design and user emotions, underscoring the profound influence of *Kansei* emotional keywords on the visual presentation of the UI. The findings of this study indicate that the UI concepts of 'radiance', 'orderliness', and 'conciseness' exert significant influence on metaverse UX as the most dominant emotional factors. This provides guidelines to designers to construct future metaverse library with a specific emotional emphasis, in which the new concept of metaverse library may induce these three emotional factors. This research is open to further investigation on how these factors contribute to the establishment of fundamental design elements in the metaverse and their influence on user perception and demand experience.

Acknowledgments. The authors would like to thank Universiti Kuala Lumpur Malaysian Institute of Information Technology and the Centre for Research and Innovation for providing financial support for this research. The authors would also like to acknowledge the Research Initiative Group for Emotion, Kansei and Design Engineering (RIG EKDE), Universiti Teknologi MARA, as well as Malaysia Association of Kansei Engineering (MAKE), for all their assistance to the research.

Disclosure of Interests. The authors have no competing interests to declare that are relevant to the content of this article.

References

1. Raj, V., et al.: Demystifying and Analysing Metaverse Towards Education 4.0. In: Proceedings of the 2023 3rd International Conference on Innovative Practices in Technology and Management (ICIPTM 2023), pp. 1–6. IEEE (2023). https://doi.org/10.1109/ICIPTM57143.2023.10118054
2. Pizzolante, R., et al.: Awe in the metaverse: designing and validating a novel online virtual-reality awe-inspiring training. Comput. Hum. Behav. **148**, 107876 (2023). https://doi.org/10.1016/j.chb.2023.107876
3. Nagamachi, M., Lokman, A.M.: Kansei Innovation: Practical Design Applications for Product and Service Development. CRC Press, Boca Raton (2015)
4. Ahmad, A.N., Abdullah, M., Lokman, A.M., Suhaimi, A.I.H.: Preliminary emotional user experience model for mobile augmented reality application design: a kansei engineering approach. Int. J. Interact. Mobile Technol. **17**(7), 32–46 (2023). https://doi.org/10.3991/ijim.v17i07.35201

5. Taharim, F., Lokman, A.M., Isa, W.A.R.W.M., Noor, N.L.M.: A relationship model of playful interaction, interaction design, kansei engineering and mobile usability in mobile learning. In: 2013 IEEE Conference on Open Systems (ICOS), pp. 22–26. IEEE (2013). https://doi.org/10.1109/ICOS.2013.6735041
6. Karim, A., Lokman, A.M., Redzuan, F.: Older adults perspective and emotional respond on robot interaction. In: 4th International Conference on User Science and Engineering (i-USER), pp. 95–99. IEEE (2016)
7. Radovanović, D., Kovačević, D.: Emotional design in digital user experience. In: Proceedings of the 1st Doctoral Colloquium on Sustainable Development, DOC-ME'2022, pp. 1–8 (2022)
8. Pyae, A., et al.: Exploring user experience and usability in a metaverse learning environment for students: a usability study of the artificial intelligence, innovation, and society (AIIS). Electronics 12(20), 4283 (2023). https://doi.org/10.3390/electronics12204283
9. Suh, K.: How users cognitively appraise and emotionally experience the metaverse: focusing on social virtual reality. Inf. Technol. People (2023). https://doi.org/10.1108/ITP-06-2022-0461
10. Ahmad, A.N., Lokman, A.M., Abdullah, M., Suhaimi, A.I.H.: Emotional kansei words for digital and pervasive product design. In: 2023 IEEE International Conference on Computing (ICOCO), pp. 386–390. IEEE (2023). https://doi.org/10.1109/ICOCO59262.2023.10397756
11. Mystakidis, S.: Metaverse. Encyclopedia 2(1), 486–497 (2023). https://doi.org/10.3390/encyclopedia2010031
12. Zhou, L., Chen, Z., Jin, X.L.: A review of the literature on the metaverse: definition, technologies, and user behaviors. Internet Res. 34(1), 129–148 (2024). https://doi.org/10.1108/INTR-08-2022-0687
13. Na, Y., Park, S.: Usability Analysis of Public Libraries' Metaverse Platform. J. Korean BIBLIA Soc. Libr. Info. Sci. 34(2), 275–294 (2023). https://doi.org/10.14699/kbiblia.2023.34.2.275
14. Uddin, M., Manickam, S., Ullah, H., Obaidat, M., Dandoush, A.: Unveiling the metaverse: exploring emerging trends, multifaceted perspectives, and future challenges. IEEE Access 11, 87087–87103 (2023). https://doi.org/10.1109/ACCESS.2023.3281303
15. Wang, Y.: Innovation and development of smart library from the perspective of Metaverse. In: 2023 International Conference on Artificial Intelligence and Computer Information Technology (AICIT), pp. 1–3. IEEE (2023). https://doi.org/10.1109/AICIT59054.2023.10277758
16. Iakovides, C., Lazarou, A., Kyriakou, P., Aristidou, A.: Virtual library in the concept of digital twin. In: 2022 International Conference on Interactive Media, Smart Systems and Emerging Technologies (IMET), pp. 1–8. IEEE (2022). https://doi.org/10.1109/IMET54801.2022.9929598
17. Van de Broek, J., Onime, C., Uhomobhi, J.O., Santachiara, M.: Evolution of user interface and user experience in mobile augmented and virtual reality applications. In: Haptic Technology - Intelligent Approach to Future Man-Machine Interaction 7(4) (2022). https://doi.org/10.5772/intechopen.103166
18. Alkhwaldi, F.: Understanding learners' intention toward Metaverse in higher education institutions from a developing country perspective: UTAUT and ISS integrated model. Kybernetes (2023). https://doi.org/10.1108/K-03-2023-0459
19. Lokman, A.M., Kadir, S.A., Hamidi, S.R., Shuhidan, S.M.: LEIQTM as an emotion and importance model for QoL: Fundamentals and case studies. Jurnal Komunikasi: Malaysian Journal of Communication 35(2), 412–430 (2019). https://doi.org/10.17576/JKMJC-2019-3502-25
20. Ahmad, A.N., Suhaimi, A.I.H., Lokman, A.M.: Conceptual model of augmented reality mobile application design (ARMAD) to enhance user experience : an expert review. Int. J. Adv. Comput. Sci. Appl. 13(10), 574–582 (2022)

21. Shamsiah, K., Anitawati, M.L., Mokhtar, M.: Identification of positive and negative emotion towards political agenda videos posted on Youtube. In: Advances in Intelligent Systems and Computing: The 7th International KEER (2018)
22. Anitawati, L., Laila, M.N., Nagamachi, M.: Kansei structure and visualization of clothing websites cluster. In: International Symposium on Information Technology, vol. 1, pp. 1–8 (2008)
23. Lokman, A.M., Aziz, A.A.: A Kansei system to support children's clothing design in Malaysia. In: IEEE International Conference on Systems, Man and Cybernetics, pp. 3669–3676. IEEE (2010)
24. Schütte, S., Lokman, A.M., Marco-almagro, L., Valverde, N., Coleman, S.: Kansei for the Digital Era. Int. J. Affect. Eng. (2023). https://doi.org/10.5057/ijae.IJAE-D-23-00003
25. Mimura, Y., Tsuchiya, T., Moriyama, K., Murata, K., Takasuka, S.: UX design for mobile application of E-commerce site by using kansei interface. In: Advances in Intelligent Systems and Computing, vol. 1202. Springer, Cham (2020)
26. Kandambi, A.A.S.M., Charles, J., Lekamge, L.S.: Kansei color concepts for E-commerce website design - a case study using designerwear websites. In: MERCon 2022 - Moratuwa Engineering Research Conference, Proceedings, pp. 1–6. IEEE (2022). https://doi.org/10.1109/MERCon55799.2022.9906156
27. Papantonopoulos, S., Karasavova, M.: A kansei engineering evaluation of the emotional appeal of product information on E-commerce product pages. In: 1st International Conference of the ACM Greek SIGCHI Chapter, pp. 1–8 (2021). https://doi.org/10.1145/3489410.3489436
28. Abdi, J., Greenacre, Z.A.: An approach to website design for Turkish universities, based on the emotional responses of students. Cogent Engineering 7(1) (2020). https://doi.org/10.1080/23311916.2020.1770915
29. Ahmad, A.N., Abdullah, M., Suhaimi, A.I.H., Lokman, A.M.: Embedding emotions in the Metaverse: the emotive keywords for augmented reality mobile library application. Int. J. Adv. Comp. Sci. Appl. **15**(5), 264–270 (2024). https://doi.org/10.14569/IJACSA.2024.0150527
30. Lokman, A.M., Harun, A.F., Noor, N.L.: Website affective evaluation: analysis of differences in. Population (English Edition) **5619**, 643–652 (2009)
31. Hussin, N., Lokman, A.M.: Kansei website interface design: Practicality and accuracy of Kansei Web Design Guideline. In: Proceedings - 2011 International Conference on User Science and Engineering, i-USEr 2011, pp. 30–35. IEEE (2011). https://doi.org/10.1109/iUSEr.2011.6150531
32. Ahmad, A.N., Hamid, N.I.M., Lokman, A.M.: Performing usability evaluation on multi-platform based application for efficiency, effectiveness and satisfaction enhancement. Int. J. Interact. Mobile Technol. **15**(10), 103–117 (2021). https://doi.org/10.3991/ijim.v15i10.20429
33. Schrepp, M.: On the usage of cronbach's alpha to measure reliability of UX scales. J. Usability Stud. **15**(4), 247–258 (2020)
34. Hadiana, A.: The importance of users' emotional factors related to design of E-learning interface using kansei analysis. J. Hunan Univ. Natural Sci. **49**(7), 83–88 (2022). https://doi.org/10.55463/issn.1674-2974.49.7.9
35. Lokman, A.M., et al.: Kansei evaluation on web-based geographic information system: a Malaysian case analysis. In: 9th International Conference on Kansei Engineering and Emotion Research. KEER2022. Proceedings, pp. 471–480. Kansei Engineering and Emotion Research (KEER) (2022)

Developing the Concept of Emotion for Rendang Packaging Design Using Kansei Engineering

Novi Purnama Sari[1], Wiwi Prastiwinarti[1(✉)], Anitawati Mohd Lokman[2], Iqbal Yamin[1], Ade Isna[1], Lytta Yennia[1], and Raditya Naufal[1]

[1] Jakarta State Polytechnic, Depok City, Indonesia
{novi.purnamasari,wiwi.prastiwinarti}@grafika.pnj.ac.id
[2] Universiti Teknologi MARA, Shah Alam, Malaysia
anitawati@uitm.edu.my

Abstract. Rendang is one of Indonesia's iconic foods. Rendang products are so popular that they are exported to foreign countries. To support product competitiveness, packaging plays a vital role. The packaging development process that involves the emotional feelings of consumers is crucial. Packaging can be optimized in design if it matches the emotional priorities of consumer preferences. So that the impression of the packaging can represent consumers' emotional needs. One method that focuses on and proves effective in interpreting emotional feelings is Kansei Engineering. This method can interpret consumer emotional feelings into packaging design concepts. This research aims to determine packaging design concepts based on consumers' emotional feeling preferences. The method used in extracting emotional feelings (Kansei words) is Term Frequency-Inverse Document Frequency (TF-IDF), while the concept determination method uses Principal Component Analysis (PCA). The results showed that 90.4% of consumers agreed to develop rendang packaging for Mutiara Kitchen UMKM. Emotional feeling in the form of consumer Kansei obtained 15 words with the highest value weight is practical, simple, and usable. The visual design insight on the packaging is the presence of a typical Padang icon with dominant black and red colors and packaging can image savory and spicy products. The packaging design concept resulting from the PCA method obtained two main components forming the concept, the first concept "Practical-Conventional" and the second concept "Luxury-Not Suitable". The determination of this concept is based on a variance value that exceeds 1, and a cumulative proportion value that exceeds 80%, namely PC 1 and PC 2 on each positive axis "Practical-Luxury".

Keywords: emotional feeling · Kansei engineering · concept design

1 Introduction

Dapur Mutiara is an Indonesian MSME that produces rendang. The superiority in terms of taste makes Rendang "Dapur Mutiara" gain good market acceptance not only at home but also abroad. Rendang became the most delicious food in the world and has become an iconic Indonesian food as proven by CNN International in 2017 and was included in the list of 50 Best Foods in the World by Taste Atlas in 2022 [1]. Rendang "Dapur Mutiara" has been exported to Europe and Turkey since 2021. Currently, Rendang "Dapur Mutiara" has received an export offer to Saudi Arabia for one ton. For Dapur Mutiara to export by maintaining its brand image, it is necessary to develop packaging through emotional feeling. Packaging development can also be an effort for MSMEs to maintain the market and increase product competitiveness.

Packaging plays a critical role in attracting consumers. According to [2, 3], packaging can utilize certain emotions that make consumers interested in the product. Psychology and culture are also factors in consumer decisions to buy products [4]. Emotional feelings that touch the heart and emotions can influence the product [5]. In addition, emotional feelings will also have an impact on consumer expectations [6]. Identification of emotional feelings will be more accurate if done on the right target segment. Involving users' emotions produces the right concept recommendation output [7, 8]. Therefore, the target consumer segment used as the subject in this study is people of Saudi Arabian and West Sumatran (Padang) who have consumed Rendang and live in Indonesia, especially in the Jakarta area.

The difference in community acculturation in acculturation in Indonesia and Saudi Arabia is one aspect of emotional feeling. One of these acculturations is the image of culinary flavors between Indonesia and Saudi Arabia towards Rendang products. Identifying emotional feelings from these two countries will be an interesting thing to learn more about as a reference for packaging design. Rendang, which is famous for its spiciness, creates different preferences regarding the level of spiciness that people in various countries can accept. Therefore, the novelty of research using the collaboration of emotional feelings from two countries, namely Indonesia and Saudi Arabia to become a reference for packaging design is new research in the field of packaging development.

Kansei Engineering is the most appropriate approach in planning and developing packaging with a focus on the emotional feeling aspect [9]. Kansei Engineering is an approach that pays attention to consumer behavior and studies their preferences for a product [10, 11]. According to [12], KE has a quick and easy understanding to find explicit and implicit needs and analyze relationships for product design. Kansei can minimize uncertainty and increase the efficiency and frequency of decision-making [13, 14].

In the packaging development process, concept design is a crucial step in determining the product strategy in creating designs [15]. Each concept has various physical attributes such as color, height, shape, and functional features [16]. However, often in the design process, only aesthetic and visual aspects are emphasized, without considering the functional aspects of packaging [17]. The importance of packaging lies not only in its visual appeal but also in its functional usefulness and ease of use. The influence of product design attributes, packaging materials, and packaging form on consumer purchase intentions has been shown to have a positive impact [18].

The process of analyzing the priority of emotional feeling is done by extracting Kansei words using the Term Frequency Inverse Document Frequency (TF-IDF) method. TF-IDF is a common weighting method used in information retrieval and text mining to assess the significance of Kansei vocabulary terms [19]. Meanwhile, the determination of packaging design concepts based on emotional feelings is carried out using the Principal Component Analysis (PCA) method. According to [20], PCA is used to determine the design concept. The application of PCA (Principal Component Analysis) provides optimal results when used on data that correlate with each other [21, 22]. The purpose of this research is to determine the packaging concept by using emotional feeling analysis as a basis for developing packaging.

2 Methods

This research uses the Kansei Engineering method, the priority emotional analysis process is carried out by weighting using the TF-IDF method. In contrast, the design concept is determined by the PCA method. The way Kansei Engineering works is to bring together human feelings and emotions to translate consumer desires into a design concept, design elements, and prototypes [23, 24].

2.1 Identifying of Problem

The initial stage of planning and developing packaging by identifying problems in objects that have complaints. Through the distribution of questionnaires to customers of Arab and Indonesian descent, especially loyal respondents who Minang descent. After that, a literature study was conducted based on related sources regarding Kansei Engineering.

2.2 Collecting of Packaging Samples

Collecting rendang packaging samples will help understand the packaging design concept. According to Nagamachi, a minimum of 20–25 packaging samples are required [25]. Samples that have been collected are then extracted based on material, shape, or in terms of design which includes color and typography, size, and the same features [26].

2.3 Determining of Respondent

The sampling technique used in this study is non-probability sampling. The size of the sample population is crucial in determining the validity of the results, with the general assumption that the larger the sample will yield higher statistical power [27]. The population is aimed at the descendants of Padang and Arab descendants in Indonesia. The respondents used to have special characteristics or requirements, namely Indonesian citizens, especially people of West Sumatra or Padang ethnicity, and people of Saudi Arabian descent who have consumed rendang products. The domicile area of the respondents who are the focus of this research is respondents spread across Jakarta. The segmentation of respondents was determined based on preliminary observations by identifying the market for rendang products which are identical to middle to upper-class food menus. As for gender, men and women were selected with an age range of 18 years to 50 years. Determination of the sampling amount is done through heterogeneous data distribution by looking at the proportion of the respondent population in the Jakarta area. The formula used is the Slovin method as follows:

$$n = \frac{N}{1 + Ne^2} \qquad (1)$$

where n is the sample size, N is the population size, e is the percentage of allowance for error (5%), and 1 is a constant value.

2.4 Extracting of Kansei Word

The statistical method used to extract Kansei words is TF-IDF [28]. This technique balances two factors: it increases a word's importance based on how often it appears in a specific document, while simultaneously reducing its significance if it occurs frequently across the entire corpus of documents. This approach helps identify words that are particularly relevant to a given document without overemphasizing common terms that appear throughout the collection. The TF-IDF process is performed with the Python programming language. The purpose of this extraction process is to get the value or weight of each Kansei word obtained in the observation so that it can be known which Kansei or emotional word has the most influence.

2.5 Determining of Concept Design

The design concept is determined by processing concept evaluation data on packaging samples using the PCA method. The evaluation process is carried out through direct interviews with respondents based on the purposive judgment sampling method [29]. The number of Kansei words obtained is generally between 50 and 600 words [30]. The questionnaire used is semantic differential to measure the emotional perception of respondents. The semantic differential scale used is seven scales, namely (−3, −2, −1, 0, 1, 2, 3). On this scale, a value of −3 indicates a result that is contrary to Kansei word. While a value of 0 indicates a neutral attitude towards Kansei Word. In this example, the Kansei Word contains the words "impractical" and "practical", a value of −3 indicates a tendency for the word "impractical", while a value of + 3 indicates a tendency for the word "practical". The aim is to capture respondents' perceptions of product design in ergonomic terms. This questionnaire is an effective method for packaging development [31]. The packaging concept design is determined using the PCA method. The results of the semantic differential questionnaire are used as input data in PCA processing to determine the design concept cluster [21].

3 Result

3.1 Identifying of Problem

The initial stage was carried out by distributing questionnaires regarding the importance of rendang packaging development. This questionnaire also aimed to find out the respondents' objective views on rendang packaging issues. The results of the Importance of Planning and Development of Rendang "Dapur Mutiara" are shown in Fig. 1.

The results of observation based on customer have shown that 90.4% of people who answered agreed. Using a Likert scale of 5, of the 72 people surveyed, 35% said it was important and 55.4% said it was very important.

3.2 Packaging Samples

Based on the results of observations via the internet, 70 samples were obtained for rendang product packaging. The packaging samples selected are samples that are tailored to the target market of rendang products. The target market for Mutiara Kitchen UMKM rendang products is products that will be exported to Saudi Arabia, so the net weight that is the basis for sample selection is a 250-g rendang package according to the majority of market demand. The packaging materials chosen are also those that can protect the product during export shipments, namely metal, rigid plastic, and flexible plastic or pouch. Samples that have been collected are then selected based on material similarity, shape, and design which includes color and typography, size, and features [26]. The results of the different samples are 40 packaging samples shown in Table 1.

3.3 Kansei Words

The Kansei words were obtained through in-depth interviews with respondents using the purposive judgment sampling technique. The number of respondents from a population of 250 was obtained with a minimum of 72 respondents to be used based on the following calculation:

$$n = \frac{250}{1 + 250(0,05)^2} = 72 \text{ people} \qquad (2)$$

expressions were then extracted using the TF-IDF method by analyzing the words Based on preliminary interviews with some Arab-descent respondents, it was found that Arabs' preferences in choosing food favored authentic Arab cuisine over rendang products. This became the basis for determining the proportion of respondents to be used. The total of 80 respondents in this study consisted of 60 people from West Sumatra (Padang) and 20 Arab respondents. The area covered by the respondents was those domiciled in Jakarta. Respondent sampling was done heterogeneously based on the areas of North Jakarta, West Jakarta, East Jakarta, South Jakarta, and Central Jakarta.

These areas were chosen based on observations that the largest Arab population in Indonesia is located in the Arab Village in the Condet area of East Jakarta. These 80 respondents were used to identify emotional feelings or Kansei words and also to evaluate the Kansei words based on packaging samples. KE can be used to incorporate user emotions into the design process [32].

The identification of Kansei words from emotional feeling expressions yielded 532 terms of words derived from respondents' expressions of organoleptic impressions of rendang products, complaints about packaging, expectations for packaging, and suggestions for packaging improvements. These emotional feelings with the highest frequency of occurrence, which would have the highest weight. The results of the evaluation of emotional expression are in two categories, namely design characteristics and organoleptic. Design characteristics are grouped into six categories: material, color, structure,

Table 1. Packaging Samples.

design elements, impression, and features. Tables 2 and 3 show the weighting results of respondents' emotional expressions.

Table 2 shows the results of respondents' emotional identification of rendang products and packaging based on the priority weight values resulting from the TIFIDF analysis. Based on observations made, consumers' emotional results can be grouped into reference insights to support decisions in terms of design characteristics of packaging elements such as recommended packaging materials ("Can" or "Plastic"), colors in packaging label designs ("Black", "Red", "Yellow", "Colored", "Dark color", or "Eccentric color"), packaging structure ("Folding box", "Standing pouch", or "Vacuumed packaging"), while in terms of design elements on the label, there is a picture of rendang products and has typical Padang characteristics. Another design element is recommended for packaging with a center layout in the label graphic design. Apart from that, there is also clear information, especially the halal logo, BPOM, and Expired date, as well as spice and spicy information. This layout gives the impression that the elements are neatly arranged and the product looks more attractive [33]. The characteristic design for the impression that consumers want is "Clear" packaging or packaging that has a label design with clear information, "Luxurious" packaging indicates luxurious packaging

Table 2. Weight of Emotional Expressions with TFIDF.

Material	Wd	Color	Wd
Can	0.3318764468	Black	1.03339943
Plastic	0.2315312495	Red	0.7756389004
Structure	Wd	Yellow	0.73541939
Folding box	0.4034888469	Colored	0.5336012682
Standing pouch	0.2911510578	Dark color	0.3292971965
Vacuumed	0.2637048134	Eccentric color	0.1465274829
Element Design	Wd	Impression	Wd
Rendang Object	1.944256346	Clear	0.5287738746
Specific Padang icon	1.488010653	Luxurious	0.1172219864
Layout	0.4654190811	Expensive	0.1041973212
Halal logo	0.4302657089	Aesthetic	0.08226104305
Herbs and spices	0.3485325352	Economis	0.05152614785
BPOM	0.06895410962	Shiny	0.01451665466
Expired information	0.05649252354	Sustainable	0.01451665466
		Hygienis	0.08226104305
Fitur		Wd	
Ziplock		1.373032515	
In accordance		0.8280846418	
Easy to open		0.4262617686	
Can be closed again		0.3635049094	
Air-tight		0.1354502304	
Window		0.111639987	
Spill-proof		0.05152614785	
Organoleptic	Wd	Organoleptic	Wd
Savory	1.362529047	Hard	0.4786391031
Spicy	1.123901249	Addiction	0.3722761255
Tasty	0.9463839086	Appetizing	0.3326314437
Soft	0.6356496574	Delicious	0.321256076
Greasy	0.612178134	Coconut milk	0.1379082192
Fibrous	0.61056816	Sweet	0.1172219864
Dry	0.5256049951	Springy	0.0366318707

because it matches the impression that rendang products in Indonesia are usually consumed by the upper middle class so that it becomes an "Expensive". According to [34], Packaging can be explained to consumers through illustrations and information to find out about the product. However, there are still some respondents who hope that rendang products are "Economical". The word "Shiny" supports the Luxurious and Expensive impression, through the application of glossy packaging so that it shines and gives a luxurious impression. The last emotional consumer impression of rendang products is that the packaging can be reused so that it can support "sustainable" packaging. On the other hand, in terms of features, consumers want packaging with "Ziplock", "Easy to open", "Can be closed again", and "Window" to see the product inside, then consumers also want packaging that "Spill-proof" so that the product doesn't spill easily. if you take it anywhere. From an emotional perspective, consumers of Rendang products produce five teen words that refer to organoleptic or the product's sense of taste. These five teen words can be used as insight into label design so that the impression of the product can be interpreted in terms of images, illustrations, and typography.

The resulting consumer emotions, apart from the characteristic designs in Table 2, also obtained consumer Kansei in the form of adjectives shown in Table 3. According to Nagamachi, the word Kansei is an adjective that reflects consumer emotions [35].

According to Mitsuo Nagamachi, explains that Kansei is described through the senses of sight, taste (tongue), touch, and interaction receptors between the respondent and the product [15]. An emotion called "perception" arises from the interaction of sensations and can be used as an assessment of what is felt [36]. The consumer Kansei obtained were fifteen adjectives which were emotional consumer preferences for Rendang packaging. Emotional priority is shown by the weight results in the TFIDF method. Words that have a high weight indicate that the word is often said by consumers so the word is considered important or something that consumers want. The Kansei emotional with the highest weight value is the "Practical" packaging, this is because of the problems with the Rendang ready-to-eat packaging object which should be made into practical packaging so that it makes it easier for consumers to use. Practical packaging can attract attention and attract consumers [37]. The second-ranking of consumer emotions is "Simple" packaging, this is also by the market segmentation of Rendang products which are usually middle and upper-class people so preferences both in terms of design and form of packaging prefer products that are simple so that they give an elegant impression. Emotional Elegant is also one of the third priorities to be used as a reference for packaging design. The next emotion that is also important is "Attractive" packaging.

3.4 Concept Design

The resulting 15 Kansei words in Table 3 were then evaluated for their suitability with 40 samples using a semantic differential questionnaire. The evaluation results from 80 respondents were used to find the average of each positive and negative weight between the Kansei words and their antonyms [38, 39]. These results were used as input data in the PCA method with R software [40]. The results of running the PCA are shown in Fig. 1.

The results of running in R software showed that the variation values greater than 1 for rendang are PC1 at 6.0803 and PC2 at 1.04388. Therefore, two main components must be maintained or selected as concept clusters [41]. According to [42], in the cumulative proportion method, the main components to be considered are the data with a total value above 80% [25]. PC1 and PC2 have a cumulative proportion total of 187.58%, meaning

Table 3. Kansei Words.

Wd	Kansei Word	Antonym	Wd	Kansei Word	Antonym
1.128895556	Practical	Impractical	0.2370849541	Disposable	Undisposable
0.8603951714	Simple	Unsimple	0.2196275049	Safety	Unsafety
0.6907125676	Elegant	Unelegant	0.1363082862	Unique	Ununique
0.5960217949	Attractive	Unattractive	0.07104362809	Efficient	Inefficient
0.4412709294	Durable	Undurable	0.06033183204	Mighty	Unmighty
0.3497649458	Informative	Uninformative	0.0551632877	Reusable	Reusable
0.3166434044	Eco-friendly	Non-Eco Friendly	0.1440718788	Easy to store	Uneasy to store
0.260493303	Eyecatching	Non-eye-catching			

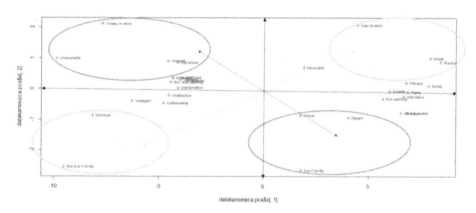

Fig. 1. Plotting of PCA Result

that PC1 can explain 92.43%, while PC2 can explain 95.15% of the information from the existing data.

4 Discussion

Based on Table 2, it is known that emotional feelings that have the highest weight have elements of rendang objects and Padang icons, black and red surface colors, ziplock features, and others. It is known that the emotional feelings with the highest weight for the organoleptic impression of the product are savory, spicy, and delicious flavors. This is a strong image of the rendang product, which must be interpreted in the packaging design. These words are used as insights into the development of the visual design of the packaging. For example, the spicy impression can later be evoked by adding a chili pepper image to the visual design, while the savory and delicious impressions can be reinforced with an image of the rendang product with a layer of spiced seasoning [43]. In addition to emotional feelings about the product impression, emotional feelings about packaging design characteristics are also known, which can be used as guidance in designing the packaging. In this study, it is known that the strongest design characteristic to be used in packaging development is having a distinctive Padang icon with dominant black, red, and yellow colors and featuring an image of rendang.

Based on Table 3, consumer Kansei is expressed through adjectives that can be evaluated by consumers through their feelings when interacting with the packaging through sensory responses, particularly sight and touch. Consumer Kansei includes practical, elegant, attractive, unique, simple, and others. In this study, it is known that the consumer Kansei with the highest weight values is practical, simple, usable, suitable, elegant, and attractive. These emotional feelings are in line with the target market segmentation for rendang consumers, namely the upper-middle class, where the lifestyle of the upper-middle class prefers simple and practical products that give an elegant impression, as evidenced in research by [44]. The upper-middle class with a modern lifestyle desires practical product. The design concept results in Fig. 1 show that two main components need to be maintained, namely PC1 and PC2. According to [42], the group of words chosen in the word scatter plot in the PCA method is a collection of words on the most positive and most negative axes of each main component (PC). PC1 on the most positive axis consists of the words practical, simple, and can be stored again. The analysis of this PC results in the concept of "practical." Meanwhile, on the negative axis of PC1, there is a collection of words that are not unique and not eco-friendly, which leads to the conclusion that the packaging concept is conventional or ordinary. On the other hand, the main component on the positive PC2 shows a collection of words such as unique, elegant, and eco-friendly. It can be concluded that the packaging concept is luxury. In contrast, on the negative side, it is known that the collection of words is not practical, cannot be reused, is not simple, and cannot be stored again. Thus, it is concluded that the concept formed on the negative PC2 is not suitable. Considering the concept results on the positive side of each PC, it is known that the practical and luxury concepts are in line with the interpretation of the emotional feelings of consumers, who are mostly from the upper-middle class, and give the impression that rendang products are synonymous with expensive and classy food products. These concept results will serve as guidance in the next stage of determining the appropriate design elements to realize the concept optimally.

5 Conclusion

The results showed that the priority of emotional feelings obtained from Arab consumers and Indonesian consumers (West Sumatra) have similar preferences, namely giving the impression that rendang products are savory and spicy. In terms of packaging the desired preferences are practical, simple, and usable. Furthermore, in terms of design characteristics of preferences, the packaging has a typical Padang icon and a dominant black color. These results follow the interpretation of concepts based on PCA analysis showing two main components, namely the concept of PC 1 "Practical-Conventional" and PC 2 "Luxury-Not Suitable".

References

1. CNN International: World's 50 Most Delicious Foods (2017). https://edition.cnn.com/travel/article/world-best-foods-readers-choice/index.html. Accessed 10 Sept 2024
2. Fauzan, R., et al.: Produk Dan Merek (2023)
3. Lokman, A.M., Kadir, S.A., Hamidi, S.R., Shuhidan, S.M.: LEIQ as an emotion and importance model for QoL: Fundamentals and case studies. J. Komun. Malaysian J. Commun. **35**(2), 412–430 (2019). https://doi.org/10.17576/JKMJC-2019-3502-25
4. Mauludin, M.S., Saputra, A.D., Sari, A.Z., Munawaroh, I., Regita, E.P.: Analisis Perilaku Konsumen Dalam Transaksi Di e-Commerce. Proc. Islam. Econ. Business, Philanthr. **1**(1), 108–123 (2022). https://jurnalfebi.iainkediri.ac.id/index.php/proceedings. Accessed 10 Sept 2024
5. Wicaksono, F.B., Yuliati, L.R., Muhyiddin, H., Hutasuhut, I.P.: Pengaruh experiential marketing terhadap experiential value serta dampaknya pada customer satisfaction coffee toffee. J. ISIP J. Ilmu Sos. dan Ilmu Polit. **19**(2), 88–101 (2023). https://doi.org/10.36451/jisip.v19i2.6
6. Schütte, S., et al.: Kansei for the digital Era. Int. J. Affect. Eng. **23**(1), 1–18 (2023). https://doi.org/10.5057/ijae.ijae-d-23-00003
7. Vilano, N., Budi, S.: Penerapan kansei engineering dalam Perbandingan Desain Aplikasi mobile marketplace di Indonesia. J. Tek. Inform. dan Sist. Inf. **6**(2), 354–364 (2020). https://doi.org/10.28932/jutisi.v6i2.2705
8. Karim, H.A., Lokman, A.M., Redzuan, F.: Older adults perspective and emotional respond on robot interaction. In: Proceedings of the 4th International Conference on User Science and Engineering (i-USEr 2016), pp. 95–99. IEEE (2017). https://doi.org/10.1109/IUSER.2016.7857941
9. Orshella, D.D.: Penerapan kansei engineering pada perancangan ulang desain kemasan produk umkm. J. Indutrial Galuh **1**(2), 80–87 (2019)
10. Nagamachi, M., Lokman, A.M.: Kansei Innovation: Practical Design Applications for Product and Service Development. CRC Press, Boca Raton (2014)
11. Redzuan, F., Lokman, A.M., Othman, Z.A., Abdullah, S.: Kansei design model for engagement in online learning: a proposed model. In: Communications in Computer and Information Science, vol. 251, pp. 64–78. Springer, Berlin, Heidelberg (2011). https://doi.org/10.1007/978-3-642-25327-0_7
12. Najib, Betanursanti, I.: Redesign produk peci menggunakan metode kansei engineering. Spektrum Ind. **15**(2), 169 (2017). https://doi.org/10.12928/si.v15i2.7551
13. Zabotto, N., da Silva, S.L., Amaral, D.C., Hornos, J.M.C., Benze, B.G.: Automatic digital mood boards to connect users and designers with kansei engineering. Int. J. Ind. Ergon. **74**, 102829 (2019). https://doi.org/10.1016/j.ergon.2019.102829
14. Sari, P., Zulkarnain, Z., Muzaki, V.A., Meilani, Y.D.: Implementasi kansei engineering dalam pengembangan kemasan minuman kopi ready to drink. J. Teknol. Ind. Pertan. **18**(1), 200–209 (2024). https://doi.org/10.21107/agrointek.v18i1.12443
15. Thamrin, S., Windiastuti, E., Halawa, M.V.: Preferensi Konsumen Pada Desain Kemasan Gula Aren dan Gula Coklat dengan Kansei Words **10**(1), 550–555 (2023)
16. Suarim, Neviyarni, N.: Hakikat belajar konsep pada peserta didik. Edukatif J. Ilmu Pendidik. **3**(1), 75–83 (2021). https://doi.org/10.31004/edukatif.v3i1.214
17. Maharani, D., et al.: Perancangan kemasan makanan pie lumpur 'pesona bakery' sebagai oleh-oleh khas sidoarjo. Petra. Ac. Id, 1–9 (2020)
18. Mufrenia, N.: Pengaruh desain produk, bentuk kemasan dan bahan kemasan terhadap minat beli konsumen (studi kasus teh hijau serbuk tocha). J. EMBA J. Ris. Ekon. Manajemen, Bisnis dan Akunt. (2016). https://doi.org/10.35794/emba.v11i1.45579

19. Wang, K.: A novel approach of integrating natural language processing techniques with fuzzy topsis for product evaluation. Symmetry **14**(1), 120 (2022). https://doi.org/10.3390/sym14010120
20. Nurdin, Muhaemin, A.: Implementasi Kansei Engineering dalam Desain Website Profile Perguruan Tinggi (Studi Kasus:STMIK Sumedang). Infoman's J. Ilmu-ilmu Manaj. dan Inform. **10**(2), 39–48 (2016). https://ejournal.stmik-sumedang.ac.id/index.php/infomans/article/view/46. Accessed 10 Sept 2024
21. Nasution, Z.: Face recognition based feature extraction using principal component analysis (PCA). J. Informatics Telecommun. Eng. **3**(2), 182–191 (2020). https://doi.org/10.31289/jite.v3i2.3132
22. Aprilia, R., Sari, N.P., Faizi, I., Wati, R.: Penerapan metode pca dalam penentuan konsep desain kemasan sekunder untuk produk X. Performa Media Ilm. Tek. Ind. **22**(2), 136 (2023). https://doi.org/10.20961/performa.22.2.80739
23. Faisal, Fathimahhayati, L.D., Sitania, F.D.: Penerapan metode kansei engineering sebagai upaya perancangan ulang kemasan takoyaki (Studi Kasus: Takoyakiku Samarinda). J. TEKNO (Civil Eng. Elektr. Eng. Ind. Eng. **18**(1), 92–109 (2021)
24. Isna, Sari, N.P., Maharani, D., Fadhillah, F.: Implementasi kansei engineering dalam menentukan konsep pengembangan kemasan rujak buah potong. J. INTECH Tek. Ind. Univ. Serang Raya **10**(1), 9–18 (2024). https://doi.org/10.30656/intech.v10i1.7832
25. Naftasha, H., et al.: Perencanaan Dan Pengembangan Kemasan Produk Umkm Kebab Gilss Menggunakan Metode Kansei Engineering. Pros. Semin. Nas. Tetamekraf **1**(2), 2022 (2022)
26. Lamalouk, I., Simanjuntak, R.A.: Re-design kemasan produk keripik tempe dengan menggunakan metode kansei engineering. J. Rekayasa Ind. **5**(1), 35–42 (2023). https://doi.org/10.37631/jri.v5i1.838
27. Lokman, A.M., Harun, A.F., Noor, N.L.: Website affective evaluation: analysis of differences in. Popul. (English Ed.) **5619**, 643–652 (2009)
28. Yuan, X., Tang, Y., Sun, W., Liu, L.: A detection method for android application security based on TF-IDF and machine learning. PLoS One **15**(9 September), 1–19 (2020). https://doi.org/10.1371/journal.pone.0238694
29. Andrade, C.: The inconvenient truth about convenience and purposive samples. Indian J. Psychol. Med. **43**(1), 86–88 (2021). https://doi.org/10.1177/0253717620977000
30. Shieh, M.D., Yeh, Y.E.: Developing a design support system for the exterior form of running shoes using partial least squares and neural networks. Comput. Ind. Eng. **65**(4), 704–718 (2013). https://doi.org/10.1016/j.cie.2013.05.008
31. Pertiwi, A., Aristriyana, E., Ningrat, N.K.: Desain Kemasan Produk Pada UMKM Berkah dengan Menggunakan Metode Kansei Engineering di Cipaku. INTRIGA (Info Tek. Ind. Galuh), J. Mhs. Tek. Ind. **1**(1), 1–8 (2023). https://doi.org/10.25157/intriga.v1i1.3593
32. Lokman, A.M., Aziz, A.A.: A Kansei system to support children's clothing design in Malaysia. In: Conference Proceedings - IEEE International Conference on Systems, Man and Cybernetics, pp. 3669–3676. IEEE (2010). https://doi.org/10.1109/ICSMC.2010.5641868
33. Hendratman: Attractive Layout using Design Principles: Includes: Align, Guidelines, Origami, Corner Technique and Logo Guide. Exotic (2023). https://books.google.co.id/books?id=K8S7EAAAQBAJ&dq=Center-aligned+Layout&lr=&source=gbs_navlinks_s. Accessed 10 Sept 2024
34. Mukhtar, S.: Muchammad: peranan packaging dalam meningkatkan hasil produksi terhadap konsumen. J. Sos. Hum. **8**(2), 181–191 (2015)
35. Soikun, M., Ibrahim, A.A.A.: Pendekatan Analisis Formalistik Dan Semiotik Visual: Mencari Faktor 'Appeal' Dalam Watak Animasi Tempatan Formalistic and Visual Semiotic Analysis Approaches: Finding 'Appeal' Factors in Local Animation Character. Gendang Alam **10**(2) (2020)

36. Fan: Desain Model Pengembangan Produk Cokelat Padat Berbasis Tipe Kepribadian Dengan Pendekatan Kansei Engineering. Universitas Sumatera Utara (2018). https://repositori.usu.ac.id/handle/123456789/9740. Accessed 10 Sept 2024
37. Rosha, Khaidir, A.: Perlindungan Konsumen Terhadap Penggunaan Plastik Berbahaya Sebagai Kemasan Pangan Dalam Upaya Meningkatkan Minat Beli. J. Manaj. Univ. Bung Hatta **14**(1), 28–36 (2019). https://doi.org/10.37301/jmubh.v14i1.13876
38. Nasional, T.I., Gadjah, U.: Pengembangan kemasan wedang uwuh menggunakan metode kansei engineering. In: Seminar Nasional Teknik Industri Universitas GADJAH MADA Yogyakarta (2023)
39. Segita, E., Putri, K.A.A., Nuryadin, R.N., Nur, Y.L., Sari, N.P.: Perencanaan konsep desain kemasan kerak telor menggunakan metode kansei engineering. Ind. Inov. J. Tek. Ind. **14**(1), 12–21 (2024). https://doi.org/10.36040/industri.v14i1.8033
40. Puspasari, A., Amelia, C., Pangestu, B.P., Sari, N.P.: Perancangan Konsep Desain Kemasan Makanan Kucing Menggunakan Metode Kansei Engineering. Performa Media Ilm. Tek. Ind. **22**(2), 143 (2023). https://doi.org/10.20961/performa.22.2.80744
41. Sari, P., et al.: Perancangan Desain Kemasan Penyedap Rasa Berbasis Kansei Engineering. Semin. Nas. Inov. Vokasi **2**(1), 1–11 (2023)
42. Coghlan, A.: A Little Book of R For Bioinformatics. (2014). http://cdn.bitbucket.org/psylab/r-books/downloads/Coghlan2014.pdf. Accessed 10 Sept 2024
43. Ramdhani, R., Rachmawati, I., A.S, Y., Kusnadi, A.S.M., Parman, S., Putra, G.M.: Perancangan Desain Kemasan Keripik Singkong Dua Saudara. J. Pengabdi. Univ. Catur Insa. Cendekia **1**(2), 1–9 (2023). https://jpucic.id/index.php/jpucic/article/view/40 https://jpucic.id/index.php/jpucic/article/download/40/27. Accessed 10 Sept 2024
44. Hasbi, M.L., Muis, I.: Segmentasi, Targeting dan Positioning Produk Neo Coffee dari Wings Food untuk Pasar Kota Bekasi. J. Mhs. Bina Insa. **5**(1), 63–72 (2020)

Exploring Emotion Classification for Children with Autism in Response to Robot Movement: A Preliminary Case Study in Malaysia

Fatin Nadhirah Zabani[1], Jo Anne Saw[2], Nur'aina Daud[1], Azhar Abd Aziz[1], and Anitawati Mohd Lokman[1](✉)

[1] College of Computing, Informatics and Mathematics, Universiti Teknologi MARA, UiTM Shah Alam, 40450 Shah Alam, Selangor, Malaysia
`{aina,azhar313,anitawati}@uitm.edu.my`
[2] Faculty of Medicine, Universiti Teknologi MARA, Sungai Buloh Campus, Jlaan Hospital, Sungai Buloh, 47000 Selangor, Malaysia
`annejosaw@uitm.edu.my`

Abstract. Children with autism struggle to comprehend and express their emotions, leading to misinterpretation and distress. Technology, particularly robots, holds promise for helping children with autism understand and communicate their emotions. However, there is limited research on emotional identification in children with autism during interactions with robots. This study aims to fill that gap by identifying emotional responses and proposing a emotion classification for children with autism during robot movement interactions. Observation and semi-structured interviews were conducted to gather data, using seven specific robot movements. Then, KJ Method was utilized to develop the classification. The result includes emotional response keywords and its classification for emotional responses in children with autism during robot movement interactions. While these findings are preliminary, the reference classification provides valuable insights into interpreting emotional responses in children with autism and can serve as a helpful guide for integrating robots into their daily activities.

Keywords: Autism · Emotion classification · Kansei · KJ Method · Robot Movement

1 Introduction

Autism is a lifelong chronic disability, and the number of reported cases is on the rise [1]. While there is no cure for autism, effective interventions based on developmental, behavioral, and cognitive approaches during early childhood have shown positive effects on reducing autism symptoms [2]. Technological interventions play a significant role in supporting children with autism, and one widely used intervention is the use of robots. Robots are employed in various assistive scenarios, addressing diverse human needs, and aiding in the rehabilitation of children with autism [3].

The adoption of robots has demonstrated positive outcomes in promoting the social and emotional development of children with autism. Social robots have proven to be an effective tool in overcoming social obstacles and engaging children with autism in interaction, gradually fostering the development and practice of social and emotional skills [4]. The use of robots has also had a positive impact on children with autism, improving their social skills, socialization, and overall quality of life [5]. However, understanding the emotions of children with autism poses challenges due to their difficulties in recognizing and expressing emotions.

This study aims to adopt the Kansei Engineering approach to identify the emotions of children with autism. Kansei Engineering primarily focuses on enhancing the quality of life, comfort, and enjoyment of individuals by focusing on the emotional elements [6]. It also has been effective in leveraging product design, including physical consumer products and IT artifacts to elicit emotional responses from users [7, 8]. An intriguing model that examines how children's emotional responses to robots as teaching mediators can influence their learning motivation. Their study emphasized the importance of understanding and leveraging emotional responses in educational settings involving robots. Building upon this research, the current study recognizes the potential of the Kansei Engineering approach in assessing emotional responses and linking them to motivational elements in robots [9].

By employing the KJ Method, this study aims to delve deeper into the emotional experiences of children with autism during their interactions with a robot. The utilization of the KJ Method in this study will enable the identification and classification of emotional responses exhibited by children with autism. Through the analysis, the study aims to shed light on the emotional landscape of children with autism during their interactions with robots. By uncovering the range of emotional responses and their connection to motivational elements, valuable insights can be gained for designing and optimizing educational interventions that effectively engage and support the learning process of children with autism. Ultimately, this research endeavors to contribute to the development of innovative approaches that enhance educational experiences and outcomes for children with autism using robots.

2 Literature Background

2.1 Autism Spectrum Disorder (ASD)

Autism is a developmental condition that affects brain development and behavior. Individuals with autism often live in their own world. This comprehensive developmental disorder on the autism spectrum was initially identified by American psychiatrist Leo Kanner. Additionally, Austrian pediatrician Asperger discovered similar clinical symptoms of autism, leading to the recognition of Asperger syndrome as another autism spectrum disorder. Autism is frequently accompanied by significant behavioral challenges. Diagnosis of autism involves observing impairments in three areas: social interaction, limited range of activities and interests, and communication abilities [10].

Individuals with autism exhibit a wide range of abilities. Some may have excellent speech skills, while others may be non-verbal. Some autistic individuals require extensive support in their daily lives, while others can function with minimal assistance.

Furthermore, children with autism often display distinct changes as they grow, such as avoiding eye contact or conversations, exhibiting intense interest in specific objects, or becoming disinterested and withdrawn from social interactions. These individuals are likely to experience "infantile" autism, which can persist into adolescence and adulthood [1]. The term "infantile" autism was introduced in the Third Edition of the Diagnostic and Statistical Manual of Mental Disorders, approximately 30 years after the publication of the first version in 1952. The initial description of "infantile" autism referred to individuals with severe impairments [11].

Autism spectrum disorder is a group of disorders that affect neurological development. While the key characteristic of autism involves difficulties in social interaction and communication, other associated features include behavioral issues, anxiety, sleep disorders, and depression. Eating disorders and attention-deficit or hyperactivity problems may also coexist. Moreover, children with autism typically experience significant delays in language acquisition, misuse words, and lack coherence in communication [12]. Autism spectrum disorder typically manifests before the age of three and can persist throughout a person's life, although symptoms may improve over time. Some children exhibit autism symptoms within their first year, while others may not display signs until they are 24 months old or older. Additionally, some children with autism may develop new skills and reach developmental milestones until around 18 to 24 months, after which they may stop acquiring new skills or lose those they have previously acquired [1].

2.2 Robot Movement Interaction

The development of robot systems that possess human-like agility, stability, and precision is imperative for effectively integrating robots into human environments [13]. There was a study examining the interaction between children with autism and either a human arm or a robotic arm model. The findings demonstrated that children with autism exhibited improved performance when primed with robotic arm movements, as indicated by shorter movement times and anticipated peak velocities. This suggests that interactions with robots influence the visuomotor priming processes of children with autism [14].

Moreover, the IROMEC (Interactive Robotic Social Mediators as Companions) is a therapeutic and educational robot toy designed to target various developmental areas in children with autism, including sensory development, communication and interaction, cognitive development, and social and emotional development. The IROMEC gradually introduces more complex movement interactions, presenting new cues to foster additional and diverse learning. It evolves alongside children with autism, assisting them in exploring and accomplishing challenging tasks related to dynamic social interaction [15].

A study proposed a framework for capturing, evaluating, and modeling three-dimensional emotional motions in an embodied game application aimed at developing social interaction abilities in children with autism. The movements are enacted by robots or other sensory-motor devices as part of a game. The emotional behavior exhibited depends on the embodiment of the robot and the game's setup. The test results revealed that children with autism reliably recognize most robot behaviors [16]. Additionally, an experiment involving the Humanoid Robot NAO and an autistic child demonstrated that the child could recognize emotions associated with NAO's body language. Therapists noted that the NAO robot exhibits positive attributes such as an appealing appearance to children with autism, good speaking and listening abilities, smooth movements, and a friendly demeanor, making it suitable for autism rehabilitation [17].

However, a study argued that while robot movement interaction in rehabilitation and therapy can aid in the treatment of children with autism, ethical considerations should be taken into account. These include examining the emotional relationship between children with autism and robots, as well as determining the appropriate extent of robot usage within the overall therapeutic process. Furthermore, the researchers believe that robots should not replace human interaction in autism therapy but rather be used as a supportive tool alongside human involvement [18].

2.3 Emotional Responses of Children with Autism Towards Robot

Children with autism exhibit distinct emotional responses compared to neurotypical children [19]. The preferences, aversions, and behavioral reactions to biomimetic stimuli are not universally consistent and depend on cognitive processes. In particular, children with autism spectrum disorder display altered sensory attraction and repulsion towards visual stimuli [20]. Simple shapes, elicit emotional responses such as preferences and behavioral reactions in children with autism [19]. Additionally, the use of robots has been found to elicit emotional responses in children with autism. Different interactions with robots evoke varied emotions in these children, and they demonstrate more distinct emotional responses when the robot engages in verbal communication or hand gestures [9].

Moreover, developing techniques to evoke and differentiate affective play and emotional response patterns induced by robots contributes to the advancement of early screening methods for autism spectrum disorder [21]. Furthermore, human-robot interaction can serve as a facilitator in the treatment of autism by improving children's social interaction skills and emotional well-being in real-life situations. However, the involvement of a child psychologist in robot-assisted intervention sessions is considered essential. It requires the knowledge, expertise, and experience of a psychologist to effectively manage unexpected situations that may arise during interactions with autistic individuals [22]. Moreover, focusing on factors such as the robot's appearance, voice, and interaction patterns, it seeks to foster positive interactions and enhance learning experiences. Just as Kansei Engineering is used to create online learning materials that engage students emotionally [23, 24], it can also be applied to design robots that appeal to children's emotions, making them more effective as tools for education and therapy.

2.4 KJ Method

The KJ Method, developed by Professor Kawakita, is a widely recognized approach in Kansei Engineering that aims to facilitate the creation of innovative and user-friendly designs that provide a pleasant user experience [25, 26]. This method was specifically devised to deconstruct the classification of design categories based on specific concepts of physical design features. It involves gathering and synthesizing Kansei Words extracted from various sources such as journals, product websites, and related studies. These Kansei Words are then subjected to expert clustering to identify their affinities and subsequently classified into specific design characteristics of the product, all while considering user requirements.

The researchers examined the difference between the understanding of wheelchair design specifications and the psychological needs of the user. Expert validation was employed to verify the affinity and classification, and a designer created a prototype based on the category classification to assess the feasibility of the design guidelines. The researchers emphasized that continuous enhancements in product design, particularly in the field of assistive devices, can be achieved by employing the KJ Method. This method allows individuals with physical disabilities and the elderly to effectively express their emotional expectations regarding assistive devices, which in turn enables expert designers to address their implicit psychological needs [27, 28].

KJ Method offers a valuable approach for improving the design process and enhancing the emotional experience of users with physical disabilities and the elderly. By utilizing this method, designers can gain insights into the users' expectations and design products that effectively cater to their psychological requirements.

3 Method

This study utilized a qualitative methodology carried out through an observational approach. Purposive sampling was employed to select participants based on criteria relevant to the study. Data collection for the qualitative analysis took place during observation sessions and semi-structured interview sessions. Figure 1 depicts the research setting.

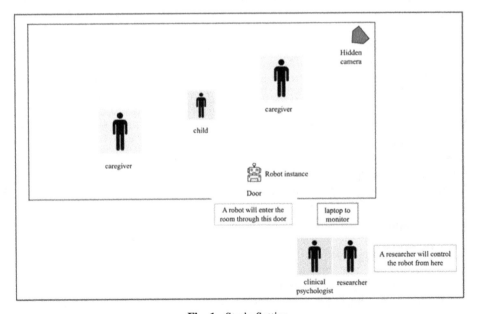

Fig. 1. Study Setting.

This study focused on children with autism at a care center, Pusat Jagaan Kanak-kanak Istimewa SAYANG in Shah Alam, Malaysia as the target population. Three children, aged four to nine years old, with a mild severity level were selected as participants for the observation session based on their ability to respond and provide basic feedback. Caregivers from the center, who had more than three years of experience with children with autism and were closest to the children, were also involved in assisting the participants. The caregivers helped understand and identify the emotional responses of the children towards robot movement.

In addition, two experts were chosen for their specialized knowledge in autism and Kansei engineering. A clinical psychologist and a Kansei expert were recruited to provide their expertise. The clinical psychologist monitored the session to prevent any undesirable or unforeseen circumstances arising from the children and offered guidance on classifying the emotional responses of children with autism during data analysis. Meanwhile, the KJ method, known for its ability to evaluate emotions, was utilized in this study. Therefore, a Kansei expert was necessary to assist in classifying the emotional responses of children with autism.

The objective of this study was to identify and categorize the emotional responses of children with autism when interacting with robot movements. Despite common perceptions that children with autism lack emotions, their difficulty in understanding and expressing emotions makes it challenging for them to convey their own feelings. Hence, the KJ Method in Kansei Engineering was employed to identify and classify the children's emotional responses during interactions with various robot movements, including circular, forward reverse, forward and left, forward and right, reverse and left, and reverse and right motions. The KJ Method was chosen due to its effective mechanism for categorizing implicit user feelings.

The analysis using the KJ Method was conducted after completing the semi-structured interview session with the caregivers. Several steps were taken to establish a reference classification of emotional responses, including categorizing the emotional responses of children with autism using sticky notes and grouping similar emotions into clusters. Each cluster was then labeled accordingly. This process successfully identified a reference classification of emotional responses. To ensure validity and eliminate bias during the evaluation, a validation process was undertaken in consultation with an autism expert. This process ensured that the final emotional response reference classification was adequate and appropriate for this study.

4 Result and Discussion

The participants selected for this observation session were aged seven and nine and had been diagnosed with autism spectrum disorder by a medical officer and therapist. Initially, three children with autism were chosen to take part in the observation session. However, during the session, one of the participants experienced allergic symptoms, which could potentially affect the validity of their emotional responses to the robot movement. Consequently, only two children with autism ultimately participated in the observation session.

The analysis of the first objective was conducted after the observation session with the children and a semi-structured interview session with the caregivers. The interview session with the caregivers yielded keywords for the emotional responses elicited by specific movements observed during the interaction between the children with autism and the robot. Despite their limitations in expressing feelings through communication and interaction, the observation method and interview session with the caregivers enabled the identification of emotional responses in children with autism during interactions with the robot. The caregivers played a crucial role in recognizing the children's emotional responses to robot movement, as they possessed an understanding of how the children express their feelings. Based on the interview sessions with Caregiver_A and Caregiver_B, both Subject_Mild1 and Subject_Mild2 exhibited positive emotions in response to the robot movement. Following the identification of emotional response keywords, the exploratory study expanded the number of emotional response keywords through expert analysis and cross-referencing with glossaries. Table 1 illustrates the emotional response keywords corresponding to each robot movement.

Table 1. Emotional Responses Keywords for each Robot Movement.

Robot Movement	Emotional Responses Keywords
Circular	Happy, Cheerful, Excited, Attracted, Confused, Weird, blur, Attentive, Interested
Forward	Curious, Confuse, Attentive, Attracted, Excited, Delighted, Accepting, Interested, Anticipating
Forward and to the Left	Surprised, Interested, Amazed, Amused, Anticipating, Weird, Adoring, Admiring, Happy, Amused, Excited
Forward and to the Right	Interested, Amazed, Amused, Anticipating, Enjoy, Joyful, Excited
Reverse and to the Left	Curious, Attentive, Enjoy, Excited, Amazed, Amused, Happy
Reverse and to the Right	Curious, Excited, Delighted, Enjoy, joyful, Anticipating, Satisfied, Amazed, Attentive, Adoring, Admiring

Once the emotional response keywords of children with autism towards robot movement were identified, the researcher, Kansei expert, and clinical psychologist expert engaged in discussions and conducted the KJ Method to categorize each emotional response keyword into distinct clusters. This session utilized online sticky notes for the process. In the implementation of the KJ Method, three crucial conditions were observed. The Kansei expert guided the researcher and clinical psychologist through the session, ensuring adherence to the following steps. Firstly, the method was carried out in silence to foster unconventional thinking and prevent any individual from dominating the process. Secondly, quick responses were encouraged to avoid overthinking. Thirdly, disagreements were resolved in a simple manner, allowing for free movement and rearrangement of words until the groupings made sense.

During the generation of cluster names, the researcher, Kansei expert, and clinical psychologist engaged in discussions to determine if the cluster adequately represented the terms within the groupings. They dedicated ample time to selecting the most appropriate emotional response keywords, aiming to minimize ambiguity and disagreements through thorough deliberation. Once all cluster names were established, a Kansei expert and clinical psychologist manually validated the cluster headers and word groupings. Finally, a clinical psychologist examined the final diagram to ascertain the suitability and appropriateness of the groupings and cluster names in the proposed reference classification of emotional responses for children with autism during interactions with robot movement. Figure 2 displays the clustered emotional response keywords using online sticky notes.

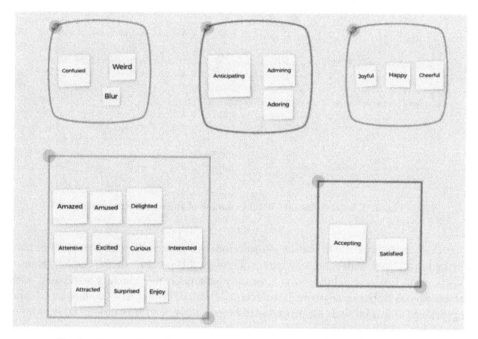

Fig. 2. Clustered Emotional Responses Keyword using Online Sticky Notes.

This study successfully identified and categorized the emotional responses exhibited by children with autism during interactions with robot movement. This conclusion was drawn based on the observation session with the children, the robot movement itself, and the interview session with the caregivers. Through the interview session, it was determined that the children expressed distinct emotions for each robot movement, with the reverse, reverse and to the left, and reverse to the right movements garnering the most interest. Furthermore, the study revealed that positive emotions such as "excited," "amazed," and "interested" were the most commonly expressed emotions during the interactions with the robot movement. Figure 3 depicts the cluster names and word groupings of emotional responses exhibited by children with autism during robot movement interactions.

Based on the obtained emotional response keywords related to robot movement, the researcher, Kansei expert, and clinical psychologist employed the KJ Method to establish

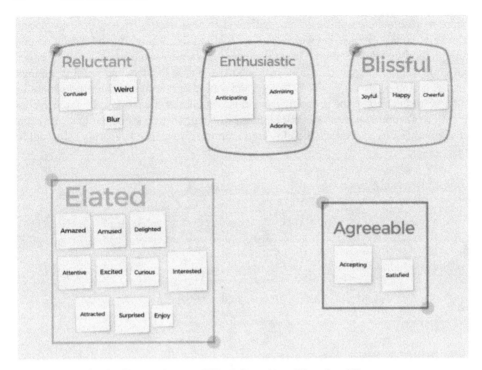

Fig. 3. Cluster Name and Word Grouping of Emotional Responses.

a proposed reference classification of emotional responses for children with autism during interactions with robot movement. Through this process, the researcher, Kansei expert, and clinical psychologist successfully generated five clusters associated with the emotional response keywords: "reluctant," "elated," "blissful," "enthusiastic," and "agreeable." Fig. 4 presents the proposed reference classification of emotional responses for children with autism during robot movement interaction.

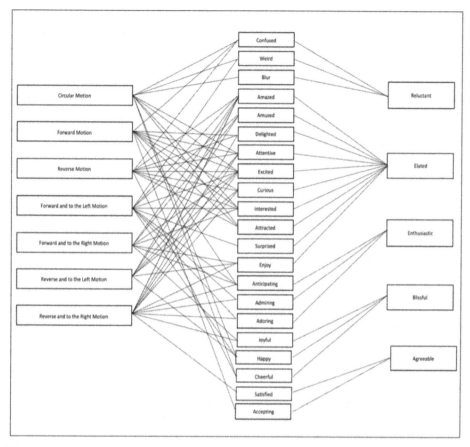

Fig. 4. The Proposed Reference Classification of Emotional Responses for Children with Autism during Robot Movement Interaction.

5 Conclusion

The primary objective of this study, which aimed to identify and categorize the emotional responses of children with autism during interactions with robot movement, has been successfully accomplished. Initially, the main concern of the study was the possibility of identifying emotional responses in children with autism, given their challenges in expressing emotions and the various factors that can trigger their emotions. To address this concern, caregivers were present throughout the observation session with the children and conducted interview sessions afterward to identify the emotional responses exhibited during the robot movement interactions. Through these methods, the emotional response keywords related to robot movement were successfully identified, revealing that the children expressed positive emotions during the interactions. This indicates that the robot's movements effectively captured their interest and attention.

Furthermore, the analysis using the KJ Method facilitated the grouping of emotional response keywords into five clusters. Consequently, this study proposes a reference classification of emotional responses for children with autism during robot movement interactions, enabling these children to express their emotions despite their impairments. This proposed reference classification holds significant benefits, not only for children with autism by helping identify and understand their emotional responses to the robot but also for parents and caregivers who can gain a better understanding of the children's emotional responses during robot interactions. The owner of the center has shown great interest in this emotional classification, and the researcher plans to extend the research results to other schools as a guide to understanding the emotional responses of children with autism during robot movement interactions.

Moreover, this reference classification of emotional responses will be valuable for future studies examining children's emotional responses during robot movement interactions. Researchers can utilize the Kansei Engineering Approach to synthesize emotional response keywords, search for potential missing keywords, or use them as a starting point when encountering difficulties in identifying appropriate language adjectives that accurately reflect the emotional state, as emotions are inherently implicit and can pose challenges for inexperienced researchers.

To further advance this study, it is strongly recommended that future researchers conduct more comprehensive investigations into subject availability, as recruiting individuals with special needs, such as children with autism, poses greater challenges compared to recruiting typically developing children. Additionally, future studies should aim for larger sample sizes of children with autism and their caregivers to minimize bias and obtain more comprehensive data. It is also advised to include children with varying severity levels of autism, such as mild, moderate, and severe, to identify a broader range of emotional response keywords to robot movement and generate multiple reference classifications of emotional responses for children with autism during robot movement interactions. Based on the analysis of the results, this study holds significant implications for understanding the emotional responses of children with autism during robot movement interactions, and it is expected to inspire further investigations, such as exploring the challenges involved in classifying the emotions of children with autism in response to robot movement, providing researchers with a clearer and more comprehensive understanding of this subject matter.

Acknowledgments. The authors would like to acknowledge the Research Initiative Group for Emotion, Kansei and Design Engineering (RIG EKDE), Universiti Teknologi MARA (grant code: 600-RMC/GPM ST 5/3 (028/2021)), as well as Malaysia Association of Kansei Engineering (MAKE), for all their assistance to the research paper.

Disclosure of Interests. The authors declare that there is no conflict of interest.

References

1. Centers for Disease Control and Prevention. Autism Spectrum Disorder (ASD) (2022). Retrieved on August 2022, from Treatment: https://www.cdc.gov/ncbddd/autism/treatment.html

2. Maw, S.S., Haga, C.: Effectiveness of cognitive, developmental, and behavioral interventions for Autism Spectrum Disorder in preschool-aged children: A systematic review and meta-analysis. Heliyon (2018)
3. Saleh, M., Hashim, H., Mohamed, N., Almisreb, A.A.: Robots and autistic children: a review. Periodicals of Engineering and natural Sciences, 1247–1262 (2020)
4. Barti-Pokorny, K., Pykala, M., Uluer, P., Barkana, D.E.: Robot-Based Intervention for Children With Autism Spectrum Disorder: A Systematic Literature Review. IEEE Access (2021)
5. Valadão, C.T., Goulart, C., Rivera, H., Caldeira, E.: Analysis of the use of a robot to improve social skills in children with autism spectrum disorder. Research on Biomedical Engineering, 161–175 (2016)
6. Lokman, A.M.: KE as affective design methodology. In: 2013 International Conference on Computer, Control, Informatics and Its Applications (IC3INA), pp. 7–13. IEEE (2013)
7. Lokman, A.M., Ishak, K.K., Razak, F.H.A., Aziz, A.A.: The feasibility of PrEmo in cross-cultural Kansei measurement. In: 2012 IEEE Symposium on Humanities, Science and Engineering Research, pp. 1033–1038. IEEE (2012)
8. Lokman, A.M., Harun, A.F., Noor, N.L., Nagamachi, M.: Website affective evaluation: Analysis of differences in evaluations result by data population. In: Human Centered Design: First International Conference, HCD 2009, Held as Part of HCI International 2009, San Diego, CA, USA, July 19-24, 2009 Proceedings 1, pp. 643-652. Springer Berlin Heidelberg (2009)
9. Karim, H.A., Lokman, A.M., Redzuan, F.: Older adults perspective and emotional respond on robot interaction. In: 2016 4th international conference on user science and engineering (i-user), pp. 95–99. IEEE (2016)
10. Ali, E.M., Adwan, F.E., Al-Naimat, Y.M.: Autism Spectrum Disorder (ASD); Symptoms, Causes, Diagnosis, Intervention, and Counseling Needs of the Families in Jordan. Modern Applied Science (2019)
11. Hyman, S.L., Levy, S.E., Myers, S.M.: Identification, Evaluation, and Management of Children With Autism Spectrum Disorder. American Academy of Pediatrics (2020)
12. Magnuson, K.M., Constantino, J.N.: Characterization of Depression in Children with Autism Spectrum Disorders. National Library of Medicine, 332–340 (2011)
13. Aziz, A.A., Moghanan, F.F.M., Mokhsin, M., Ismail, A., Lokman, A.M.: Humanoid-robot intervention for children with autism: a conceptual model on FBM. In: Soft Computing in Data Science: First International Conference, SCDS 2015, Putrajaya, Malaysia, September 2–3, 2015, Proceedings 1, pp. 231–241. Springer Singapore (2015)
14. Pierno, A.C., Mari, M., Lusher, D., Castiello, U.: Robotic movement elicits visuomotor priming in children with autism. Neuropsychologia, 448–454 (2008)
15. Ferrari, E., Robins, B., Dautenhahn, K.: Therapeutic and educational objectives in Robot Assisted Play for children with autism. RO-MAN 2009-The 18th IEEE international (2009)
16. Barakova, E.I., Lourens, T.: Expressing and interpreting emotional movements in social games with robots. Personal and ubiquitous computing, 457–467 (2010)
17. Miskam, M.A., et al.: Study on Social Interaction between Children with Autism and Humanoid Robot NAO. Applied mechanics and materials, 573–578 (2013)
18. Ntaountaki, P., Lorentzou, G., Lykothanasi, A., Anagnostopoulou, P.: Robotics in Autism Intervention (2019)
19. Belin, L., Henry, L., Destays, M., Hausberger, M., Grandgeorge, M.: Simple Shapes Elicit Different Emotional Responses in Children with Autism Spectrum Disorder and Neurotypical Children and Adults. Frontiers in Psychology (2017)
20. Ben-Sasson, A., et al.: A Meta-Analysis of Sensory Modulation Symptoms in Individuals with ASD. J. Autism Develop. Disorders, 1–11 (2009)

21. Boccanfuso, L., et al.: Emotional robot to examine different play patterns and affective responses of children with and without ASD. 2016 11th ACM/IEEE International Conference on Human-Robot Interaction (HRI), pp. 19–26. IEEE (2016)
22. Redzuan, F., Mohd. Lokman, A., Ali Othman, Z., Abdullah, S.: Kansei design model for engagement in online learning: a proposed model. In Informatics Engineering and Information Science: International Conference, ICIEIS 2011, Kuala Lumpur, Malaysia, November 12-14, 2011. Proceedings, Part I, pp. 64–78. Springer Berlin Heidelberg (2011)
23. Redzuan, F., Lokman, A.M., Othman, Z.A., Abdullah, S.: Kansei design model for e-learning: A preliminary finding. In: Proceeding of the 10th European Conference on e-Learning (ECEL-2011), pp. 685–696 (2011)
24. Pour, A.G., Taheri, A., Alemi, M., Meghdari, A.: Human–robot facial expression reciprocal interaction platform: case studies on children with Autism. Int. J. Soc. Rob. 179–198 (2018)
25. Lokman, A.M., Kadir, S.A., Hamidi, S.R., Shuhidan, S.M.: LEIQ™ as an emotion and importance model for QoL: fundamentals and case studies. Jurnal Komunikasi: Malaysian Journal of Communication **35**(2), 412–430 (2019)
26. Lokman, A.M., Kadir, S.A., Noordin, F., Shariff, S.H.: Modeling factors and importance of happiness using KJ method. In: Proceedings of the 7th International Conference on Kansei Engineering and Emotion Research 2018: KEER 2018, 19–22 March 2018, Kuching, Sarawak, Malaysia, pp. 870–877. Springer Singapore (2018)
27. Lokman, A.M., Ismail, M.N., Abdullah, N.A., Omar, A.R.: Kansei Wheelchair Design based on KJ Method. J. Computat. Theoretical Nanoscience, 4349–4353 (2017)
28. Ismail, M.N., Lokman, A.M., Abdullah, N.A.S.: Formulating Kansei concept of assistive device for people with physical disabilities. In: 2014 3rd International Conference on User Science and Engineering (i-USEr), pp. 30–35. IEEE (2014)

Implementation of the K-Means Genetic Algorithm to Determine the Design Concept for Toast Bread Packaging by Kansei Engineering

Novi Purnama Sari[✉], Wiwi Prastiwinarti, Rachmadita Dwi Pramesti, Lytta Yennia Putri, and Rafi Ramdan Permana

Jakarta State Polytechnic, Depok City, Indonesia
novi.purnamasari@grafika.pnj.ac.id

Abstract. Indonesian's bread sales in 2021 became the highest in Southeast Asia, reaching USD 18.7 billion and continuing to increase until 2023. This condition is an opportunity factor for the development of processed bread businesses such as toast which is growing rapidly in Indonesia. However, the packaging used by toast products currently still has many shortcomings so it is necessary to make improvements according to the needs and emotions of consumers through the development of Kansei Engineering. Packaging with a clear concept can facilitate the packaging development process. The stage of determining the packaging concept is an important thing that must be done. The purpose of this research is to determine the concept of packaging design quantitatively based on consumer emotions. The appropriate method used to identify consumer emotions is Kansei Engineering supported by the K-Means method optimized by Genetic Algorithm to determine design concepts. The result of the concept obtained is "practical-unique" because it has a greater range than the second cluster. The resulting concept will serve as guidance in carrying out the next process, namely the determination of packaging design elements in the packaging development process.

Keywords: Kansei Engineering · K-Means Cluster · Genetic Algorithm · Packaging Design Concept

1 Introduction

Bread consumption has increased every year in Indonesia. This increase in consumption within a week occurred from 2021 to 2023, according to the Indonesian Central Statistics Agency (BPS) was IDR 1,008,231.00 to IDR 1,164,522.20. Along with the rise in public spending on consuming bread, it can be a business opportunity for processed bread products, one of which is toast. Toast is considered to have high nutritional value, which can make it a viable option for sustainable baked food production [1]. Based on this, it is unsurprising that the toast business continues to experience development, one of which is the Bandung Toast MSME in Indonesia.

The current development of Bandung's toast culinary business has not been matched by the right packaging. Toast packaging only uses recycled paper which has no information and is less aesthetically pleasing, so Bandung Toast MSMEs have not utilized the function of packaging as brand awareness that can attract consumer attention.

Whereas packaging is one of the promotions that can increase sales and indirectly protect consumers through information about packaged products [2].

An important factor in making attractive packaging comes from consumer preferences [3]. The existence of consumer preferences is an opportunity to get a combination of specific packaging design ideas by extracting consumer emotions. so that consumer preferences can be a strategy in designing concepts to create attractive packaging designs [4]. The packaging development process is expected to determine specific design concepts that can influence consumer emotions and behavior.

One development method that uses a consumer emotion approach is Kansei Engineering [5]. This method is done by translating consumer emotions into words called Kansei words as parameters in conducting product development and improvement [6]. The emotional keywords are used as the measure of strength of the emotional responses [2]. Concept determination is the first step in Kansei Engineering before determining the design elements in packaging development. Kansei engineering is considered capable of transforming the psychological and affective aspects of consumers towards a packaging design into the form of packaging specifications [3]. Several methods can be used in the process of determining this concept.

Previous research shows the ability of the Kansei Engineering method to determine packaging design concepts that are in accordance with consumer emotions [4], processing data using the Principal Component Analysis (PCA) method produces the concepts of "Modern-Practical" and "User Friendly-Untenable" for secondary packaging of skincare products. Similar Kansei Engineering research also produced the concepts of "Useable & Safety" and "Attractive & General" for cat food packaging [5]. As well as reusable and functional concepts for fried meatball packaging [6]. The same research was conducted [7], in determining the concept of egg crust product packaging with the Kansei Engineering approach resulting in the concepts of User Friendly-Unsustainable" and "Standard-Simple". The Kansei Engineering method has proven effective in translating consumer emotions into clearer packaging design concepts.

The advantage of the Kansei Engineering method is that it can convert qualitative data in the form of consumer emotions (Kansei words) into quantitative data to determine packaging design concepts. The use of statistical methods is usually done to support this. Some commonly used statistical methods are Principal Component Analysis (PCA), Factor Analysis, and K-Means Cluster. These methods are able to cluster data to facilitate the process of extracting Kansei words into concepts. Each method has its own advantages and disadvantages. One of them is the K-Means Cluster method still has weaknesses in determining the centroid so that optimization needs to be done. One of the efforts in the process of optimizing the K-means cluster method by combining using the Genetic Algorithm method. According to research [8], there are differences in the results of data cluster centers from the K-Means method without the Genetic Algorithm and the Genetic Algorithm optimization K-Means method. K-Means with a Genetic Algorithm can produce Centroid or Cluster center data that is more optimal than just using the K-Means method without a Genetic Algorithm [9]. This is a novelty in this research to provide an alternative to a more optimal concept determination process. This research aims to determine the packaging design concept of Bandung toast bread in accordance with consumer emotions. Through this research, it is hoped that it can provide alternative concepts as a strategy for developing Bandung toast bread packaging to increase selling power and be able to compete.Subsequent paragraphs, however, are indented.

2 Method

The method in developing Bandung toast bread packaging uses the Kansei Engineering method. The following are the stages as shown in Fig. 1.

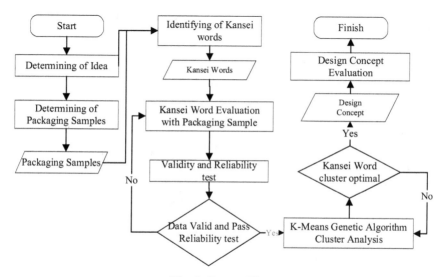

Fig. 1. Process Flow.

2.1 Determining of Packaging Samples

Sample collection is based on segmenting, targeting, positioning, and product background. The packaging samples used are samples of similar toast products. The minimum sample collection is 20–25, so the Kansei Engineering process can be carried out [10]. The collected packaging samples can provide an overview of the design concept [11]. The sample collection used was obtained through an online survey on social media. Sample measurement is carried out in a questionnaire with a numerical scale in each question.

2.2 Indentifing of Kansei Words

The Kansei Words identification process is carried out through direct and indirect interviews with respondents. Respondents were selected using purposive sampling [12]. Kansei Words collection used stimuli by packaging samples and videos to describe the product more emotionally [13]. In addition, the use of packaging samples in the questionnaire can also explore the emotions in explaining the expectations and problems of the product [14]. Kansei words collected generally vary in number between 50–600 words [10]. The purpose of this Kansei word is to be used as input data for processing methods in finding packaging concept categories.

2.3 K-Means Cluster Analysis

The K-Means Cluster analysis process is carried out with data from the Semantic Differential questionnaire results that have been tested for validity and reliability. The K-Means Cluster process is determined by the centroid value.

$$Vij = \frac{1}{Ni} \sum_{k=0}^{Ni} Xkj \qquad (1)$$

where Vij is Cluster average centroid, Ni is number of Cluster member data, k is Index of data, j is Index of variables.

The next step is to calculate the distance between the center points of the Cluster (centroid) which will form a group of objects.

2.4 K-Means Optimization with Genetic Algorithm

K-Means can cluster complex data, but the resulting data can be optimized with the help of Genetic Algorithm. K-Means is considered to be very dependent on the centroid value so the results obtained in the form of a cluster center are not optimal [15]. The utilization of GA is to map the problem into a chain of chromosomes, this can cause differences in the results of the Silhouette Coefficient value. After obtaining the Optimal Cluster Plot, the results of the selected kansei words for the concept were discussed through expert panelists.

3 Result and Discussions

3.1 Packaging Development Ideas

The identification of this research was carried out by conducting survey observations to consumers. It was obtained that 97% of 67 respondents stated the need to develop Bandung toast packaging.

3.2 Packaging Samples

From the results of the Observation Survey to 67 Respondents, 97% stated the need for packaging development for Bandung Toast. 75 packaging samples for toast were obtained which were then screened independently and with the help of experts. Thus, it can be seen in Table 1 that 33 packaging samples were obtained for Bandung Toast products.

Table 1. Packaging samples.

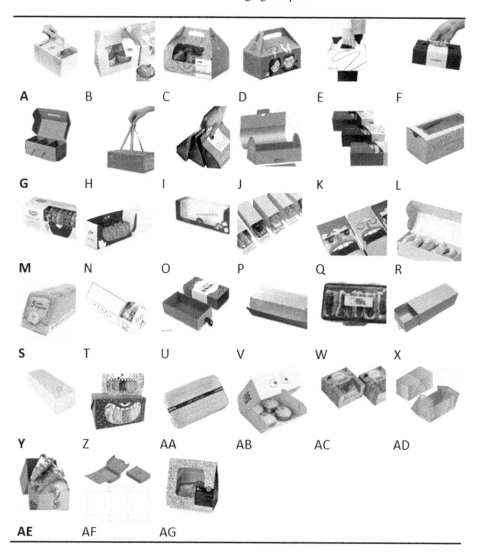

3.3 Kansei Words

The results of the representation of respondents' wishes for the toast packaging to be developed are called Kansei words [10]. KW can be used as a representation of consumers' emotional responses [16]. From the results of the packaging samples in Table 1, the Kansei Word collection process continued through a survey to 67 respondents and obtained 488 Kansei Words. The Kansei words collected generally vary in number between 50–600 words [10]. Next, Kansei Word is selected based on the similarity of word meaning, so that 40 pairs of Kansei Word and their antonyms are obtained as shown in Table 2 below:

Table 2. Kansei Words.

No.	Kansei Word	Antonym	No.	Kansei Word	Antonym
1	Paper material	Not paper material	21	Color Reflects Toast	Color does not Reflect Toast
2	Sturdy	Not sturdy	22	Efficient	Inefficient
3	Informative Design	Uninformative design	23	Handle feature	Non handle feature
4	Depicts toast	Not depicts Toast	24	Window feature	Non window feature
5	Practical	Impractical	25	Simple	Complex
6	Eco friendly	Not environmentally friendly	26	Traditional	Not traditional
7	Eating utensils are available	eating utensils not available	27	Aesthetics	Not Aesthetics
8	Slide form	Not slide form	28	Heat resistance	Not heat resistant
9	Oil resistant	Not Oil resistant	29	Ergonomic	Unergonomic
10	With cover	Without cover	30	Coating	Non coating
11	Tight lock	Loose lock	31	Air Circulation Features	Not Air Circulation Features
12	With Flavor variants	Without flavor variants	32	Unique Packaging Design	Not Unique Packaging Design
13	Food Grade	Non-Food Grade	33	Full Design	Non full Design
14	Depicts a visual of toast	Not Depicts a visual of toast	34	Typical	Non typical
15	Strong	Not strong	35	Attractive	Unattractive

(continued)

Table 2. (*continued*)

No.	Kansei Word	Antonym	No.	Kansei Word	Antonym
16	Hygienic	Unhygienic	36	Qualified	Unqualified
17	Easy to open and close	Not easy to open and close	37	Effective	Ineffective
18	Box shaped	Not box shaped	38	Flexible	Non flexible
19	Able to protect toast	Unable to protect toast	39	Functional	Non functional
20	Size fits toast	Size not fit toast	40	Layer	Non Layer

An evaluation of 40 pairs of Kansei words was conducted using a semantic differential questionnaire by approaching the extent to which the 33 samples in Table 1 approached the Kansei word. The questionnaire used 7 scales for 42 purposive sampling respondents.

3.4 Design Concept

Analysis of packaging design concepts using K-Means optimization Genetic Algorithm processed using R software. Input data for the data running process is in the form of numeric scale results obtained through questionnaires when determining the suitability of samples obtained with Kansei Word in Excel format. Through the input data, this method will cluster words according to the performance level of the method that can support the aesthetics of packaging design. From this process, the concept category for packaging design will be obtained. To obtain more accurate results, the running process is carried out twice for comparison between the running coding of the K-Means method and the Genetic Algorithm optimization K-Means method. Figure 2 shows the cluster results of the K-Means. The Kansei words obtained by each cluster are described in Table 1. In the two K-Means Clusters, there are 38 Kansei words in Cluster 1 and 2 Kansei words in Cluster 2 (Table 3).

In Fig. 2, shows the results of two groups of Kansei words based on the results of running R software. The Kansei words obtained by each cluster are described in Table 1. In the two K-Means Clusters, there are 38 Kansei words in Cluster 1 and 2 Kansei words in Cluster 2 (Table 4).

The cluster results with K-Means will then be compared with the cluster results using the K-Means Genetic Alghoritma to prove how optimal the cluster results are with the K-Means Genetic Alghoritma. Figure 3 shows the clusters formed with the K-Means Genetic Algorithm which also produces two clusters. However, the cluster results have a wider distribution of points than the centroid position due to the exploration ability of the GA implementation and the sensitivity of K-Means. Another difference is the Silhouette Coefficient value which can determine the level of excellence. The Silhouette coefficient value is part of the matrix value that can evaluate clustering performance. Following are the Silhouette Coefficient values for both methods in Fig. 4.

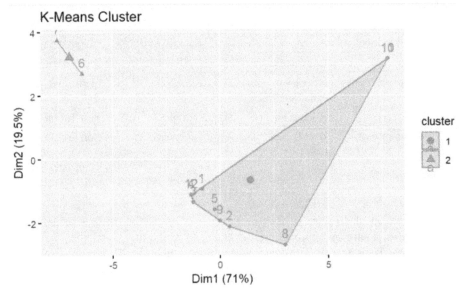

Fig. 2. Cluster by K-Means.

Table 3. Kansei words of k-means results.

Cluster 1					Cluster 2	
1	Paper Material	22	Efficient		6	Eco Friendly
2	Sturdy	23	Handle Features		7	Eating utensils available
3	Informative Design	24	Window Features			
4	Depict Toast	25	Simple			
5	Practical	26	Traditional			
8	Slide form	27	Aesthetic			
9	Oil resistant	28	Heat resistant			
10	With cover	29	Ergonomic			
11	Tight Lock	30	With Coating			
12	Flavour Variant	31	Air circulation features			
13	Food Grade	32	Unique Packaging Design			
14	Depicts a visual of toast	33	Full Design			

(*continued*)

Table 3. (*continued*)

Cluster 1				Cluster 2	
15	Strong	34	Typical		
16	Hygienic	35	Attractive		
17	Easy to open and closed	36	Qualified		
18	Box shaped	37	Effective		
19	Able to protect toast	38	Flexible		
20	Size fits toast	39	Functional		
21	Color reflects toast	40	Layer		

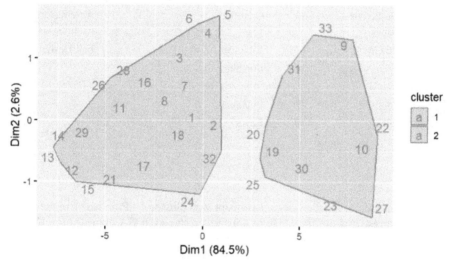

Fig. 3. Cluster by K-Means GA Optimization.

The K-Means Cluster results which are considered better have a higher Silhouette Coefficient value [17]. In this study, the cluster chosen for the Bandung toast design concept was the GA optimization K-Means Cluster with a Silhouette Coefficient value of 0.4817299, while the K-Means Cluster had a value of 0.4132317. Conclusions on the GA K-Means Cluster optimization results were obtained based on discussions with 5 expert panelists. The expert panelists in question are expert panelists in the field of graphic design who have at least 5 to 10 years of experience in that field. The panelists used were lecturers who taught in the graphic design study program. Based on expert panelist input, Cluster 1 was concluded as a "Practical-Unique" concept, and Cluster 2 was concluded as a "Safety-Aesthetic" concept. The concept obtained based on expert

Table 4. Kansei words of GA optimization k-means results.

Cluster 1					Cluster 2	
1	Paper material	18	Box Shaped	9	Oil resistant
2	Sturdy	21	Color reflects toast	10	With cover
3	Informative design	24	Window Features	19	Able to protect toast
4	Depicts toast	26	Traditional	20	Size fits toast
5	Practical	28	Heat resistant	22	Efficient
6	Eco friendly	29	Ergonomic	23	Handle Feature
7	Eating utensils available	32	Unique packaging design	25	Simple
8	Slide Form	34	Typical	27	Aesthetic
11	Tight lock	35	Attractive	30	With coating
12	With flavour variants	36	Qualified	31	Air circulation features
13	Food Grade	37	Effective	33	Full Design
14	Depicts Visual toast	38	Flexible		
15	Strong	39	Functional		
16	Hygienic	40	Layer		
17	Easy to open and close				

panelists concluded that the selected concept was cluster 1 "Practical-Unique", based on the widest results from the range of clusters produced in GA optimization K-Means data processing.

```
> # Visualize the K-Means clustering results using 'fviz_cluster'
> fviz_cluster(final, data = data, main = "K-Means Cluster")
>
> # Add the cluster assignment as a new column to the data frame
> data <- data.frame(data, cluster = final$cluster)
>
> # Calculate the silhouette coefficient
> silhouette_coef <- silhouette(final$cluster, dist(data))
> cat("Silhouette Coefficient:", mean(silhouette_coef[,
+                                             "sil_width"]), "\n")
Silhouette Coefficient: 0.4132317
>
```

(a)

```
Optimal number of clusters (K) suggested by GA: 2
>
> # Calculate and print the silhouette coefficient
> silhouette_coef <- silhouette(kmeans_result$cluster, dist(data))
> cat("Silhouette Coefficient:", mean(silhouette_coef[,
+                                             "sil_width"]), "\n")
Silhouette Coefficient: 0.4817299
>
> # Step 6: Visualize the clustering results using fviz_cluster
> library(factoextra)
>
> # Extract only the numeric columns for visualization
> numeric_data <- data[, sapply(data, is.numeric)]
```

(b)

Fig. 4. Comparison of Silhouette Coefficient: (a) Silhouette Coefficient K-Means, (b) Silhouette Coefficient K-Means GA Optimation

4 Conclusions

Based on the results of data analysis and processing in the development of Bandung toast packaging design, it is concluded that the application of Kansei Engineering is successful in translating consumer preferences into packaging design concepts. In this research, the selected K-Mean method with GA optimization is used which has a better Silhouette Coefficient value, which is 0.4817299 compared to the Silhouette Coefficient value of the K-Mean method without optimization, which is 0.4132317. Concept translation based on the Kansei word in the selected cluster was obtained with the help of experts. The selected concept in this development is "Practical-Unique" which is obtained based on the widest range in the cluster results of the K-Means GA method.

References

1. Ribeiro, A.E.C., Oliveira, A.R., da Silva, A.C.M., Caliari, M., Júnior, M.S.S.: Physicochemical quality and sensory acceptance of toasts with partial replacement of wheat flour by maize biomass flour. J. Food Sci. Technol. **57**(10), 3843–3851 (2020). https://doi.org/10.1007/s13197-020-04416-3
2. Bidin, S.A.H., Lokman, A.M.: The emotional keywords are used as the measure of strength of the emotional responses. In: KEER 2018. KEER (2018)
3. Sari, N.P., Zulkarnain, Z., Muzaki, V.A., Meilani, Y.D.: Implementasi kansei engineering dalam pengembangan kemasan minuman kopi ready to drink. ARGOINTEK **18**(1), 200–209 (2024). https://doi.org/10.21107/agrointek.v18i1.12443

4. Aprilia, I.R., Sari, N.P., Faizi, I., Wati, R.: Penerapan Metode PCA dalam Penentuan Konsep Desain Kemasan Sekunder untuk Produk X. Performa: Media Ilmiah Teknik Industri **22**(2), 136 (2023). https://doi.org/10.20961/performa.22.2.80739
5. Puspasari, R.A., Amelia, C., Pangestu, B.P., Sari, N.P.: Perancangan Konsep Desain Kemasan Makanan Kucing Menggunakan Metode Kansei Engineering. Performa: Media Ilmiah Teknik Industri **22**(2), 143 (2023). https://doi.org/10.20961/performa.22.2.80744
6. Sari, N.P., Rizwan, R., Hafidah, E., Andriyani, S.Z.P.: Perancangan Desain Kemasan Bakso Goreng (Basreng) dengan Metode Kansei Engineering. Performa: Media Ilmiah Teknik Industri **22**(2), 109 (2023). https://doi.org/10.20961/performa.22.2.80674
7. Putri, K.A.A., Segita, N.E., Nuryadin, R.N., Nur, Y.L., Sari, N.P.: Perencanaan Konsep Desain Kemasan Kerak Telor Menggunakan Metode Kansei Engineering (2024)
8. Ezar, M., Rivan, A., Sonaru, R.A., Kunci-Algoritma, K.: Perbandingan Metode K-Means Dan GA K-Means Untuk Clustering Dataset Heart Disease Patients hasil intra cluster dari GA K-Means lebih baik dibandingkan dengan K-Means dan untuk inter cluster sangat kecil perbedaannya, dimana rata-rata inter cluster metode K-Means sedikit lebih baik daripada GA K-Means (2022). http://jurnal.mdp.ac.id
9. Agarwal, N., Sikka, G., Awasthi, L.K.: WGSDMM+GA: a genetic algorithm-based service clustering methodology assimilating dirichlet multinomial mixture model with word embedding. Futur. Gener. Comput. Syst. **145**, 254–266 (2023)
10. Nagamachi, M., Lokman, A.M.: Kansei Innovation: Practical Design Applications for Product and Service Development. CRC Press (2015)
11. Isna, Sari, N.P., Maharani, D., Fadhillah, F.: Implementasi Kansei Engineering dalam Menentukan Konsep Pengembangan Kemasan Rujak Buah Potong. Jurnal INTECH Teknik Industri Universitas Serang Raya **10**(1), 9–18 (2024). https://doi.org/10.30656/intech.v10i1.7832
12. Andrade, C.: The inconvenient truth about convenience and purposive samples. Indian J. Psychol. Med. **43**(1), 86–88 (2021). https://doi.org/10.1177/0253717620977000
13. Sari, P.: Perencanaan Dan Pengembangan Kemasan: Kansei Engineering. PNJ Press (2018)
14. Putri, A.A., Segita, N.E., Nuryadin, R.N., Nur, Y.L., Sari, N.P.: Perencanaan Konsep Desain Kemasan Kerak Telor Menggunakan Metode Kansei Engineering. Industri Inovatif (2024)
15. Taslim, F.: Penerapan algoritma k-mean untuk clustering data obat pada puskesmas rumbai (2016)
16. Mohd Lokman, A., Harun, A.F., Md Noor, N.L., Nagamachi, M.: Website Affective Evaluation: Analysis of Differences in Evaluations Result by Data Population (2009)
17. Ananda, Yamani, A.Z.: Penentuan Centroid Awal K-means pada proses Clustering Data Evaluasi Pengajaran Dosen. Masa Berlaku Mulai **4**(3), 544–550 (2020)

Design Guide for Emotional Evocative Student's Leadership Program

Zaiha Ahmad[1], Zuraeda Ibrahim[2], Nordiana Ibrahim[3], Nur Nafishah Azmi[4], Mohd Sazili Shahibi[5], and Anitawati Mohd Lokman[5(✉)]

[1] Faculty of Communication and Media Studies, Universiti Teknologi MARA, Shah Alam, Malaysia
`zaiha964@uitm.edu.my`

[2] Faculty of Accountancy, Universiti Teknologi MARA Selangor, Puncak Alam, Malaysia
`zurae229@uitm.edu.my`

[3] Faculty of Plantation and Agrotechnology, Universiti Teknologi MARA Melaka, Jasin, Malaysia
`nordiana@uitm.edu.my`

[4] College of Creative Arts, Universiti Teknologi MARA Selangor, Puncak Perdana, Malaysia
`nafishah2610@uitm.edu.my`

[5] College of Computing, Informatic, and Mathematic, Universiti Teknologi MARA Selangor, Puncak Perdana, Malaysia
`{mohdsazili,anitawati}@uitm.edu.my`

Abstract. This paper presents an in-depth study on the emotional experiences induced by student leadership programs. Utilizing the Kansei Engineering approach, the research examines three leadership programs, dissecting their design components and corresponding attributes. A comprehensive evaluation process involving 150 students and a Kansei checklist of 28 words was used to synthesize students' emotions. The collected data was then analyzed using Factor Analysis (FA), Principal Component Analysis (PCA), and Partial Least Squares (PLS) to identify the attributes from the program components that significantly influence emotional experiences. The results reveal that the emotional structure towards student leadership programs is predominantly characterized by two factors: Excitement and Confidence. Further, the PLS analysis discerns the program components and attributes that significantly influence student emotions. Importantly, these findings have led to the creation of a framework for Emotionally Evocative Leadership Program Design. This framework can become invaluable reference for educators, program designers, and leadership coaches, providing clear direction on how to incorporate significant emotional elements into program design to enhance student experience and engagement effectively. By prioritizing elements that foster a sense of excitement and confidence, the framework helps to create more impactful and emotionally resonant leadership programs. Its usability lies in its systematic approach, making it easy to apply across various leadership program contexts. Ultimately, this work underscores the significant role that emotions play in educational settings, contributing to a more holistic approach to developing future leaders.

Keywords: Program Design · Emotion · Kansei Engineering · Student's Leadership program

1 Introduction

Higher Education Institutions (HEIs) are pivotal in shaping the leaders of tomorrow in the contemporary educational landscape. This responsibility extends far beyond promoting academic and technical excellence. It also involves nurturing emotionally intelligent leaders, a task that is increasingly being recognized as crucial in today's rapidly changing world (McKay et al., 2024). This commitment aligns seamlessly with the broader educational objectives outlined in the Malaysia Education Blueprint 2015–2025 and the Ministry of Higher Education's Eight Soft Skills Characteristics (Ministry of Higher Education, 2021).

In the traditional educational paradigm, the focus was primarily on intellectual or rational learning, with emotions often relegated to the background (McKay et al., 2024). This perspective saw emotions as secondary to the cognitive processes involved in learning. However, more recent research and understanding have shown that emotions hold a central position in the learning process, significantly influencing learning outcomes (Kremer et al., 2019, Shafait et al., 2021). Despite this understanding, there is a notable gap in the literature and practice. Emotional intelligence frequently gets overlooked in discussions surrounding successful student leadership development programs. This is concerning, given that emotional intelligence is a critical factor for effective leadership and contributes significantly to an individual's ability to succeed in various aspects of life (Ohiku, 2021).

This paper seeks to address this research gap by exploring the complex interplay between emotional responses and program design parameters. The study employs the Kansei Engineering approach, a methodology that focuses on the emotional impact of product design, and applies it to the field of educational program design. The aim of this research is to comprehend how program design can effectively utilize emotions to enhance student outcomes. It seeks to understand how the integration of emotional considerations into the design of leadership programs can contribute to their effectiveness and impact. In doing so, it hopes to provide valuable insights that can inform the development of student leadership programs.

The study recognizes the importance of fostering emotional intelligence in students, given its impact on leadership abilities and overall success (Ohiku, 2021). It seeks to contribute to the ongoing discourse on effective leadership development, with a specific focus on the role of emotional intelligence in this process. The insights gained from this exploration will inform the development of student leadership programs, thus contributing to the formation of future leaders. It emphasizes the need for a more holistic approach to education, one that recognizes the importance of emotional intelligence alongside academic and technical skills. It is hoped that this research will contribute to the ongoing efforts to enhance the effectiveness of leadership development programs and, ultimately, shape emotionally intelligent leaders who are well-equipped to navigate the challenges of the future.

2 Theoretical Background

This section reviews relevant past publications related to the research. The review establishes a theoretical basis for the research endeavor, aiming to uncover the design of a student leadership program and the emotional experiences students feel throughout the program.

2.1 Student's Leadership Program in Higher Education

The concept of leadership in Higher Learning Institutions (HEIs) encompasses leading by example, influencing others, and fostering ethical relationships. Northouse (2010) defines leadership as the capacity to influence a specific group towards a common goal. Crawford and Kelder (2019) emphasize that leaders are subject to tighter parameters for accountability and higher expectations of performance and outcomes. Leadership in HEIs is pivotal for shaping the direction of educational institutions. Effective leaders are knowledgeable, dedicated, and committed, playing crucial roles in strengthening planning and management within these institutions. Quality leadership garners support from followers and navigates challenges adeptly. The comprehensive engagement of leaders with diverse stakeholders is essential for developing future leaders with well-rounded identities—intellectually, spiritually, emotionally, and physically (Malaysia National Education Philosophy). This holistic approach ensures that leaders can consistently make positive impacts on their institutions, society, and the nation as a whole.

The understanding of leadership has evolved over the past decades, transitioning from a focus on individual leaders to a more process-oriented approach. This contemporary notion of leadership development emphasizes the interaction between the leader and the environment (Fiedler, 2006). Iles and Preece (2006) note that this shift has led to the emergence of relational models of leadership, which prioritize the construction of social capital over individual heroism. Kezar et al. (2006) Observe that new theories in general leadership literature are increasingly applied in higher education environments. This evolving perspective allows for a more dynamic and inclusive understanding of leadership that is well-suited to the diverse and complex nature of HEIs.

2.2 Emotion and Student's Leadership Program

Emotions play a critical role in student leadership development programs at HEIs. Well-designed programs promote collaboration, self-efficacy, character development, and personal growth. Pascarella and Terenzini (2005) suggest that engaging in leadership-related activities enhances students' skills and knowledge, boosting self-efficacy and civic engagement. Emotional skills are crucial for enhancing leadership capabilities, as positive emotions improve cognitive strategies and overall well-being (Li et al., 2020). Leadership programs that incorporate emotional intelligence training are more effective in preparing students for real-world leadership challenges.

The emotional experiences of students in leadership programs have significant implications for their personal and professional growth. Positive emotions contribute to deeper learning, critical thinking, and problem-solving abilities. Conversely, addressing negative emotions is vital for developing resilience and emotional intelligence (Gruicic and

Benton, 2015). Silva and Almeida (2023) advocate for integrating social and emotional skills into training programs to enhance performance and well-being. By fostering emotional intelligence, students can better navigate interpersonal relationships and leadership challenges.

Understanding and addressing emotional elements in leadership programs are essential for ensuring holistic student development. Yip and Côté (2013) highlight the importance of emotions in the knowledge acquisition process, suggesting that emotional engagement enhances learning outcomes. Crossman (2007) emphasizes the need for teachers to balance objectivity with empathy, considering students' emotional responses during assessments. These insights underscore the need for HEIs to integrate emotional skills training into their leadership programs to develop well-rounded leaders. Additionally, a supportive and emotionally engaging learning environment can motivate students to actively participate and apply their knowledge and skills effectively (Redzuan et al., 2011a).

2.3 Factors Influencing Student's Emotion in Leadership Programs

Emotional experiences significantly influence how students perceive their environment, interact with others, and shape their behavior. Emotions directed towards goals, values, and interactions affect students' ability to achieve personal and organizational objectives (McEnrue et al., 2009). Recognizing the intricate nature of emotions, it is crucial to consider them when developing leadership programs to enhance learning, teaching, and student outcomes (Schutz et al., 2006). Emotions such as enthusiasm, pride, and joy can enhance student engagement and motivation, leading to better learning experiences and outcomes.

Positive emotions during leadership development enhance cognitive strategies, well-being, and the development of social and emotional skills. Students experiencing positive emotions are more likely to engage in deep learning, critical thinking, and effective application of leadership concepts (Harvey et al., 2019). Creating a supportive and emotionally engaging learning environment motivates students to actively participate and apply their knowledge and skills effectively (Redzuan et al., 2011a). Moreover, positive emotions can foster creativity and innovation, which are essential for effective leadership.

Addressing negative emotions is equally important for student development. Leadership programs can be challenging, and students may experience self-doubt, anxiety, or frustration. Providing a safe space for expressing and processing emotions helps develop resilience and emotional intelligence (Tekerek and Tekerek, 2017). Furthermore, fostering an environment where students feel safe to take risks and make mistakes can lead to significant growth and learning. By acknowledging and addressing these emotional challenges, leadership programs can help students build the necessary skills to navigate difficult situations effectively.

3 Methodology

This section explains the methods and processes used to examine students' emotional preferences in the context of leadership programs. The study employs the Kansei Engineering method and utilizes its analysis for multivariate statistical analysis. The research is structured into three phases, aiming to offer an exhaustive exploration of students' emotional reactions related to the leadership program. The phases, which includes Instrumentation, Kansei Evaluation, and Kansei Analysis, is depicted in Fig. 1 and further detailed in the subsequent subsections.

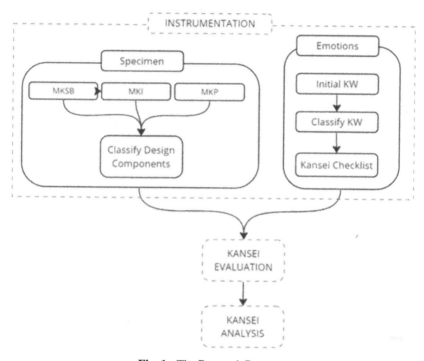

Fig. 1. The Research Process.

3.1 Instrumentation

Specimen - When practicing KE, it is important to choose specimens from the field domain that effectively capture the many features and characteristics pertinent to the intended design goals (Abdi and Greenacre, 2020; Lokman et al., 2009; Ismail et al., 2016). The scope of this research is specifically directed towards the Student's Leadership Program, with the objective of investigating student's emotion within the program design context. Henceforth, this study was provided with three programs, namely Survival Management Module (MKSB), Professional Management Module (MKP), and Institutional Management Module (MKI). The specimen design components were then classified based on each specimen and formed a matrix for specimen vs design components, to be used in the analysis process.

Emotion The identification of relevant Kansei Words (KWs) for use in the research context is another important aspect of KE investigations. The initial KWs in this study were developed by a team of the program experts based on KJ Method. The experts involving 28 participants, including facilitators and students from leadership programs, provided the range of emotional experiences encountered. The KJ Method was used to classify these emotions into similar groups. This categorization provided a nuanced understanding of the diverse range of emotions from the leadership programs. The process allowed participants to express their emotions and reflect on their experiences, providing valuable insights into the overall emotional landscape of the programs. The subsequent step involves converting the KWs into the semantic differential (SD) scale structure to form a Kansei checklist. This study utilizes all sixty KWs obtained from the previous step, using a five-level scale pattern that spans from 1 to 5. Additionally, this scale employs a negatively phrased statement accompanied by a "Not" prefix; with the presence of the prefix indicating the lowest score (scale 1) and the absence indicating the greatest score (scale 5). The SD Scale was then transformed into a Google Form to facilitate the process of gathering data online.

3.2 Kansei Evaluation

150 undergraduate students were selected to be involved as test subjects. The instruments used for the evaluation consisted of three specimens and 28 KWs in SD scale form. The Kansei checklist was used to evaluate emotional experiences of students participating in a leadership program. This approach is useful in gaining insights into the emotional experiences of the target audience, which can inform the future development of programs that better meet their needs and preferences. During the evaluation process, the students were asked to recall their experience with the programs, module by module. The resulting data was then processed, and the average was calculated. This dataset was the input for the following phase.

3.3 Kansei Analysis

After the experimental data was collected, multivariate statistical analysis was performed on the averaged data. This analysis was a crucial part of the research process as it helped to uncover patterns in the data and identify relationships between variables that might not be immediately apparent. The analysis was carried out using a statistical technique known as Factor Analysis (FA) AND Partial Least Squares (PLS) analysis. FA was used to identify the significant emotional factors relevant to the student leadership program. To further understand the data, PLS analysis was conducted in this research process to analyze the relationships between the identified emotional factors and the various components of the program design. This analytical process provided valuable insights into how different aspects of the program design could influence the emotional responses of the students. The knowledge derived from this process could then be utilized to optimize the program design to evoke desirable emotional responses, thereby enhancing the effectiveness and success of the student leadership program.

4 Results and Discussion

This section provides the analysis results and discussion of the findings. The details are further elaborated in the subsequent sub-sections.

4.1 Classification of Program Design Components

The study uncovered a total of 15 program components, which were then classified into 41 design attributes. Table 1 shows the excerpt of the design component matrix.

Table 1. Excerpt from the Program Components.

Module	Patriotic Themes	Icon Themes	Community Themes	Indoor Activity-Bench Marking	Indoor Activity-Team Building	Affective Skill (Internal)-Self - Confidence	Affective Skill (Internal) - Creativity	Affective Skill (Internal) – Trust
MKSB	✓				✓			✓
MKI		✓		✓		✓		
MKP			✓		✓		✓	

The table is a comparison of different modules, namely MKSB, MKI, and MKP, across various parameters. These parameters include Patriotic Themes, Icon Themes, Community Themes, Indoor Activity-Bench Marking, Indoor Activity-Team Building, Affective Skill (Internal)-Self-Confidence, Affective Skill (Internal)-Creativity, and Affective Skill (Internal)-Trust. A tick under a specific module and parameter represents the presence of that particular parameter in the module. For instance, if there's a tick under the MKSB module and the Patriotic Themes parameter, it signifies that the MKSB module includes elements or activities related to Patriotic Themes. Similarly, a tick under the MKI module and the Indoor Activity-Team Building parameter denotes that team-building activities conducted indoors are a part of the MKI module.

This result will be utilized in the Partial Least Squares (PLS) analysis process as an input matrix to identify the interrelationships among the emotions and design components. The presence or absence of various parameters in different modules can significantly influence the results of the PLS analysis. Understanding these relationships aids in design guide development and can help identify which components are key to enhancing the emotional experience of each module.

4.2 Significant Emotion

The resulting cluster of emotions was used to develop an emotion evaluation checklist, named IKP Leadership Emotion Checklist. This checklist is used in tandem with Kansei Engineering methodology to identify significant emotional experiences of a targeted group of consumers, which in this case are students participating in a leadership program. This approach is useful in gaining insights into the emotional experiences of the target audience, which can inform the development of programs that better meet their needs and preferences.

The developed checklist was distributed to 150 students who attended leadership programs. During the evaluation process, the students were asked to recall their experience with the programs, module by module. The resulting data was then processed, and the average evaluation result was calculated. It is noteworthy that the checklist had a significant impact on the students' perception of the programs, as they were able to provide detailed and thoughtful feedback on each module. This feedback was instrumental in improving the programs and ensuring that future students have a more emotion enriching experience.

Table 2 presents the results of a Factor Analysis (FA) and lists the significant emotional factors identified from the study. The emotional experiences of students are categorized into two distinct factors: Excitement and Confidence. Each factor is accompanied by a set of variables expressing the specific emotional states associated with it. For instance, the Excitement factor includes emotions like feeling "tested," "exciting," "overwhelming," "shy," "happy," "motivating," "joyful," and "energetic." On the other hand, Confidence involves emotions such as feeling "confident," "challenging," "determined," "optimistic," "pleasant," and "sympathetic." The numbers next to the variables represent the factor loadings, which indicate the correlation between the observed variables and the factor. A high factor loading (close to 1) means that the variable is significantly associated with the factor. The analysis of these factors and their associated emotions provides valuable insights into the emotional dynamics within the leadership program. These findings can guide program designers in creating activities that foster these emotional states, thereby enhancing the overall effectiveness of the leadership program.

Table 2. Emotional Factors

Factor	Variables
Excitement	Tested (0.99), Exciting (0.99), Overwhelming (0.99), Shy (0.99), Happy (0.98), Motivating (0.98), Joyful (0.96), Energetic (0.96), Grateful (0.94), Stressful (0.92), Spirited (0.91), Tired (0.86), Confused (0.83), Sympathetic (0.82), Sad (0.78)
Confidence	Confident (0.99), Challenging (0.99), Determined (0.99), Optimistic (0.98), Pleasant (0.96), Sympathetic (0.96), Spirited (0.88), Grateful (0.84), Energetic (0.82), Joyful (0.81), Motivating (0.76), Exciting (0.66), Tested (0.64), Overwhelming (0.54), Shy (0.51)

4.3 Emotional Evocative Program Design

The study utilized Partial Least Square (PLS) analysis to explore the relationship between program components (x) and emotion (y), with a focus on identifying the influence of each program component on specific emotional responses, as well as the best and worst fit for each program component and the impact of each sample on the elicited emotional responses. PLS analysis was chosen for its ability to handle a large number of x variables and tens of y variables.

In the earlier phase of the research, 15 program components and 41 attributes were identified and converted into dummy variables for PLS analysis. The PLS coefficient score was calculated and analyzed to explore the relationship between program components and emotion. This allowed the researchers to identify the impact of program component combinations on eliciting specific emotional responses.

Table 3 shows an excerpt for the coefficient score calculated by PLS analysis. The research analyzed the result of PLS coefficient score to discover relations between emotion and program components. The following sub-sections describe how the use of these scores enables the identification of how the combinations of program components influence emotional responses.

The use of PLS analysis to analyze the relationship between program components and emotion can provide insights into the combinations of program components that influence specific emotional responses. This methodology can be applied to a variety of product design and development contexts to ensure that products meet the needs and desires of consumers.

The study calculated the PLS range for each emotion to determine the influence of program components on emotion. By calculating the range, the researchers were able to identify the design influence, both positive and negative, of each program component. The range was calculated by subtracting the minimum PLS score from the maximum PLS score. The mean of the range was then calculated for each emotion. Design influence is identified based on the range score. When the mean PLS score for program component is greater than the average range, the item is considered to have a positive influence on the design. A attribute range for each program component that is greater than the average attribute implies the best fit attribute, which heavily influences user emotion in the leadership program. Table 3 shows excerpt of the result.

Table 3. Excerpt on PLS Score for Program Components.

Program componets	PLS score	Range
Patriotic Themes	0.0451	0.0839
Icon Themes	−0.0388	
Community Themes	−0.0063	
Outdoor Activity-Explorace	0.0451	0.0839
Outdoor Activity-Build Tower	−0.0388	
Outdoor Activity-Field Study	−0.0063	
Indoor Activity-Bench Marking	−0.0388	0.0776
Indoor Activity-Team Building	0.0388	
Time-Work Days	−0.0063	0.0126
Time-Weekends	0.0063	

The table compares different program components based on their respective Partial Least Squares (PLS) scores and ranges. Each row signifies a different program component. There are three columns: one for the program component's name, one for its PLS score, and one for its range. The PLS score signifies the weight or significance of that program component within the overall program, whereas the range offers an insight into the variability or difference in PLS scores for that specific component. The calculated average range is 0.07789. A program component's range that exceeds this average is deemed significant. For instance, the program component "Patriotic Themes" has a PLS score of 0.0451 and a range of 0.0839. This suggests that "Patriotic Themes" holds a moderate significance in the program, with some variability in this significance.

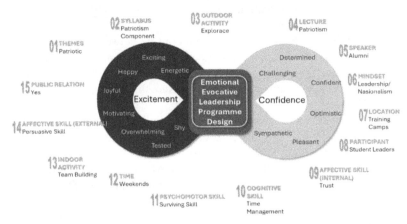

Fig. 2. Emotional Evocative Leadership Program Design Framework.

Figure 2 demonstrate a diagram for Emotional Evocative Leadership Program Design Framework, which illustrates the interplay between various components of a leadership program designed to evoke and manage emotions, specifically excitement and confidence, to enhance emotional experience in leadership programs. The framework presents a comprehensive approach to designing leadership programs by focusing on emotional experiences, particularly excitement and confidence, that students express. It illustrates how various design components significantly influence these emotional experiences. Each emotional concept is associated with a set of emotional keywords that represent the spectrum of feelings students experience. Excitement includes emotions such as happy, joyful, motivating, energetic, exciting, overwhelming, shy, and tested. Confidence encompasses emotions such as determined, confident, optimistic, pleasant, challenging, sympathetic, and confident. These keywords reflect the major variances in the emotional experience students have during leadership programs. They signify the positive emotional states that are crucial for engaging students and enhancing their learning outcomes.

The outer layer of the diagram outlines 15 design components, ranked from highest to lowest influence on students' emotional experiences. Each component plays a significant role in shaping how students feel and respond to the leadership program, with the most influential attribute indicated for each component. The following describes the design components and attributes in descending order of influence:

1) Themes: Patriotic Themes. Patriotic themes inspire a sense of pride, unity, and motivation among participants, fostering a deep emotional connection to the program's objectives.
2) Syllabus: Syllabus with Patriotism Component. Integrating a patriotism component into the syllabus ensures that the program is well-organized and comprehensive, providing clear guidance and enhancing participants' sense of purpose and belonging.
3) Outdoor Activity: Outdoor Activity - Explorace. Explorace activities engage students physically and mentally, promoting teamwork, adventure, and a sense of achievement, which boosts excitement and confidence.
4) Lecture: Lecture on Patriotism. Lectures focused on patriotism provide foundational knowledge and inspire intellectual curiosity and pride in national identity, fostering emotional engagement.
5) Speaker: Speaker Among Alumni. Alumni speakers can share relatable experiences and success stories, motivating current students by showing tangible examples of leadership and achievement.
6) Mindset: Mindset on Leadership/Nationalism. Cultivating a mindset focused on leadership and nationalism encourages students to embrace responsibilities and aspire to contribute to their community and country.
7) Location: Training Camps. Training camps provide a dedicated, immersive environment that enhances learning, bonding, and focus, creating a supportive backdrop for activities and emotional growth.
8) Participant Criteria: Student Leaders. Selecting student leaders as participants ensures that those with potential and experience can further develop their skills, creating a dynamic and inspiring peer group.
9) Affective Skill (Internal): Trust. Building trust among participants fosters a safe and supportive environment, essential for open communication, collaboration, and emotional growth.
10) Cognitive Skill: Time Management. Emphasizing time management skills helps students balance various tasks efficiently, reducing stress and increasing confidence in their ability to manage responsibilities.
11) Psychomotor Skill: Surviving Skill. Practical exercises in survival skills provide hands-on experience that enhances students' confidence in their ability to face and overcome challenges.
12) Indoor Activity: Team Building. Indoor team-building activities foster collaboration, trust, and a sense of belonging, which are crucial for creating a supportive and engaging learning environment.
13) Affective Skill (External): Persuasive Skill. Training in persuasive skills enhances participants' ability to influence and motivate others, which is vital for effective leadership and communication.
14) Public Relation: Have Public Relation Element. Incorporating public relations elements helps students develop strong communication skills and manage their public image, which is essential for leadership and public engagement.

15) Time: Program During Weekends. Scheduling programs during weekends ensures that participants can fully engage without conflicting with their weekday commitments, maintaining high energy and focus.

This framework serves as a guide for designing leadership programs that evoke positive emotional experiences and foster the development of effective leaders. Each design component is carefully chosen to maximize its impact on students' excitement and confidence, ensuring a comprehensive and engaging program.

5 Conclusion

This study sheds light on the crucial role of emotional experiences in leadership development programs. Key emotional factors identified are 'Excitement' and 'Confidence', which aligns with previous research emphasizing the significance of emotional intelligence in enhancing learning outcomes and leadership capabilities (McKay et al., 2024; Ohiku, 2021).

The findings reinforce the necessity for a holistic educational approach that incorporates emotional skills with academic and technical abilities (Kremer et al., 2019; Shafait et al., 2021; Taharim et al., 2016). The research contributes to the discourse on effective leadership development by offering a design guide for Emotionally Evocative Leadership Program Design, serving as a practical tool to create supportive and emotionally engaging learning environments (Redzuan et al., 2011a; Redzuan et al., 2011b; Redzuan et al., 2014). These findings are consistent with previous research, underlining the importance of developing emotionally intelligent leaders (Li et al., 2020; Gruicic and Benton, 2015). The emotional factors identified resonate with other studies that highlight the central role of emotions in boosting learning outcomes. This study supplements the growing body of literature emphasizing a holistic educational approach that recognizes the importance of emotional intelligence alongside academic and technical skills. The findings also corroborate studies suggesting that emotional skills are vital for enhancing leadership capabilities. The study supports the integration of social and emotional skills into training programs to enhance performance and well-being (Silva and Almeida, 2023). The results align with assertions on the importance of emotions in the learning process (Yip and Côté, 2013) and recognize the need to balance objectivity with empathy, considering students' emotional responses during assessments (Crossman, 2007).

Building on these findings, the study develops a framework for Emotionally Evocative Leadership Program Design. This framework aligns with the view on creating a supportive and emotionally engaging learning environment for effective learning (Redzuan et al., 2011a). The study contributes to our understanding of the complex interplay between emotional responses and program design parameters, providing a practical framework for designing emotionally evocative leadership programs. The framework offers a novel approach to understanding the intricate relationship between emotional responses and program design parameters. It lays the groundwork for future research to investigate other emotional factors and their impact on program design and student outcomes, contributing to the broader understanding of leadership development in the context of emotional intelligence.

Acknowledgments. Authors acknowledge the Universiti Teknologi MARA (UiTM) for funding under the Geran Penyelidikan MyRA (600-RMC/GPM SS 5/3 (084/2021)) and the Student's Affairs Division of UiTM. The authors would also like to acknowledge Malaysia Association of Kansei Engineering (MAKE), and RIG Kansei and Design Engineering (RIG EKDE) for the support to the research activities.

Disclosure of Interests. The authors have no competing interests to declare.

References

Crawford, J.A., Kelder, J.-A.: Do we measure leadership effectively? Articulating and evaluating scale development psychometrics for best practice. Leadersh. Q. **30**(1), 133–144 (2019). https://doi.org/10.1016/j.leaqua.2018.07.001

Crossman, J.: The role of relationships and emotions in student perceptions of learning and assessment. High. Educ. Res. Dev. **26**(3), 313–327 (2007)

Fiedler, F.E.: The contingency model: a theory of leadership effectiveness. In: Levine, J.M., Moreland, R.L. (eds.) Small Groups, pp. 369–381. Psychology Press (2006)

Gruicic, D., Benton, S.: Development of managers' emotional competencies: mind-body training implication. European J. Train. Develop. **39**(9), 798–814 (2015). https://doi.org/10.1108/EJTD-04-2015-0026

Harvey, M., Baumann, C., Fredericks, V.: A taxonomy of emotion and cognition for student reflection: introducing emo-cog. High. Educ. Res. Dev. **38**(6), 1138–1153 (2019)

Iles, P., Preece, D.: Developing leaders or developing leadership? The Academy of Chief Executives' programmes in the North East of England. Leadership **2**(3), 317–340 (2006). https://doi.org/10.1177/1742715006066

Ismail, A., Kadir, S.A.S.A., Aziz, A., Mokshin, M., Lokman, A.M.: ITourism travel buddy mobile application. In: 2016 10th International Conference on Next Generation Mobile Applications, Security and Technologies (NGMAST), pp. 82–87. IEEE (2016)

Kezar, A.J. (ed.): Rethinking leadership in a complex, multicultural, and global environment: New concepts and models for higher education. Taylor & Francis (2023)

Kremer, T., Mamede, S., Martins, M.A., Tempski, P., van den Broek, W.W.: Investigating the impact of emotions on medical students' learning. Health Professions Education **5**(2), 111–119 (2019)

Li, L., Gow, A.D.I., Zhou, J.: The role of positive emotions in education: A neuroscience perspective. Mind Brain Educ. **14**(3), 220–234 (2020)

Lokman, A.M., Harun, A.F., Md. Noor, N.L., Nagamachi, M.: Website affective evaluation: Analysis of differences in evaluations result by data population. In: Human Centered Design: First International Conference, HCD 2009, Held as Part of HCI International 2009, pp. 643–652. Springer Berlin Heidelberg, San Diego, CA, USA (2009)

McEnrue, M.P., Groves, K.S., Shen, W.: Emotional intelligence development: Leveraging individual characteristics. J. Manage. Develop. **28**(2), 150–174 (2009)

McKay, A., MacDonald, K., Longmuir, F.: The emotional intensity of educational leadership: A scoping review. Int. J. Leaders. Edu. 1–23 (2024)

Ministry of Education Malaysia: Malaysia National Education Philosophy. Principles of education (2006)

Ministry of Higher Education: Malaysia Education Blueprint 2015–2025 (2021)

Northouse, P.G.: Leadership: Theory and practice, 6th edn. Sage Publications (2010)

Ohiku, P.A.: The Impact of Emotional Intelligence on Leadership Styles and Leadership Outcomes in College and University Students in Leadership Roles: A Quantitative Study. Doctoral dissertation, The Chicago School of Professional Psychology (2021)

Pascarella, E.T., Terenzini, P.T.: How College Affects Students: A Third Decade of Research, vol. 2. Jossey-Bass, An Imprint of Wiley, Indianapolis (2005)

Redzuan, F., Lokman, A.M., Othman, Z.A., Abdullah, S.: Kansei design model for e-learning: a preliminary finding. In: Proceedings of the 10th European Conference on e-Learning (ECEL-2011), pp. 685–696 (2011a)

Redzuan, F., Lokman, A.M., Othman, Z.A.: Kansei semantic space for emotion in online learning. In: 2014 3rd International Conference on User Science and Engineering (i-USEr), pp. 168–173. IEEE (2014)

Redzuan, F., Mohd. Lokman, A., Ali Othman, Z., Abdullah, S.: Kansei design model for engagement in online learning: a proposed model. In: Informatics Engineering and Information Science: International Conference, ICIEIS 2011, Kuala Lumpur, Malaysia, November 12-14, 2011. Proceedings, Part I, pp. 64–78. Springer, Berlin Heidelberg (2011b)

Schutz, P.A., Hong, J.Y., Cross, D.I., Osbon, J.N.: Reflections on investigating emotion in educational activity settings. Educ. Psychol. Rev. **18**(4), 343–360 (2006). https://doi.org/10.1007/s10648-006-9030-3

Shafait, Z., Khan, M.A., Sahibzada, U.F., Dacko-Pikiewicz, Z., Popp, J.: An assessment of students' emotional intelligence, learning outcomes, and academic efficacy: a correlational study in higher education. PLoS ONE **16**(8), e0255428 (2021)

Silva, A.J., Almeida, N.: Can engagement and performance be improved through online training on emotional intelligence? A quasi-experimental approach. Int. J. Educ. Manag. **37**(2), 449–464 (2023). https://doi.org/10.1108/IJEM-03-2022-0092

Taharim, N.F., Lokman, A.M., Hanesh, A., Aziz, A.A.: Feasibility study on the readiness, suitability and acceptance of M-learning AR in learning history. In: AIP Conference Proceedings, 1705(1). AIP Publishing (2016)

Tekerek, M., Tekerek, B.: Emotional intelligence in engineering education. Turkish Journal of Education **6**(2), 88–95 (2017)

Yip, J.A., Côté, S.: The emotionally intelligent decision maker: emotion-understanding ability reduces the effect of incidental anxiety on risk taking. Psychol. Sci. **24**, 48–55 (2013). https://doi.org/10.1177/0956797612450031

Extraction of Toast Packaging Design Elements Using Long Short Term Memory-Neural Network with Kansei Engineering Approach

Wiwi Prastiwinarti(✉), Novi Purnama Sari, Rafi Ramdan Permana, and Lytta Yennia

Jakarta State Polytechnic, Universitas Indonesia, Jl. Prof. DR. G.A. Siwabessy, Kukusan, Kecamatan Beji, Kota Depok, Jawa Barat 16425, Indonesia
{wiwi.prastiwinarti,novi.purnamasari}@grafika.pnj.ac.id,
{rafi.ramdan.permana.tgp21,
lytta.yennia.putri.tgp21}@mhsw.pnj.ac.id

Abstract. Packaging is one of the factors that can increase consumer satisfaction. Packaging performance can be realized through the right combination of design elements based on design concepts that are in accordance with consumer preferences. The process of extracting design elements quantitatively is an important thing to do in minimizing subjectivity. The purpose of this research is to determine the optimal packaging design elements quantitatively with the Kansei Engineering approach. Artificial Neural Network (ANN) and Long-Short Term Memory Neural Network (LSTM-NN) ore some method that is able to predict packaging design elements. The novelty of this research is comparing the accuracy of the two methods so that it can be a reference in selecting the appropriate element extraction method. Design elements prediction was carried out by determining factors such as packaging shape (X1), packaging material (X2), packaging features (X3), image elements (X4), design style (X5), design surface (X6), and lock opening (X7). Design elements determining using ANN method for Practical-Unique concept is obtained X1.2 is Horizontal Beam, X2.5 is Paper and Polymer, X3.2 is Window, X4.2 is None, X5.1 is Trendy, X6.1 is Direct, X7.4 is Perforation. While the design elements for the Practical-Unique concept using the LSTM-NN method obtained X1.2 is Horizontal Beam, X2.2 is Brown Kraft, X3.4 is Cutlery & Handle, X4.2 is None, X5.1 is Trendy, X6.3 is None, X7.3 is Die cut. It is concluded that the results of the LSTM-NN method show better element design due to high training accuracy and having complete variables compared to ANN.

Keywords: ANN · LSTM-NN · Kansei Engineering · Element Design Concept

1 Introduction

Packaging is one of the factors that can increase consumer satisfaction when buying a product. In an effort to increase consumer satisfaction, the packaging of toast products must also be considered in order to meet consumer needs. Toast packaging on the market does not have design elements that are in accordance with consumer needs and preferences. Design elements play an important role as a means of promotion and visual communication of products. Elements of toast packaging have also not been able to support the packaging function well in making it easier for consumers to use. The lack of information on the packaging is also a problem that can reduce consumer attractiveness. Packaging not only functions as a protector, but also a means of communication for consumers in product purchasing decisions. Packaging needs to apply appropriate design elements to the packaging created, such as text, images, colors. Appropriate design elements are able to get consumer attention by bringing verbal messages closer to consumers. Problems with toast packaging are the main factor in packaging development to improve packaging quality. Packaging quality, such as ease of opening packaging, and packaging shape have a significant influence on consumer buying decisions [1]. Relevant features if added to the packaging can also attract potential customers and can improve the user experience [2, 3]. Analysis of packaging elements is important to do in the packaging development process. The packaging concept that has been obtained will be connected to the packaging design elements as a reference for product packaging.

The Kansei Engineering method is one of the methods that can be used to fulfill consumer preferences, based on sensory responses. This method is used to develop products that match the user's feelings and increase their satisfaction [4, 5]. The Kansei Engineering method allows product development that is customer-oriented and can increase the chances of the product being sold on the market [6]. The process of extracting design elements can be carried out using several methods, such as Artificial Neural Network (ANN) and Long-Short Term Memory Neural Network (LSTM-NN). These two methods will be compared to get better results based on suitability with Kansei Kata which will be new in research. Several studies have proven the effectiveness of LSTM and ANN. For example, time series calculations produce a stable error value of 20% for LSTM and 40%, which is greater using ANN [7], then prediction research on electricity in a building with an MSE value for LSTM gets a figure of 17.3% while ANN only 12% [8]. This research also provides new knowledge, especially in the field of logic-based packaging and Kansei Engineering, because predictions of packaging design elements using LSTM and ANN have not yet been discovered. Packaging development is carried out not only to get better packaging but also to achieve usability including efficiency, effectiveness, and consumer satisfaction [9]. This research is new because it can provide additional reference knowledge for further research to choose the right design element extraction method. This research aims to determine optimal packaging design elements using the LSTM and ANN methods.

2 Method

This method uses a database in the form of packaging samples and respondent data used to train Artificial Intelligence in determining packaging design elements. The stages of the method flow are shown in Fig. 1. The LSTM method is a type of neural network that imitates human long-term and short-term memory so that it can remember core information and discard useless information using gates. Meanwhile, ANN is a method that imitates the neural network of the human brain by using nodes (points) which function as a simulation of human brain neurons for data processing [10]. This method has good results and fast processing and can handle complex variables while maintaining the accuracy of the results [11, 12]. In this research, the two methods will be compared to determine the results of design specifications based on the concept. The ANN method has good accuracy results for large linear datasets and runs over long periods of time [13]. Meanwhile, LSTM-NN can process long-term linear data stored in artificial neural network memory and is able to solve sequential information problems. Meanwhile, ANNs more often process linear data and in the process sometimes bias occurs.

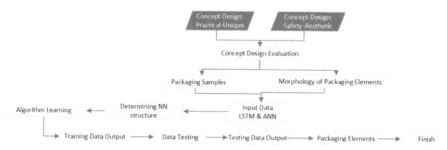

Fig. 1. Process Flow.

2.1 Concept Design Evaluation

The packaging concept for Roti Bakar Bandung was obtained using the K-Means Genetic Algorithm method, resulting in the concepts of "Practical-Unique" and "Safety-Aesthetic" based on the Kansei Engineering method. The concepts obtained were then evaluated to determine the suitability of the packaging samples to the concepts. The questionnaire used is Likert with a scale of 1–7 points. Respondents used in the form of purposed judgment sampling with the characteristics of frequent consumption of toast, with a minimum of 30 respondents. The data results will be used as test data in NN training.

2.2 Morphology Analysis

This stage is carried out by analyzing the morphology of the samples that have been collected. Packaging design elements are obtained by analyzing packaging samples such as: Lid, shape, seal, design style, and material. The identification of design elements is assisted by expert panelists who have at least 2 years of experience in their field [15].

2.3 Determining NN Structure

The structure is designed by determining the input layer, hidden layer, and output layer. As an illustration of the flow of data that will be learned by the NN algorithm [16] A hidden layer will perform a nonlinear transformation from the input into the network. Hidden layers vary depending on the function of the NN network and the associated weights. In determining the hidden layer, using three is sufficient to achieve test convergence [17].

2.4 Algorithm Learning

The algorithm learning process is carried out using the respondent data that has been obtained. The training process is carried out continuously until it gets the smallest error result. The following is the general equation used in the LSTM method.

$$f_t = \sigma_g(W_f x_t + U_f h_{t-1} + b_f) \qquad (1)$$

$$i_t = \sigma_g(W_f x_t + U_f h_t + b_i) \qquad (2)$$

$$o_t = \sigma_g(W_o x_t + U_o h_{t-1} + b_o) \qquad (3)$$

$$c_t = fg°c_{t-1} + i_t°\sigma_c(W_c x_t + U_c h_t + b_c) \qquad (4)$$

$$h_t = \sigma_g°\sigma_h(c_t) \qquad (5)$$

where, it is input gate, ht is hidden state, ct is cell state, and ft is forget gate. This method allows information to enter through the input gate and insert it into the memory cell [18]. Then the forget gate will decide which information will be discarded from the memory cell. After that, the memory will exit in the form of output to be forwarded to the next state.

2.5 Data Testing

The data that has been trained as a learning database to be used as a data pattern for testing. Testing was carried out using a morphological analysis table. The results of the test are in the form of ANN graphs and LSTM values. They will be used to determine packaging design elements. When using Machine Learning (ML), if you use the same dataset, ML will produce good output, but cannot recognize new data patterns [8]. This research divides the dataset into 2, namely respondent data as training data and morphological classification data as data test.

2.6 Determining Design Element

Design elements are determined using graphs from the output results in the previous stage. The ANN method in the research uses backpropagation which is supervised learning, where Artificial Intelligence will learn using target data (Goal) to predict the output for each new input entered [19]. LSTM-NN is a variant of the Recurrent Neural Network (RNN) cycle [20]. This method allows artificial intelligence programmed in Python to analyze design elements based on a given morphological dataset.

3 Results

3.1 Morphology

Determination of packaging morphology elements is done through discussion with experts who have experience in the field [15]. Table 1 shows the morphology design element.

Table 1. Morphology Analysis.

Design Element	Packaging Shape (X1)	Packaging Material (X2)	Packaging Feature (X3)	Design Element (X4)	Design Style (X5)	Design Surface (X6)	Lock (X7)
Type 1	Vertical Box (X1.1)	Ivory (X2.1)	Handle (X3.1)	Vector Illustration (X4.1)	Trendy (X5.1)	Direct (X6.1)	Slide (X7.1)
Type 2	Horizontal Box (X1.2)	Brown Craft (X2.2)	Window (X3.2)	None (X4.2)	Minimalist (X5.2)	Sticker/Label (X6.2)	Telescope (X7.2)
Type 3	Triangular Prism (X1.3)	Board (X2.3)	Handle & Window (X3.3)		Traditional (X5.3)	None (X6.3)	Die Cut (X7.3)
Type 4	Rounded Beam (X1.4)	Paper & Polymer (X2.4)	Eating Utensils & Handle (X3.4)		Funny (X5.4)		Perforation (X7.4)
Type 5		Duplex (X2.5)	None (X3.5)		Plain (X5.5)		Glue End (X7.5)

3.2 NN Structure

The NN structure requires layers in the process. Among them are input layer, hidden layer, and output layer. The input layer is the input data that will be entered into the neural network process. Hidden layer is the hidden layer between the input and output layers. The determination of the hidden layer is not limited and standardized. Some problems need fewer hidden layers, some complex problems need more hidden layers. The function of the hidden layer is to transfer data and train data from one layer to another.

3.3 Data Training and Data Testing

Data training is input data that will be used to create and build ANN models. The training data used in this study are the results of a questionnaire to 33 respondents using a Likert scale from 1–7. The results of the data in the table are then entered into the Neural Network to serve as a reference for the training model. Test data is data that will be tested using a reference model based on training data that has been created. In this study, 33 packaging samples with codes A to AG were used. Each sample is then identified by its design elements based on the morphology analysis data in Table 1. Examples of sample identification results are shown in Table 2 below.

Table 2. Element Design Identification of Packaging Samples.

Factor\Sample	A	B	...	AG
X1	2	3	....	3
X2	1	1	....	2
X3	3	3	....	2
X4	1	2	....	1
X5	2	5	....	1
X6	2	3	....	1
X7	2	3	....	5

Table 2 shows seven factors from X1-X7 based on the morphology of Fig. 1. In sample A, namely Horizontal Beam, Ivory Material, Window Feature, Vector Illustration, Minimalist, Label, Telescope Lock.

3.4 Packaging Concept

The packaging concept was obtained by conducting K-Means analysis in the previous research process. Which is used to find data clusters and optimized using genetic algorithm to find a better centroid point. The results obtained are two clusters or two packaging concepts. In cluster 1 with practical-unique results, and cluster 2 concept safe-aesthetic.

3.5 ANN Data Training

Before the data is trained, it is important to construct the NN network first, the construction can be seen in Fig. 2.

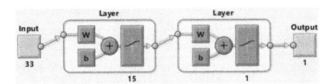

Fig. 2. NN Construction.

The network used uses 33 inputs from the amount of sample data, 15 hidden layers for hidden layer inputs as test data and an output value of 1. A hidden layer of 15 was used after going through several tests and getting a hidden layer value of 15 with good results. The learning process set is 1000 epoch. The packaging design concept is trained using the Levenberg-Marquardt algorithm until it produces a small error value. The algorithm was chosen because of its effectiveness and speed of convergence, where this method is a nonlinear optimization during backpropagation error correction to find adjusted weights [21]. The results of the training will produce performance and linear regression. The training performance results can be seen in Fig. 3.

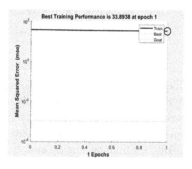 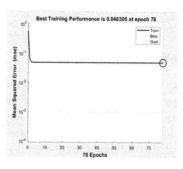

a) Performance State Practical-Unique   b) Performance State Safety-Aesthetic

Fig. 3. Performance State Diagram.

Performance State. The performance stage in ANN is a value based on the correlation coefficient (R) with the mean square error (MSE) value [22]. The MSE (Mean Squared Error) value in the Fig. 2 caption shows that the "Safe-Esthetic" concept (a) gets a value of 0.048305 which is achieved at epoch 78, and the "Practical-Unique" concept (b) gets a value of 0.48758 which is achieved at epoch 4. The results of the two networks have reached the desired target because they have reached the smallest MSE value and are below the predetermined maximum epoch limit, which is 1000 epochs or 1000 training times, so the training process can be stopped (Fig. 4).

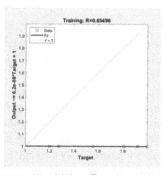 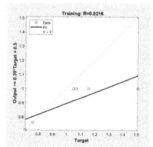

a) Regression Value Practical-Unique   b) Regtession Value Safety-Aesthetic

Fig. 4. Linear Regression Diagram.

Linear Regression. Linear regression is a straight line analysis method used to identify input factors and response factors [23]. The training graph data shows that the "Safe-Esthetic" concept (a) gets a value of 0.8216, and the "Practical-Unique" concept (b) is 0.65496. The Safe-Esthetic concept data produces a regression value close to 1, meaning that the data is close to the target to be achieved in making design elements. While the "Practical-Unique" concept data has lower results. This can happen because in the training process there are data habits that produce outputs with incomplete variables. After the network training process, the network is stored as a database of learning results.

3.6 ANN Data Testing

Testing is carried out based on training data that has been inputted previously as a learning database and test data as testing. The running results in Fig. 5 are then analyzed for correlation of design elements with the desired packaging concept. The selected bar is the longest bar, the higher the direction, the stronger the correlation with the design concept. The following are the results of ANN data testing for both packaging concepts. The longest, the higher the direction, the stronger the correlation with the design concept.

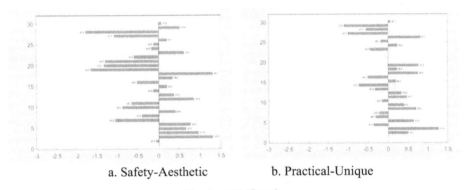

a. Safety-Aesthetic   b. Practical-Unique

Fig. 5. ANN Result.

After testing, the 'Practical-Unique' design concept was selected to proceed to the mock-up stage. This is because in the K-Means cluster that has been optimized by Genetic Algorithm shows a wider area in the "Practical-Unique" cluster. The results of the test produce design elements that can be used to design toast packaging. The graph is obtained after all series of training data and test data are successfully run and reach the smallest MSE error value. Based on the morphological analysis shown in Table 1, the results of the selected design elements for Practical-Unique are as follows: Horizontal beam (X1.2), Paper Material (X2.3), Handle Feature (X3.1), Image Element None (X4.2), Minimalist Design Style (X5.2), Label Design Surface (X6.2), Slide lock opening (X7.1).

3.7 LSTM-NN Data Processing

The data processed in the LSTM-NN process is the same as that of the NN method. The data is trained until it reaches the smallest loss to achieve good running. The training results are carried out for the training data model, in this case the packaging morphology. The training results show a range of 2–5% in most epochs, indicating that the training results are quite good. The accuracy graph also shows a range of 90–100% which means accurate. After the training process is done and the train data is ready to be used, then the determination of packaging elements can be done. The training graph can be seen in Fig. 6.

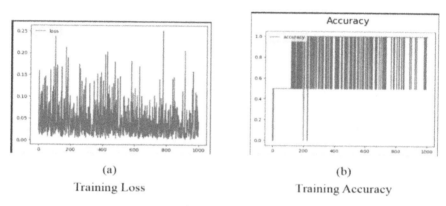

(a) Training Loss

(b) Training Accuracy

Fig. 6. Training Diagram.

In LSTM-NN there is no concept of right and left sides, and only focuses on the positive axis, namely the right side which means the "Unique" axis. The test results for the selected elements are as follows: Horizontal Beam (X1.2), Brown kraft material (X2.2), Handle and window features (X3.3), None Design Element (X4.2), Trendy design style (X5 .1), Surface Design Direct (X6.1), Die Cut Open/Lock (X7.3).

3.8 Mockup Design

From the two methods that have been run, there are significant differences in results. Of the two mockups, the LSTM result has a better design and leads to a unique design. With the window feature shaped and printed directly, it is easier for consumers to see the contents of the product (Fig. 7).

Fig. 7. Mockup Packaging Design.

4 Conclusion

Based on the research conducted using ANN and LSTM-NN methods, it aims to find out the best packaging design elements that can be applied to toast packaging design based on the packaging concept that has been obtained previously using the K-Means Genetic Algorithm method. Determination of packaging design elements is to fulfill and increase consumer satisfaction with toast products. The output obtained for the "Practical-Unique" concept in the ANN method is: Horizontal beam (X1.2), Paper Material (X2.3), Handle Feature (X3.1), Image Element None (X4.2), Minimalist Design Style (X5.2), Label Design Surface (X6.2), Slide lock opening (X7.1). While the output of LSTM-NN: are: Horizontal Beam (X1.2), Brown kraft material (X2.2), Handle and window feature (X3.3), None Design Element (X4.2), Trendy design style (X5.1), Surface Design Direct (X6.1), Die Cut Opening/Lock (X7.3). From the research and trials carried out, ANN processing is still often biased and a lot of trial and error is carried out to get good results, but the regression value is still low, namely 0.6−0.8 under of 1.0. Meanwhile, LSTM is able to produce accuracy values ranging from 0.95–1.0.

References

1. Nemat, B., Razzaghi, M., Bolton, K., Rousta, K.: The potential of food packaging attributes to influence consumers' decisions to sort waste. Sustain **12**(6) (2020). https://doi.org/10.3390/su12062234
2. Araminta, S., Sari, N.P.: Analisis Elemen Kemasan Sekunder Pembalut Dengan Metode Rough Set. **9**(2), 157–163 (2024)
3. Redzuan, F., Lokman, A.M., Othman, Z.A.: Kansei Design Model for Engagement in Online Learning: A Proposed Model. pp. 64–78
4. Yang, M., Lin, L., Chen, Z., Wu, L., Guo, Z.: Research on the construction method of kansei image prediction model based on cognition of EEG and ET. Int. J. Interact. Des. Manuf. (2020). https://doi.org/10.1007/s12008-020-00651-2
5. Lokman, A.M., Harun, A.F., Noor, N.L.: Website Affective Evaluation: Analysis of Differences in, pp. 643–652
6. Sari, N.P., Zulkarnain, Z., Muzaki, V.A., Meilani, Y.D.: Implementasi kansei engineering dalam pengembangan kemasan minuman kopi ready to drink **18**(1), 200–209 (2024). https://doi.org/10.21107/agrointek.v18i1.12443

7. Song, X., Liu, Y., Xue, L., Wang, J., Zhang, J., Wang, J.: Time-series well performance prediction based on Long Short-Term Memory (LSTM) neural network model. J. Pet. Sci. Eng. **186**, 106682 (2020). https://doi.org/10.1016/j.petrol.2019.106682
8. Byeongmo, S., Yeobeom, Y., Kwang, H.L., Soolyeon, C.: Comparative Analysis of ANN and LSTM Prediction Accuracy and Cooling Energy Savings through AHU-DAT Control in an. (2023)
9. Azlina, N., et al.: Performing Usability Evaluation on Multi-Platform Based Application for Efficiency. Effectiveness and Satisfaction Enhancement **15**(10), 103–117
10. Wu, Feng, J.: Development and Application of Artificial Neural Network. Wirel. Pers. Commun. (2017). https://doi.org/10.1007/s11277-017-5224-x
11. Mao, et al.: Comprehensive comparison of artificial neural networks and long short-term memory networks for rainfall-runoff simulation. Phys. Chem. Earth. 123 (2021). https://doi.org/10.1016/j.pce.2021.103026
12. Guo, Tian, K., Ye, K., Xu, C.: MA-LSTM: A Multi-Attention Based LSTM for Complex Pattern Extraction, pp. 3605–3611 (2021)
13. Wentz, V., Maciel, N.J., Ledesma, G.J.J., Hideo, O., Junior, A.: Solar Irradiance Forecasting to Short-Term PV Power: Accuracy Comparison of ANN and LSTM Models, pp. 1–23 (2022)
14. Inthachot, Boonjing, V., Intakosum, S.: Artificial Neural Network and Genetic Algorithm Hybrid Intelligence for Predicting Thai Stock Price Index Trend (2016)
15. Wu, Y., Chen, Y.H.: Factors affecting consumers' cognition of food photos using Kansei engineering. Food Sci. Technol. **42**, 1–9 (2022). https://doi.org/10.1590/fst.38921
16. Di Franco, G., Santurro, M.: Machine learning, artificial neural networks and social research. Qual. Quant. **55**(3), 1007–1025 (2020). https://doi.org/10.1007/s11135-020-01037-y
17. Shen, Yang, H., Zhang, S.: Neural network approximation: three hidden layers are enough. Neural Networks **141**, 160–173 (2021). https://doi.org/10.1016/j.neunet.2021.04.011
18. Kumar, Goomer, R., Singh, A.K.: Long short term memory recurrent neural network (LSTM-RNN) based workload forecasting model for cloud datacenters. Procedia Comput. Sci. **125**, 676–682 (2018). https://doi.org/10.1016/j.procs.2017.12.087
19. Lee, Oh, S., Kim, C., Bae, J., Lee, J.: Hardware-based spiking neural network architecture using simplified backpropagation algorithm and homeostasis functionality. Neurocomputing. (2020). https://doi.org/10.1016/j.neucom.2020.11.016
20. Solgi, Loáiciga, H.A., Kram, M.: Long short-term memory neural network (LSTM-NN) for aquifer level time series forecasting using in-situ piezometric observations. J. Hydrol. **601** (2021). https://doi.org/10.1016/j.jhydrol.2021.126800
21. Cheema, N., et al.: Intelligent computing with Levenberg – Marquardt artificial neural networks for nonlinear system of COVID-19 epidemic model for future generation disease control, vol. 123. Springer Berlin Heidelberg (2020). https://doi.org/10.1140/epjp/s13360-020-00910-x
22. Ocampo, López, R.R., Camacho-León, S., Nerguizian, V., Stiharu, I.: Comparative evaluation of artificial neural networks and data analysis in predicting liposome size in a periodic disturbance micromixer. Micromachines **12**(10) (2021). https://doi.org/10.3390/mi12101164
23. Kadnár, M., et al.: Comparison of linear regression and artificial neural network models for the dimensional control of the welded stamped steel arms. Machines **11**(3), 376 (2023)

Assistive Tool for Emotion and Importance Quadrant LEIQ™

Nur Batrisyia Damia Mustafa[1], Saidatul Rahah Hamidi[1(✉)], Surya Sumarni Hussein[1], and Shuhaida Mohamed Shuhidan[2]

[1] College of Computing, Informatics and Mathematics, Universiti Teknologi MARA, 40450 Shah Alam, Selangor, Malaysia
saida082@uitm.edu.my

[2] Center for Research in Data Science, Computer and Information Sciences Department, Universiti Teknologi PETRONAS, Perak, Malaysia

Abstract. In today's rapidly growing digital world, effective and user-centered experiences rely on integrating emotions into the design of digital products. This study focuses on investigating the LEIQ™ paradigm and creating computer-aided assistive tools for enhancing the evaluation of emotion-driven user experience (UX) in web-based solutions. Given the increasing incorporation of technology into our everyday routines, it is critical to understand the impact of emotions on user interactions. Emotional experiences play a crucial role in the efficacy of online design and development, as they have a substantial impact on user engagement, contentment, and product loyalty. The study acknowledges the limitations of existing methods for evaluating emotions, which include challenges in managing and analyzing the large datasets generated by frameworks like LEIQ™. Thus, this paper aims to develop a prototype assistive tool specifically designed for LEIQ™ emotion measurement, with the intention of addressing these shortcomings. This advancement reduces the need for a human workforce by providing the opportunity to quickly and accurately acquire information using automated technology. Potential gains for LEIQ™ practitioners, IT artifact designers, developers, and product organizations are envisaged by the research. The ensuing solution, which goes beyond LEIQ™ data consolidation, has the potential to completely alter emotional UX research by going deeper and encompassing all areas of LEIQ™ data processing. Finally, the assistive tool's usefulness may improve by making data interpretation in the digital age easier across a range of prospects and disciplines.

Keywords: Assistive Tool · Emotional UX · Lokman's Emotion and Importance Quadrant (LEIQ)™

1 Introduction

1.1 Research Background

Researchers have examined comprehensive frameworks for evaluating emotional user experiences (UX) in order to properly deconstruct the intricate interplay among design components, emotional experiences, and user viewpoints. The LEIQ™ model has become increasingly popular among the many frameworks available [1]. By separating the UX space into four quadrants according to the perceived relevance of a touchpoint and the emotional valence (positive or negative), the LEIQ™ model provides a methodical approach. This methodical classification makes it easier to build products that resonate with consumers on an emotional level and allows for a detailed analysis of emotional reactions and their impact on user behavior. The growing significance of emotional user experience (UX) in web-based solutions makes it imperative to study the capabilities and appropriateness of the LEIQ™ framework in this research. This gives valuable information regarding how users' perceptions, decision-making pro-cesses, and general level of enjoyment while dealing with web-based items are influ-enced by their emotional experiences [2].

The creation of good emotional user experiences (UX) involves a variety of design elements, including visual design, interactive features, user interfaces, and content display. These aspects are important in eliciting emotions and shaping user experiences [1]. However, previous research has mostly focused on using the LEIQ™ frame-work in traditional ways—that is, through publications and in-person interactions—which may provide challenges due to a lack of technological support for implementation [3]. Similarly, Hussein and colleagues [4] argue that effectively integrating business and technology, as well as aligning people and information and communication technology (ICT), are essential factors for achieving success in ICT strategic planning initiatives.

The goal of this study is to gather feedback from LEIQ™ practitioners in order to address these issues. With a focus on web-based development, the goal is to provide design components for computer-aided assistive tools that significantly improve data handling when evaluating emotions. The research aims to create a prototype of useful tools for analyzing emotions produced in any subject of study using the LEIQ™ technique. The project's goal is to improve the effectiveness and accuracy of users' emo-tional evaluations by speeding up data collection, processing, and analysis using au-tomation and an internet application. This assistive tool has the potential to transform emotional research by providing researchers and practitioners with fast, actionable information on users' emotional experiences.

2 Research Work

2.1 Assistive Tools

Assistive tools, often known as assistive technology (AT), include devices, software, or equipment designed to aid people with disabilities in performing tasks that would be difficult or impossible for them otherwise [5]. These tools are intended to help people strengthen, maintain, or improve their functional skills, thereby promoting independence, well-being, and involvement [6]. According to Mallin and Carvalho [7], assistive equipment with style, formal, functional, and emotional characteristics can enhance creative processes by linking the designer's ideas with the actual needs of users, who face constant challenges due to improvisation and poor design. Aside from that, assistive tools are items or gadgets that allow people to study, work, or live more productive lives. Collaboration with experts, such as occupational therapists or assistive technology specialists, is critical to selecting and implementing the most appropriate and effective solutions for each individual's specific circumstances and accessibility requirements [8, 9].

Although assistive technology seeks to identify and resolve human emotions based on personal preferences and needs, a systematic study conducted by Musri and colleagues [10] identified a number of challenges that prevent assistive ambient technology from being widely adopted. These include issues related to technology, user difficulties, organizational or societal limits, and budgetary limitations. To improve the usability and acceptance of assistive technologies among people with disabilities or limitations, these barriers must be removed. A product that is designed and constructed to be used by the widest range of people—including those who are less tech-savvy—is considered accessible [11]. This technique prioritizes inclusivity and diversity throughout the design process. To do this, each feature or design must be carefully studied with accessibility in mind, allowing their experience to be tailored to their specific needs [12].

2.2 Emotional UX

Numerous studies show emotions play an important role in UX design; for example, excellent emotional experiences improve users' perceptions of usefulness and overall satisfaction [13]. The significance of an emotional user experience extends beyond simple user satisfaction. According to Forlizzi and Battarbee [14], emotional experiences can create rapport and long-term product engagement. Positive emotional interactions with a digital interface increase the likelihood of meaningful relationships and brand loyalty among users. There is an increasing body of literature highlighting the critical role emotions play in UX design [15]. Nyagadza and colleagues [16] conducted a study to determine the relationship between user happiness and emotions when it comes to internet banking. Their findings revealed a significant link between high levels of client satisfaction and positive emotions such as surprise and joy. This emphasizes the need for understanding and responding to emotions, especially in banking and lifestyle, which affect people from all walks of life and socioeconomic backgrounds.

According to Bidin and colleagues [17], it is crucial to incorporate emotional responses that reflect users' values and perspectives while utilizing Kansei robotic technology. This is particularly crucial when designing interactions for elderly users. The organization prioritizes the integration of local cultural elements into events and products, recognizing their impact on community attitudes and behaviors, as well as the reception of innovative initiatives[18]. Besides, Redzuan and colleagues [19] also highlight the significance of using emotion in e-learning to enhance user experiences through the integration of interface, content, and interaction design. Thus, a successful website must go beyond only giving functional utility; it should also promote comfortable and efficient interaction between the user and the site [13, 20, 21].

Similarly, Lomas and colleagues [22] investigated how consumers' emotions influenced their perceptions of artificial intelligence (AI) chatbots. According to their findings, customers' positive opinions of chatbots were heavily influenced by pleasant emotions such as trust and empathy. These advancements hold enormous potential for improving mental health and wellness because they provide better tools and platforms. Using AI technologies to better understand emotions not only improves user experiences but also enables the development of more efficient tools that encourage sympathetic and emotionally sensitive interactions, promoting positive emotional experiences and, ultimately, improving people's overall well-being when using digital platforms. Thus, by creating emotionally charged experiences, UX practitioners can impact user decision-making and encourage desired user behaviors. Users' emotional experiences have been shown to have an impact on e-commerce, marketing, and other businesses. Lombart and colleagues [23] found that emotionally engaging design characteristics in online shopping environments increase consumer trust, purchase intentions, and brand impressions. Emotional appeal in digital advertising has a significant impact on consumers' perceptions and intentions toward the provided commodity or service [24]. These findings highlight the need to incorporate emotive design aspects into the user experience (UX) in order to achieve marketing objectives and increase user engagement.

2.3 Lokman's Emotions and Importance Quadrant (LEIQ™)

Kansei Engineering (KE) is a technology that merges the realms of human emotions with engineering to integrate emotional responses into product design [25]. KE has been effectively applied across various product categories, including physical consumer goods and IT artifacts, to enhance their emotional appeal. Hence, Lokman's Emotion and Importance Quadrant, or LEIQ™ for short, is an approach for categorizing and evaluating emotions based on their intensity and the perceived significance of associated tasks based on Kansei Engineering [26, 27]. This paradigm has evolved into an essential framework, particularly in research on the relationship between subjectively significant appraisal and emotions in the context of quality of life (QoL). It aids in the identification of feelings, the factors that influence them, and the importance of these emotions to people's overall health [1, 26]. There are four overall quadrants: positive emotion in the Important quadrant, positive emotion in the Not Important quadrant, negative emotion in the Important quadrant, and negative emotion in the Not Important quadrant.

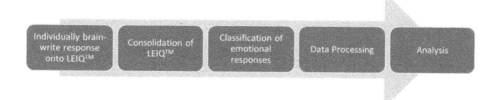

Fig. 1. Process for LEIQ [28]

As shown in Fig. 1, the LEIQ™ technique process consists of several critical components. This study's primary focus was the integration of stages 1–2, which are required for the participants' verbal self-report, was the primary focus of this study. First, participants were given two different colored cards (blue and pink) on which to write topics that elicited positive and negative emotions in their minds. The cards were then placed in the relevant LEIQ™ template quadrants. Because the questions were open-ended, participants could freely express their emotions. Researchers can use the model's four quadrants to quantify the impact of users' emotional reactions and preferences on product and service design and development. Researchers can design emotionally appealing and meaningful research by considering the strong and lasting influence that specific emotions can have on people's decision-making [29]. Nevertheless, the LEIQ™ model provides a solid foundation for exploring feelings and applying various life domains, making it a viable option for analyzing user emotions. Furthermore, the model provides a framework for researching emotions and their importance in a variety of domains, including effective strategies for developing organizational skills. More research and implementation are needed to investigate the precise application of LEIQ™ in analyzing user experiences with an assistive tool in the context of this study.

3 Research Design

3.1 Research Model

The LEIQ™ serves as the foundation for comprehending the appraisal of emotions and is the basis of the conceptual framework of this study. These well-established theories serve as a framework for knowing how variables interact and influence the creation of research questions and hypotheses. Based on the LEIQ™ paradigm, this study endeavors to increase our understanding of how emotions are valued and considered important by providing new insights into the area. Figure 2 depicts the conceptual framework graphically.

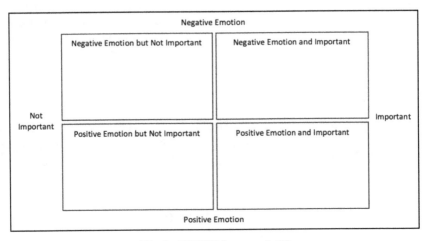

Fig. 2. LEIQ™ framework [1]

According to Barbara L. Fredrickson [30], a proponent of the broaden-and-build theory of positive emotions, positive emotions expand an individual's thought-action repertoire, leading to increased cognitive, social, psychological, and physical resources over time. Fredrickson's theory posits that positive emotions are crucial for well-being and resilience. This theory proves relevant to this research as it aids in comprehending users' emotions when interacting with established products or services [30]. By integrating this theory into the LEIQ™ model, decision-makers can gain deeper insights into people's emotional responses and the importance of elements influencing their well-being, thereby leading to more effective methods for enhancing quality of life.

Incorporating Emotional UX and Kansei Engineering into this research conceptual framework allows for a comprehensive examination of various perspectives, providing a thorough grasp of the research issue. The primary concepts under study are emotions and perceived importance, representing the foundational elements necessary to address the research challenge. Visual representations aid in understanding the theoretical structure and guide the data analysis and interpretation process, indicating the specific outcomes the research is expected to yield.

4 Findings

4.1 Assistive Tools: The Integration of Miro

After completing the requirement analysis phase, the researcher designed and built the Lo-Fi prototype for the web-based LEIQ™ assistive tool using the Miro platform, an online interactive collaboration board. The researcher successfully designed and executed a functional prototype for the LEIQ™ assistive tool. The technique entails creating an interactive digital platform using the LEIQ™ to measure users' emotions in assessments on a variety of themes or titles. The completed prototype serves as a rigorous proof of concept, demonstrating the instrument's practical application.

The suggested assistive tool for web-based LEIQ™ evaluation consists of four panels, each dedicated to a separate activity. It is built on the Miro platform, an online workspace that promotes collaboration among distant teams. Miro's inclusion in the web-based prototyping design improves communication between remote and scattered teams. The design demonstrates how acquired requirements were transformed into a simple mock-up, with an emphasis on user-friendly and intuitive interfaces. The following part delves into the layout of the assistive tool prototype. Notably, the web-based tool's design goes beyond a single assessment to accommodate a wide range of measuring criteria. The web-based tool's adaptability makes it useful for measuring emotional user experience using the LEIQ™ framework. As a result, the prototype of assistive tools is designed to aid the research of emotions elicited across a variety of topics, demonstrating its versatility and utility.

4.1.1 Design for Task Completion

This subsection includes five (5) boards intended to duplicate tasks in analyzing the use of technology within the LEIQ™ framework for empowerment design.

Figure 3 shows the Task 1 board, which is an introspective process of reflecting on personal experiences connected to the desired topics. Participants are asked to link their experiences with certain feelings. Color-coded sticky notes on the screen board make the process easier to complete. Participants are encouraged to select the hue that corresponds to their experiences and emotions, creating a visual representation of their reflections. The sticky note function allows for seamless placement within the boards,

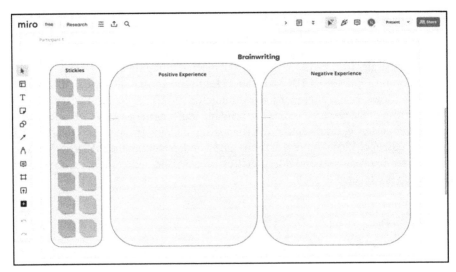

Fig. 3. Prototype Screen for Task 1

and participants can drag and drop them as needed. This feature provides an intuitive way for participants to express their thoughts and emotions in a structured manner.

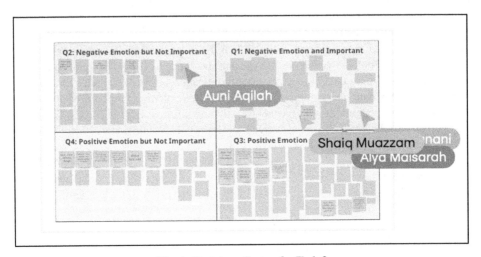

Fig. 4. Prototype Screen for Task 2

Figure 4 shows Task 2, which involves categorizing events using the LEIQ™ model. Participants divide their encounters into one of four quadrants based on experiences related to desired themes. Notably, participants are guided by each event's relevance, determining whether it is positive or negative and carries significant importance. The screen board has designated areas for each quadrant (Q1 through Q4), providing a structured layout for players to strategically position cards representing their experiences.

Participants are encouraged to consider the detailed details of each experience and use the drag-and-drop functionality to arrange the cards in the appropriate quadrant. This feature allows for a dynamic portrayal of the classification process, which supports a full emotion assessment and the importance assigned to each experience.

Task 3 involves the consolidation of experiences, which requires participants to analyze the collected experience notes and arrange them based on similarities as shown in Fig. 5, Participants are instructed to begin the clustering procedure by selecting any card and clustering them using the 'Cluster Objects' tool, ensuring that those with comparable meanings are in their appropriate quadrants. This iterative clustering process continues until all cards are properly grouped. When certain cards do not match with others, players are requested to create a separate cluster for them. Following the clustering process, participants are tasked with assigning a thematic appellation to each cluster that represents the overarching subject, such as "process," "technology," or "knowledge." This feature offers a hierarchical organization of experiences, which improves the interpretability and analysis of clustered data. To improve aggregated cluster clarity and coherence, participants should ensure that the group name is visible at the summit of the LEIQ™ quadrant.

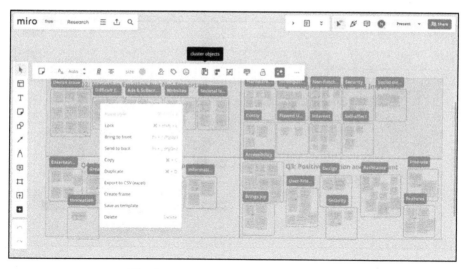

Fig. 5. Prototype Screen to Execute Task 3

Task 4, as shown in Fig. 6, entails participants designating significance levels for clustered themes inside the LEIQ™ quadrant. The numbering method, which ranges from 1 to 3, corresponds to the significance level of each cluster based on its location in Q1, Q2, Q3, and Q4. This step gives a visual representation that aids in decision-making, allowing the team to focus on areas of greatest impact. Using the colorful 'Tag' function, the three levels—less significant (1), significant (2), and highly significant (3)—serve as descriptors for the assigned numbers, improving the visual guide's interpretability. This feature contributes to a full understanding of the value assigned to various topic clusters, enabling informed decision-making during the analytical process.

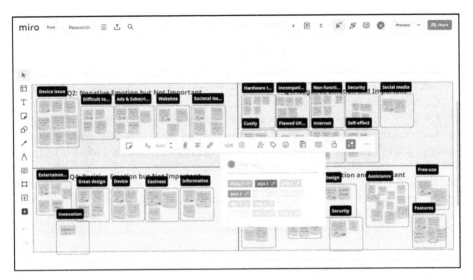

Fig. 6. Prototype Screen to Execute Task 4

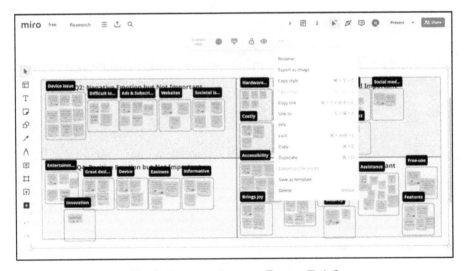

Fig. 7. Prototype Screen to Execute Task 5

For the final task in Fig. 7, the researcher uses a simple method of exporting the outcomes of all completed jobs by participants using the 'Export to CSV (Excel)' tool. This streamlined solution eliminates the need for manual transcribing, increasing efficiency and accuracy. The exported data is cleaned and aligned using the LEIQ™ template before being analyzed in Excel. This features not only speeds up the analysis process but also guarantees that the data is seamlessly integrated into the chosen template, allowing for a thorough examination.

Simultaneously, the supportive function previously performed by many facilitators in the manual technique is converted into tutorial films, which serve as essential resources for guiding participants through and understanding the assessment process. This carefully crafted blend of written instructions and educational films provides a seamless and instructive user experience, emulating the manual support structure while harnessing the benefits of digital mediums for improved participant assistance and comprehension.

4.1.2 Data Analysis of Manual Testing

The manual assessment phase, characterized by the traditional use of pens and sticky notes, revealed inherent limitations that warrant attention. Although the traditional method facilitated participant engagement, it was observed that the pace of data collection was significantly hindered by the manual writing process. Participants, utilizing pens and sticky notes, generated information at a measured rate, leading to a limitation in the quantity of information gathered within the stipulated time frames. Each of the four tasks, allocated 30 min each, exhibited constrained data output due to this inherent constraint. Upon completing task 4, the transfer of comprehensive datasets into Excel for subsequent analysis encountered substantial procedural impediments. The manual transcription of handwritten information from sticky notes demanded a noteworthy temporal investment, averaging over 30 min per group. Additionally, the essential process of data cleaning, imperative for alignment with the Excel template, required an additional 10 min per group.

The cumulative commitment for the tasks of data transfer and cleaning resulted in an extended analysis timeframe, approximating 30–40 min per group. Refer Table 1 for comparison time taken per group to complete all task by using manual vs via assistive tool. The initial step in the process involves meticulous organization of the collected data, encompassing written responses, and results derived from the completed activities. In the context of discussions, the written responses were systematically arranged, to preserve the temporal categorically based on recurring topics. This systematic organization lays the foundation for a comprehensive understanding of the qualitative data, facilitating subsequent stages of analysis and interpretation. An analysis of the task performance data, extracted from user activities during these sessions, is meticulously presented not only the outcomes of the tasks but also the completion times for each task.

4.1.3 Data Analysis of Web-Based Testing

Participants indicated that web-based testing collected more data than manual testing. The digital interface, combined with drag-and-drop functionality, made it easier to use computers and type quickly, resulting in efficient task completion. This significant gain over manual testing was attributed to the easy nature of the web-based approach, which increased user satisfaction. Furthermore, consumers who participated in web-based testing consistently demonstrated shorter completion times for specified tasks. This implies a significant reduction as compared to manual testing, highlighting the efficiency of the web-based technique. While web-based testing provided a more pleasant user experience, several participants initially struggled to use the digital interface. In contrast, the web-based examination made use of new methods to increase data collection efficiency.

This method sped up the data collection process, allowing participants to provide more information within the same time constraints, resulting in a larger collection of user inputs. Figures 8 and 9. Show task performance data extracted from user behaviors during these sessions.

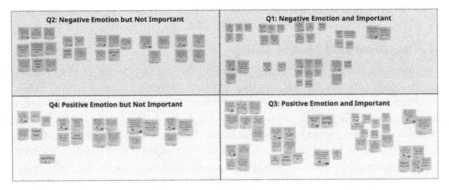

Fig. 8. Web-Based Task Completed by Group 1

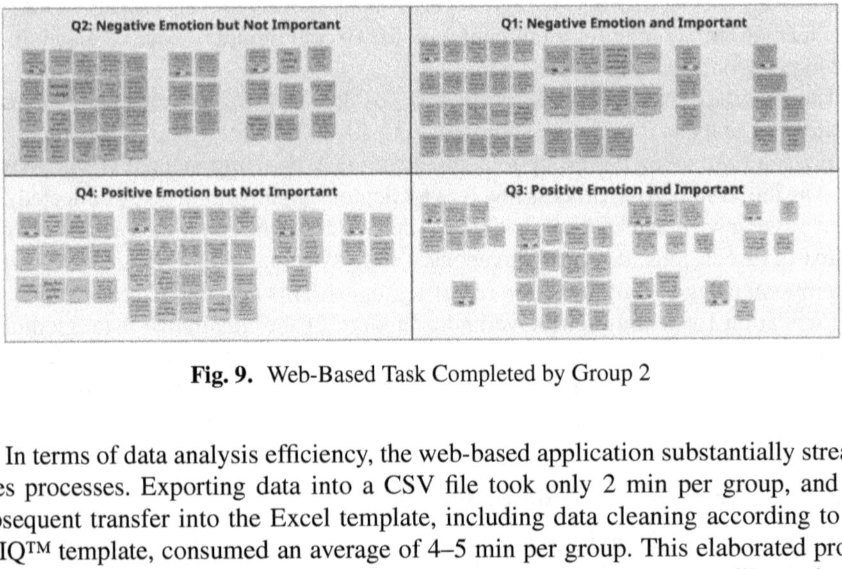

Fig. 9. Web-Based Task Completed by Group 2

In terms of data analysis efficiency, the web-based application substantially streamlines processes. Exporting data into a CSV file took only 2 min per group, and the subsequent transfer into the Excel template, including data cleaning according to the LEIQ™ template, consumed an average of 4–5 min per group. This elaborated procedure significantly decreased the time investment for data processing, totalling only 6–7 min per group. In summation, the web-based application not only proved more efficient in data collection but also revealed a significant advantage in accelerating the data analysis phase (Refer Table 1 for summary comparison of time taken). These findings demonstrate the potential benefits of adopting technology-driven approaches in data-intensive research efforts.

The analysis is affirmed and validated by an experienced LEIQ™ practitioner, verifying the validity and reliability of the findings. This comparison provides for a comprehensive evaluation of the efficiency and effectiveness of the web-based LEIQ™ prototype

Table 1. Time Taken Per Group Comparison to Complete 5 Tasks.

Group Code	Time Taken (min)						Total Time (M)	Total Time (AT)
	Task 1	Task 2	Task 3	Task 4	Task 5 (M)	Task 5 (AT)		
G1	30:00	30:00	30:00	30:00	38:31	06:21	159:31	126:20
G2	30:00	30:00	30:00	30:00	35:16	04:10	155:16	124:10

(M) - Manual; (AT) – Assistive Tool using Web Based

(M) - Manual; (AT) – Assistive Tool using Web Based.

in handling huge datasets, delivering vital insights into its potential advantages over traditional paper-based approaches. The result discussed above leverages the study analysis, indicating that the resulting prototype delivers efficiency in processing complex information as a proof of concept.

5 Conclusions

This study addresses the final research objective by comparing the performance of the web-based LEIQ™ prototype and existing paper-based approaches in the data gathering process. The analysis is verified by an experienced LEIQ™ practitioner, confirming the reliability of the findings. This comparison evaluates the efficiency of the web-based LEIQ™ prototype in managing large data sets and highlights its advantages over traditional approaches.

The findings indicate that the web-based assistive tool prototype improves efficiency, offering faster completion times, a higher volume of results, and fewer user errors. The web-based LEIQ™ tool expedited data collection, drag-and-drop capability, and user-friendly interface contribute to these benefits. Participants utilizing the digital tool reported faster task completion with structured and intuitive digital interface contributes to a reduction in errors generally associated with manual or semi-manual procedures, consequently boosting the overall accuracy of data collection. This comparative analysis emphasizes the enormous advantages presented by the built prototype, positioning it as a more efficient and accurate instrument for measuring emotional user experiences in comparison to traditional approaches.

In conclusion, this research meets its primary objectives by providing a realistic and scalable computer-aided assistive tools inside web-based environments, enhancing the assessment of emotions in user interactions through the integration of technology into the LEIQ™ assessment process. The prototype demonstrates clear benefits, not only contributes to the knowledge of intricate emotional user interactions, but also simplifies the transformation of raw emotional data into usable insights in a more efficient and speedy manner. While this research effectively meets its current aims, it also serves as a fundamental reference for future design refinements and user-centric adaptations to fully automate the LEIQ™ evaluation process. The implications extend to researchers, practitioners, and developers, underlining the advantages of technology-driven techniques. This research enriches the understanding of emotional user experience assessment, providing significant guidance for future attempts in this expanding field.

Acknowledgments. The authors would like to acknowledge the Research Initiative Group for Emotion, Kansei and Design Engineering (RIG EKDE), Universiti Teknologi MARA, and Malaysia Association of Kansei Engineering (MAKE), for all their assistance to the research paper.

Disclosure of Interests. The authors declare that there is no conflict of interest.

References

1. Lokman, A.M., Kadir, S.A., Hamidi, S.R., Shuhidan, S.M.: LEIQ as an emotion and importance model for QoL: Fundamentals and case studies. J. Komun. Malaysian J. Commun. **35**, 412–430 (2019)
2. Brown, K.S., Lee, R.: Adapting to change: strategies for implementing differentiated learning in phase D English courses. J. Curric. Instr. **15**, 48–65 (2021)
3. Lee, S.B., Liu, S.H., Maertz, C., Jr.: The relative impact of employees' discrete emotions on employees' negative word-of-mouth (NWOM) and counterproductive workplace behavior (CWB). J. Prod. Brand. Manag. **31**, 1018–1032 (2022)
4. Hussein, S.S., Alfiansyah, M.W., Daud, R., Ya'acob, S., Lokman, A.M.: Developing an effective ICT strategic framework for higher education institutions: a case of Mataram University. In: International Conference on Knowledge Management in Organizations, pp. 210–221. Springer (2023)
5. Bouck, E.C.: Assistive technology. Sage Publications (2015)
6. Buning, M.E., Hammel, J., Angelo, J., Schmeler, M., Doster, S., Voelkerding, K., et al.: Assistive technology within occupational therapy practice. AJOT Am. J. Occup. Ther. **58**, 678–681 (2004)
7. Mallin, S.S.V., de Carvalho, H.G.: Assistive technology and user-centered design: emotion as element for innovation. Procedia Manuf. **3**, 5570–5578 (2015)
8. Layton, N., Smith, R.O., Smith, E.M.: Global outcomes of assistive technology: what we measure, we can improve. In: Assistive Technology, p. 673. Taylor & Francis (2022)
9. Soares, L.F.L., Mollo Tormin, L., Carvalho, K.S., de J. Alves, A.C.: Assistive technology for Para-badminton athletes: the application of the matching person and technology theoretical model in occupational therapy. Disabil. Rehabil. Assist. Technol. **19**, 1170–1177 (2024)
10. Musri, N.I.F., Osman, R., Hamid, N.H.A., Razak, F.H.A.: Adoption barriers of assistive ambient technology: a systematic literature review. In: International Visual Informatics Conference, pp. 200–208. Springer (2023)
11. Waller, R.: Graphic literacies for a digital age: the survival of layout. Inf. Soc. **28**, 236–252 (2012)
12. Adhithya, K.: Designing accessible products, https://uxdesign.cc/designing-accessible-products-e8aa79b55ebc

13. Bidin, S.A.H., Lokman, A.M.: Enriching the comfortability emotion on website interface design using Kansei engineering approach. In: Proceedings of the 7th International Conference on Kansei Engineering and Emotion Research 2018 (KEER 2018), 19–22 March 2018, pp. 792–800. Springer, Kuching, Sarawak, Malaysia (2018)
14. Forlizzi, J., Battarbee, K.: Understanding experience in interactive systems. In: Proceedings of the 5th conference on Designing interactive systems: processes, practices, methods, and techniques, pp. 261–268 (2004)
15. Lokman, A.M., Ishak, K.K., Razak, F.H.A., Aziz, A.A.: The feasibility of PrEmo in cross-cultural Kansei measurement. In: 2012 IEEE Symposium on Humanities, Science and Engineering Research, pp. 1033–1038. IEEE (2012)
16. Nyagadza, B., Mazuruse, G., Muposhi, A., Chuchu, T., Makoni, T., Kusotera, B.: Emotions' influence on customers'e-banking satisfaction evaluation in e-service failure and e-service recovery circumstances. Soc. Sci. Humanit. Open. **6**, 100292 (2022)
17. Bidin, S.A.H., Lokman, A.M., Mohd, W.A.R.W., Tsuchiya, T.: Initial intervention study of kansei robotic implementation for elderly. Procedia Comput. Sci. **105**, 87–92 (2017)
18. Karim, H.A., Lokman, A.M., Redzuan, F.: Older adults perspective and emotional respond on robot interaction. In: 2016 4th International Conference on User Science and Engineering (i-USEr), pp. 95–99. IEEE (2016)
19. Redzuan, F., Lokman, A.M., Othman, Z.A., Abdullah, S.: Kansei design model for e-learning: A preliminary finding. In: Proceeding of the 10th European Conference on e-Learning, pp. 685–696 (2011)
20. Redzuan, F., Lokman, A.M., Othman, Z.A., Abdullah, S.: Kansei design model for engagement in online learning: a proposed model. In: Informatics Engineering and Information Science: International Conference, ICIEIS 2011, Kuala Lumpur, Malaysia, November 12–14, 2011. Proceedings, Part I, pp. 64–78. Springer (2011)
21. Hussin, S.N., Lokman, A.M.: Kansei website interface design: practicality and accuracy of Kansei Web Design Guideline. In: 2011 International Conference on User Science and Engineering (i-USEr), pp. 30–35. IEEE (2011)
22. van der Maden, W., Lomas, D., Hekkert, P.: A framework for designing AI systems that support community wellbeing. Front. Psychol. **13**, 1011883 (2023)
23. Lombart, C., Millan, E., Normand, J.-M., Verhulst, A., Labbé-Pinlon, B., Moreau, G.: Consumer perceptions and purchase behavior toward imperfect fruits and vegetables in an immersive virtual reality grocery store. J. Retail. Consum. Serv. **48**, 28–40 (2019)
24. Kim, Y.H., Kim, D.J., Wachter, K.: A study of mobile user engagement (MoEN): Engagement motivations, perceived value, satisfaction, and continued engagement intention. Decis. Support. Syst. **56**, 361–370 (2013)
25. Lokman, A.M., Harun, A.F., Noor, N.L.M., Nagamachi, M.: Website affective evaluation: analysis of differences in evaluations result by data population. In: Human Centered Design: First International Conference, HCD 2009, Held as Part of HCI International 2009, San Diego, CA, USA, July 19–24, 2009. Proceedings 1, pp. 643–652. Springer (2009)
26. Shuhidan, S.M., Lokman, A.M., Hamidi, S.R., Kadir, S.A., Syahirah, S., Alam, M.M.: Unfolding emotions for creating happiness and quality of life in Malaysia's low-income community using text mining. J. Community Appl. Soc. Psychol. **33**, 1078–1098 (2023)
27. Lokman, A.M.: KE as affective design methodology. In: 2013 International Conference on Computer, Control, Informatics and Its Applications, pp. 7–13. IEEE (2013)
28. Lokman, A.M., Kadir, S.A., Noordin, F., Shariff, S.H.: Modeling Factors and Importance of Happiness Using KJ Method. In: Lokman, A.M., Yamanaka, T., Lévy, P., Chen, K., and Koyama, S. (eds.) Proceedings of the 7th International Conference on Kansei Engineering and Emotion Research 2018 (KEER 2018), pp. 870–877. Springer, Singapore (2018)

29. Lokman, A.M., Mohamed Shuhidan, S., Hamidi, S.R., Kadir, S.A., Syahirah, S.: Exploring indicators for happiness and its effect to people's emotion using LEIQ™. In: 9th International Conference on Kansei Engineering and Emotion Research KEER2022 Proceedings, pp. 315–322. Kansei Engineering and Emotion Research (KEER) (2022)
30. Fredrickson, B.L.: The role of positive emotions in positive psychology: The broaden-and-build theory of positive emotions. Am. Psychol. **56**, 218 (2001)

Key Determinants of Value-Based Management for ICT Project Delivery: Insights from Kansei Engineering and Emotion Research

Surya Sumarni Hussein[1](✉), Nur Hanis Solehah Mohd Rosli[1], and Azran Ahmad[2]

[1] College of Computing, Informatics, and Mathematics, Universiti Teknologi MARA, Shah Alam, Malaysia
`suryasumarni@uitm.edu.my`
[2] College of Computing and Informatic, Universiti Tenaga Nasional, Putrajaya, Malaysia
`azran@uniten.edu.my`

Abstract. This study explores the key determinants of Value-Based Management (VBM) in Information and Communication Technology (ICT) project delivery, incorporating insights from Kansei Engineering and emotion research. VBM emphasizes optimizing project outcomes to align with organizational strategic goals and stakeholder value. However, traditional approaches often overlook the emotional and psychological aspects influencing user satisfaction and project success. By integrating Kansei Engineering, which focuses on translating users' emotional responses into design elements, the study employed a qualitative method with interview sessions to collect information. Five questions were formulated based on a few relevant works of literature to obtain feedback from the targeted population and a thematic analysis was conducted to analyse the interview data. At the end of this study, the objectives of this research were achieved as the eleven key determinants of value-based management and its contribution toward the successful delivery of ICT projects respectively were identified and practical guidance to improve value-based management in ICT projects was provided. The results suggest that a holistic understanding of user emotions and their translation into project deliverables can lead to more successful and value-driven ICT projects. This study contributes to the field by bridging the gap between emotion research and value-based management, proposing practical strategies for enhancing project delivery in the ICT sector.

Keywords: Value-based Management · ICT Projects · Kansei Engineering

1 Introduction

The field of ICT has seen a rise in interest in value-based management in recent years. Value-based management is a management philosophy that emphasizes the significance of providing value to stakeholders such as consumers, shareholders, workers, and society as a whole [1]. This approach focuses on delivering business value and allocating resources to activities that provide the most value for the organization.

ICT projects are particularly well-suited for value-based management since they are frequently complicated, costly, and have high stakes [2]. Effective project delivery demands meticulous planning, management, and execution, as well as a thorough grasp of stakeholders' requirements and expectations. Despite the importance of value-based management in ICT projects, there is still a lack of clarity on what constitutes best practices and how to effectively execute them [1].

To address this knowledge gap, researchers have explored the key determinants of value-based management in ICT projects. Some studies have concentrated on specific areas of value-based management, such as risk management or stakeholder involvement, while others have taken a more comprehensive approach.

This research identifies and analyse the key determinants of VBM in the context of ICT project delivery. By integrating Kansei Engineering with Value-Based Management in ICT project delivery, organizations can enhance user satisfaction and project success. This integration ensures that both the functional and emotional needs of users are addressed, leading to higher perceived value and better alignment with organizational goals. Understanding and applying the principles of Kansei Engineering in VBM can thus play a crucial role in delivering ICT projects that are both emotionally resonant and strategically valuable.

2 Literature Review

This section will discuss the previous research studies that are relevant and related to the research study. Based on the research study, there are three main concepts or criteria to focus on in this study, which are value-based management (VBM), key determinants of VBM, and theories of implementing VBM in ICT project delivery.

2.1 Value-Based Management

Definition. Value-based management (VBM) is a business model that attempts to increase long-term shareholder value by aligning managerial actions with stakeholder interests [3, 4]. VBM's major goal is to improve organizational performance by emphasizing value generation and strategic alignment in decision-making processes [5, 6]. According to Corazza [7], in the context of VBM, value in information and communication technology (ICT) projects refers to the ability to strategically leverage ICT resources to improve business performance, gain a competitive advantage, and maximize returns on investment, all while ensuring alignment with organizational objectives and stakeholder interests. Corraza's study focuses on the idea of project complexity as it relates to ICT projects and value management, emphasizing the importance of aligning project objectives with organizational goals and stakeholders' interests.

In literature, VBM has garnered a lot of attention, with several authors giving frameworks and best practices for its application. Beck and Brit [8], on the other hand, define VBM as a value-driven strategy that prioritises stakeholder satisfaction, strategic alignment, and continual development. They contend that VBM may provide value by increasing operational efficiency, boosting customer happiness, and optimising resource allocation.

The application of VBM has been proven to improve organisational performance. Beck [9] discovered that VBM can improve financial performance, product quality, and employee happiness in her study. Corazza [7] discovered that VBM may lead to increased profitability, enhanced customer happiness, and improved innovation. To add on, Nichol [10] mentioned that implementing value-based management ensures that strategic programs achieve their value.

However, implementing VBM is not without difficulties. According to Beck and Brit [8], successful VBM adoption necessitates a shift in organisational culture, good communication, and a clear knowledge of stakeholder interests. They also contend that VBM necessitates an organised approach to performance measurement, effective technology utilisation, and stakeholder involvement.

Overall, VBM is a frequently debated subject in the literature that has the potential to aid organisations looking to develop long-term value. VBM deployment demands a systematic approach to performance assessment, stakeholder involvement, and strategy alignment [11]. However, the obstacles to adopting VBM cannot be overlooked, and organisations must take these into account while establishing their VBM strategy.

VBM in ICT Project. In the field of information and communication technology (ICT) project management, VBM is a concept that is becoming more and more significant. VBM is a management approach that stresses a strong emphasis on generating value for every stakeholder involved in a project, including clients, shareholders, employees, and suppliers. VBM seeks to maximize value by coordinating organizational goals with project outcomes. VBM is a response to the shortcomings of conventional cost-based management strategies, which emphasize minimizing costs at the expense of other elements that contribute to value creation [12, 13].

ICT projects, which are frequently complicated and demand major time, money, and resource inputs, might greatly benefit from VBM. VBM is a framework that enables organizations to prioritize value creation and match project activities to those priorities. Organizations may improve their chances of completing projects and satisfying the demands of all involved stakeholders by putting a strong emphasis on value creation [14]. Despite the increased interest in VBM in ICT projects, there is continuous discussion over the best methods and strategies for putting this approach into practice. This can be a result of the challenges associated with estimating value in ICT projects. VBM demands businesses to consider a variety of intangible characteristics, such as customer satisfaction, brand reputation, and employee morale, in contrast to the conventional cost-based management approach [15]. Effective VBM implementation can be difficult since these elements are sometimes hard to evaluate and quantify.

Despite these difficulties, there is mounting evidence that VBM may be a useful strategy for managing ICT projects. For instance, research by Mendes et al. [16] discovered that businesses that implement VBM practice are more likely to succeed with projects and have greater levels of client satisfaction. This implies that businesses that put value on creation first may be better able to satisfy the demands of their stakeholders and complete their projects on time.

The ability of VBM to assist organizations in focusing on outputs rather than inputs is one possible advantage in the context of ICT projects. In other words, VBM encourages organizations to consider the goals of their projects rather than only concentrating on the inputs (such as the time, money, and resources needed to accomplish the project) [17]. This might be crucial while working on ICT projects since they frequently include rapidly changing stakeholder requirements and technological advancements.

Overall, the research points to VBM as a useful project management method for ICT projects. To find the best methods and approaches for putting this strategy into practice, additional study is necessary. Despite these obstacles, businesses that place a high priority on value creation may be better able to fulfill their stakeholders' expectations and complete projects.

2.2 Key Determinants of Value-Based Management

The key factors that influence VBM have received a great deal of attention in the literature, with various writers offering frameworks and best practices for its implementation. To begin, Norman [18] suggests that one of the primary factors of VBM is the alignment of managerial actions with the organization's strategic objectives. This entails establishing specific goals and ensuring that all administrative choices are made with these goals in mind. This necessitates a comprehensive grasp of the organization's strategic objectives and how each choice impacts the attainment of these objectives. As a result, organisations must be strategic and forward-thinking in their approach to VBM.

Second, as Piero [19] points out, effective performance assessment is another significant factor of VBM. To properly adopt VBM, organisations must have a systematic approach to performance assessment, with defined metrics that fit with the organisation's strategic objectives. This will assist in guaranteeing that performance is regularly measured and that a clear relationship exists between performance and the attainment of organisational goals.

Third, as emphasised by Firk et al., [20], a culture of continuous improvement is important to the effective deployment of VBM. This necessitates emphasis on continuing learning and growth, as well as a dedication to increasing performance and generating improved outcomes for stakeholders. Stakeholder participation and effective use of technology are also vital in developing a culture of continuous improvement.

Fourth, Beck [9] highlights the contradiction between short-term and long-term goals as a key barrier to VBM deployment. Organisations must strike a balance between short-term financial goals and long-term value generation. This calls for a strategic approach to decision-making that considers the long-term implications of actions as well as the interests of all stakeholders.

Finally, Malmi and Ikäheimo [21] emphasise the need for good communication and participation in the successful implementation of VBM, particularly in complex organisations with various stakeholders. To guarantee buy-in and support, organisations must effectively explain their VBM plan to all stakeholders and engage them in the process. This involves clear and concise communication, as well as effective engagement approaches that consider the requirements and interests of numerous stakeholders.

Overall, the major factors in VBM are complicated, necessitating a deliberate, forforward-thinking approach. Effective performance assessment, a culture of continuous improvement, and effective communication and participation are all required for VBM to be implemented successfully. Organisations must, however, balance short-term and long-term goals and consider the interests of all stakeholders in their decision-making processes. Organisations may develop long-term shareholder value and deliver better outcomes for all stakeholders by doing so.

2.3 Theories of Implementing VBM in ICT Project Delivery

Several theories have been proposed for implementing VBM in ICT project delivery. These theories provide insights into how organizations can effectively adopt VBM practices and align their project activities with the creation of value.

The resource-based view (RBV) theory is another theory that has been put forth for adopting VBM in the execution of ICT projects. This approach focuses on the significance of utilising an organization's unique assets and skills to generate value. The RBV theory states that organisations are more likely to acquire a sustainable competitive advantage if they have a special mix of resources and capabilities. The RBV theory can help organisations recognize and use their special resources and skills to add value in the context of delivering ICT projects. Malmi and Ikäheimo [21], for instance, developed a methodology for assessing the value creation potential of ICT projects based on the organization's particular resources and competencies using the RBV theory.

Kansei Engineering (KE) sees the Kansei concept applied to product design. Consumer's feelings and overall impressions are regarded as perceptual design elements and, along with the product or service's physical attributes, through a reductionist approach, are broken down and prioritized. Kansei Engineering is a method that translates customers' emotions and feelings (Kansei) into product design elements [27, 28]. It was developed in Japan and is widely used in product development to ensure that the design aligns with the emotional needs and preferences of users. The process involves gathering and analyzing users' emotional responses to different design features and using this information to guide the design process [22, 23].

In summary, theories like Kansei engineering and RBV theory can offer helpful insights into how organisations can successfully adopt VBM practices in ICT project delivery. Organisations may boost the chance of project success and long-term value creation by considering the demands and interests of all stakeholders and utilizing their unique resources and competencies.

3 Research Methodology

This section will outline the research design and its essential elements. The methodology for the research is identified first, followed by the research setting, sample, data collection method, type of data being collected and data analysis.

3.1 Research Design

This study employs a qualitative approach as a research design methodology. The primary data is collected based on interview sessions with ICT project executives, while secondary data is gathered through UiTM online databases. The setting of the study is focused on completed ICT projects in Klang Valley and the research has been conducted from November 2023 to January 2024. Table 1 below depicts the details of the research design for this research.

Table 1. Research Design.

Research Questions	Research Objectives	Method/Activities	Deliverable
RQ1: What is the definition and importance of value-based management in the context of ICT projects?	RO1: To identify the key determinants of value-based management in ICT projects	Identify the definition of VBM in the context of ICT projects through: 1. Theoretical study 2. Literature review	Research background and problem statement were determined Key determinants of VBM were identified. (LR) RO1 achieved

RQ2: What are the key determinants of value-based management in ICT projects, and how do they contribute to project success?	RO2: To analyze the key determinants of value-based management in ICT projects	Identify and analyze the key determinants of VBM by:
		1. Interview: Project Manager and PMOs of ICT projects
		  a) Tool: Interview question
		  b) Technique: Purposive/Judgemental Sampling
		2. Participant: Sampling Size (5 PMs and PMOs)
			Key determinants of VBM were identified. (Interviews)
			RO2 achieved

RQ3: How can project managers and organizations improve their value-based management practices in ICT projects, and what are the implications for project success and overall business value?	RO3: To recommend practical guidance for project managers and organizations seeking to improve their value-based management practices in ICT projects based on the research findings	Documenting the participants' feedback a) Tool: Thematic Analysis	Practical recommendations for project managers and organizations will be listed RO3 achieved

Data Collection. A semi-structured interview has been conducted to identify the key determinants of VBM in ICT project delivery. For the interviews, five questions have been formulated based on research by Beck [9] and Mielcarz et al. [24]. All interviews were conducted online via Google Meet. This type of interview is chosen because it is more powerful than other types of interviews for qualitative research as it allows researchers to acquire in-depth information and evidence from interviewees while considering the focus of the study. It also allows flexibility and adaptability for researchers to hold their track as compared to an unstructured interview, where its direction is not fully considered

[25]. The session was conducted in Bahasa Malaysia and English and lasted between 15 to 45 min. The interviews occurred from November to December 2023. Five participants have participated in this research study, two of them are from the government sector (P1 and P2), while the other three are from the private sector (P3, P4, and P5). Table 2 below summarizes the demographic information for all participants.

Table 2. Demographic Information.

Position	PMO	PMO	Project Manager /Head of Sales	Project Manager /Business Development Director	Project Manager
Participant ID	Participant 1 (P1)	Participant 2 (P2	Participant 3 (3)	Participant 4 (4)	Participant 5 (P5)
Agency/ Company	Agency 1	Agency 1	Agency 1	Agency 1	Agency 1
Category	Management & Professional	Management & Professional	Middle Management	Top Management	Middle Management
Working Experience	15 years	8 years	16 years	17 years	22 years
Experience in ICT Project	12 years	5 years	13 years	10 years	28 years
Interview Information	29th Nov 2023	3rd Dec 2023	5th Dec 2023	8th Dec 2023	16th Dec 2023

Data Analysis. After the interview sessions are completed, a thematic analysis is performed. Thematic analysis is a research approach for analysing written or spoken communication. It can aid in the organisation and simplification of complicated data into useful and manageable codes, categories, and topics [26]. The thematic analysis offers six stages of data gathering and analysis that provide an approachable way to manage the difficult procedures of shifting between specific descriptions and abstract judgments influenced by the literature.

4 Findings and Discussions

This section represents collected information on key determinants of value-based management and its contribution toward the successful delivery of ICT projects and practical guidance to improve value-based management in ICT projects based on a defined research method.

4.1 Key Determinants of VBM

From the literature review that was done during the research study, the researcher was able to identify eight key determinants of VBM in ICT projects. The key determinants are the strategic alignment of managerial actions and business goals, a solid project plan, effective performance assessment, systematic stakeholder management, supportive team members, a culture of continuous improvement, short-term and long-term conflict management, and effective communication. However, through the semi-structured interview, three key determinants emerged making the findings into a total of eleven key determinants of VBM. Table 3 below depicts the analysed results of the participants' interviews.

Table 3. Participants Input on Key Determinants of VBM.

Themes	Attribute	Participant's Input				
		P1	P2	P3	P4	P5
Key determinants of VBM	The needs of the project	/				
	Fulfill the requirement	/		/	/	
	Stakeholder management		/	/		/
	Good project plan			/		
	Wise use of resources			/		
	Compatibility technology			/		
	Good change management plan			/		
	Be customer-centric			/		
	Supportive team			/		
	Mutual understanding			/	/	
	Constant communication					/

4.2 Kansei Engineering to Improve VBM in ICT Projects

Kansei Engineering is a method that translates customers' emotions and feelings (Kansei) into product design elements. It was developed in Japan and is widely used in product development to ensure that the design aligns with the emotional needs and preferences of users [27]. The process involves gathering and analyzing users' emotional responses to different design features and using this information to guide the design process [22, 28]. Although we are moving towards increased digitization, humans remain at the centre of product design and development. The human aspects must reflect the full range of human society, particularly the exciting prospect of Society 5.0 input. Three recommendations for improving VBM in ICT projects with Kansei insight were identified [29, 30]. It has mapped with the Kansei engineering process. The process involves gathering and analyzing users' emotional responses to different design features and using this information to guide the design process as discussed below:

i. Requirement Analysis: When conducting a requirement analysis, Kansei Engineering can be used to find characteristics that will elicit favorable emotional responses from consumers. This is consistent with VBM's emphasis on locating and delivering project features that are driven by value.
ii. Design and Prototyping: To increase the possibility of project success and value delivery, Kansei Engineering may guide the design and prototyping stages of ICT projects by ensuring that the design aspects fit with users' emotional demands.
iii. User Testing and Feedback: When Kansei Engineering is used for user testing, emotional feedback may be gathered and utilised to improve project features and make sure they provide the intended value. Through continuous alignment of project outputs with value maximization, this iterative approach enables VBM.

Table 4 below depicts the analysed results of the participants' interviews that extract the attribute of project success with Kansei insight.

Table 4. VBM Attributes of Project Success with Kansei Insight.

Themes	Attribute	Participant's Input				
		P1	P2	P3	P4	P5
A practical guide to improve VBM in ICT project	Requirement Analysis			/	/	/
	Design and Prototyping			/		/
	User testing and feedback			/	/	/

5 Conclusions

This study has successfully identified key determinants of VBM and its contribution to the project's success, also providing practical guidance to project managers and organizations seeking to improve VBM practices in ICT projects. The researcher was able to identify eleven key determinants of VBM in ICT projects from this research. The key determinants are the strategic alignment of managerial actions and business goals, a solid project plan, effective performance assessment, systematic stakeholder management, supportive team members, a culture of continuous improvement, short-term and long-term conflict management, and effective communication. Understanding and applying the principles of Kansei Engineering in VBM can thus play a crucial role in delivering ICT projects that are both emotionally resonant and strategically valuable. By integrating Kansei Engineering with Value-Based Management in ICT project delivery, organizations can enhance user satisfaction and project success. This integration ensures that both the functional and emotional needs of users are addressed, leading to higher perceived value and better alignment with organizational goals. Understanding and applying the principles of Kansei Engineering in VBM can thus play a crucial role in delivering ICT projects that are both emotionally resonant and strategically valuable. By implementing these practices, ICT projects can enhance overall management, stakeholder satisfaction, and the delivery of value aligned with organizational objectives.

Acknowledgments. The authors would like to acknowledge the Research Initiative Group for Emotion, Kansei and Design Engineering (RIG EKDE), Universiti Teknologi MARA, as well as Malaysia Association of Kansei Engineering (MAKE), for all their assistance in the research paper.

Disclosure of Interests. The authors have no competing interests to declare that are relevant to the content of this article.

References

1. Wobst, J., Tanikulova, P., Lueg, R.: Value-based management: a review of its conceptualizations and a research agenda toward sustainable governance. Journal of Accounting Literature (2023)
2. Burcovich, L.: Value-Based Management: Performance Indicators of Value Creation. Alternative Measures to Assess Shareholder Returns (2021)
3. Nowotny, S., Hirsch, B., Nitzl, C.: The influence of organizational structure on value-based management sophistication. Management Accounting Research **56** (2022). https://doi.org/10.1016/j.mar.2022.100797
4. Pratomo, T.P., Suhartati, W.S.: Identification of AKHLAK Values-Based Management's Determinant in Indonesia State-Owned Enterprises Ecosystem. In: The 2021 12th International Conference on E-business, Management and Economics, pp. 610–615 (2021)
5. Alsolami, B.M.: Identifying and assessing critical success factors of value management implementation in Saudi Arabia building construction industry. Ain Shams Eng. J. **13**(6), 101804 (2022)
6. Firk, S., Richter, S., Wolff, M.: Does value-based management facilitate managerial decision-making? An analysis of divestiture decisions. Manage. Acc. Res. **51** (2021). https://doi.org/10.1016/j.mar.2021.100736
7. Corazza, G.: Value Based Management Systems and Firm Performance: An Analysis of the Literature (2020). https://doi.org/10.26493/978-961-6832-68-7.8
8. Beck, V., Britzelmaier, B.: A critical review on surveys of value-based management. Int. J. Manage. Cases 270–286 (2011). www.ijmc.orgwww.circle-international.co.uk
9. Beck, V.: The effects of the implementation of value-based management. Int. J. Eco. Sci. App. Res. **7**(2) (2014). http://ssrn.com/abstract=2535924
10. Nichol, P.: Value Management as an Organizational Capability Full Paper (2020). https://doi.org/10.13140/RG.2.2.34402.56002
11. Mavropulo, O., Rapp, M.S., Ueva, I.A.: Value-based management control systems and the dynamics of working capital: Empirical evidence. Management Accounting Research **52** (2021). https://doi.org/10.1016/j.mar.2021.100740
12. Gaponenko, T., Dovbysh, V., Filin, N., Bulatova, R.: Building a value based human resource management system. In: E3S Web of Conferences, Vol. 273, p. 08010. EDP Sciences (2021)
13. Zakharchenko, V., Yermak, S.: Directions for improving the organisational system of value-based management in a high-tech company. Economics & Education **7**(4), 1319 (2022)
14. Mandić, V., Rodríguez, P., Kuvaja, P., Oivo, M.: Value-Management Challenges: Experiences from Research Projects with FinnishICT Industry (2011). http://www.sbl.tkk.fi/vaspo
15. Oane-Marinescu, C.M., Smoląg, K., Marinescu, E.S., Szopa, R.: Value-based management as the innovating paradigm of contemporary governance: a theoretical approach. Polish J. Manage. Stud. **12**(1), 106–120 (2015)

16. Mendes, E., Rodriguez, P., Freitas, V., Baker, S., Atoui, M.A.: Towards improving decision making and estimating the value of decisions in value-based software engineering: the VALUE framework. Software Qual. J. **26**, 607–656 (2018)
17. Bijlsma, M., Jongebreur, L.P.W., Winthagen, S.: Application of System Dynamics in Value Based Management-an approach based on a real-life case (2002)
18. Normann-Tschampel, C.: Value-based management (VBM) in Mittelstand–the relevance of VBM to specifically identified areas of management (Strategic decision-making, objectives, attitudes) (2019)
19. Piero, M.: Quality, a Key Value Driver in Value Based Management (2019). https://doi.org/10.13132/2038-5498/9.4.1970
20. Firk, S., Schmidt, T., Wolff, M.: CFO emphasis on value-based management: Performance implications and the challenge of CFO succession. Manag. Account. Res. **44**, 26–43 (2019). https://doi.org/10.1016/j.mar.2018.11.001
21. Malmi, T., Ikäheimo, S.: Value based management practices – some evidence from the field. Manag. Account. Res. **14**(3), 235–254 (2003). https://doi.org/10.1016/S1044-5005(03)00047-7
22. Schütte, S., et al.: Kansei for the Digital Era. Int. J. Affect. Eng. **23**(1), 1–18 (2023)
23. Lokman, A.M.: Kansei/affective engineering and Web design. Kansei/affective engineering, 227–251 (2011)
24. Mielcarz, P., Osiichuk, D., Wnuczak, P.: An Inquiry into Determinants of Value Creation in the ICT Industry. IMPACT, 71 (2016)
25. Ruslin, R., Mashuri, S., Rasak, M.S.A., Alhabsyi, F., Syam, H.: Semi-structured Interview: a methodological reflection on the development of a qualitative research instrument in educational studies. IOSR J. Res. Method in Edu. (IOSR-JRME) **12**(1), 22–29 (2022)
26. Peel, K.L.: A beginner's guide to applied educational research using thematic analysis. Pract. Assess. Res. Eval. **25**(1), 2 (2020)
27. Redzuan, F., Mohd. Lokman, A., Ali Othman, Z., Abdullah, S.: Kansei design model for engagement in online learning: a proposed model. In: Informatics Engineering and Information Science: International Conference, ICIEIS 2011, Kuala Lumpur, Malaysia, November 12-14, 2011. Proceedings, Part I, pp. 64–78. Springer Berlin Heidelberg (2011)
28. Lokman, A.M., Harun, A.F., Md. Noor, N.L., Nagamachi, M.: Website affective evaluation: Analysis of differences in evaluations result by data population. In: Human Centered Design: First International Conference, HCD 2009, Held as Part of HCI International 2009, San Diego, CA, USA, July 19-24, 2009 Proceedings 1, pp. 643–652. Springer Berlin Heidelberg (2009)
29. Hussin, S.N., Lokman, A.M.: Kansei website interface design: Practicality and accuracy of Kansei Web Design Guideline. In: 2011 International Conference on User Science and Engineering (i-USEr), pp. 30–35. IEEE (2011)
30. Lokman, A.M., Ishak, K.K., Razak, F.H.A., Aziz, A.A.: The feasibility of PrEmo in cross-cultural Kansei measurement. In: 2012 IEEE Symposium on Humanities, Science and Engineering Research, pp. 1033–1038. IEEE (2012)

Exploring Research Direction for Human Metaverse Interaction

I Made Bambang Ariawan[1], Shamsiah Abd Kadir[2,3], Afiza Ismail[4], and Anitawati Mohd Lokman[4](✉)

[1] PT Berlian Cranserco Indonesia, Balikpapan, Kalimantan Timur 76115, Indonesia
[2] Centre for Research in Media and Communication, Faculty of Social Sciences and Humanities, Universiti Kebangsaan Malaysia, 43600 Bangi, Malaysia
shamkadir@ukm.edu.my
[3] Komunikasi Kesihatan (Healthcomm)-UKM Research Group, Universiti Kebangsaan Malaysia, 43600 Bangi, Malaysia
[4] College of Computing, Informatics and Mathematics, Universiti Teknologi MARA, UiTM Shah Alam, 40450 Shah Alam, Selangor, Malaysia
{afiza025,anitawati}@uitm.edu.my

Abstract. The term Metaverse has recently gained popularity, primarily due to the rise of virtual reality and social media. Metaverse is a shared virtual environment that allows for real-time interactions between people and digital objects. As metaverse technology evolves, it is expected to have a significant impact on social interactions, reality concepts, and cultural norms. Understanding the implications of Metaverse necessitates a thorough examination of previous research on human-metaverse interactions. The goal of this study is to conduct a bibliometric analysis on Human Metaverse Interaction (HMI), with an emphasis on identifying key concepts, authors, journals, and trends in this emerging field. Primary bibliographic data is gathered from the Scopus database using keyword searches. The data is then quantitatively analyzed using bibliometric techniques in VOSviewer software to reveal patterns and trends. The analysis looks at publication volume, content themes, country affiliations and frequency of keywords to map the intellectual structure of HMI research. The study reveals rising trends in HMI research output, with key focus areas including virtual reality, augmented reality, and metaverse taxonomy. Gaps are identified in areas such as human-robot interaction and social virtual reality. The study sheds light on the current state of knowledge and future directions for HMI research.

Keywords: Human Metaverse Interaction · Bibliometric Analysis · Virtual Reality · Augmented Reality · Emerging Technologies

1 Introduction

The research conducted a bibliometric analysis of the Human Metaverse Interaction (HMI) field, aiming to systematically review and analyze existing academic literature. The study identifies key concepts, authors, journals, and trends within the field, highlighting gaps and limitations in current research. By taking a quantitative approach, the research aims to offer a comprehensive understanding of HMI, which is increasingly relevant as Metaverse technology evolves and impacts human interaction, cultural norms, and our perception of reality.

The significance of this research lies in its potential to inform the development of Metaverse technology in a socially responsible manner. By identifying the main themes and trends in HMI, the study provides insights into areas requiring further investigation, thereby contributing to the creation of future research agendas. This will aid academics, technology companies, industry practitioners, policymakers, and the general public in understanding the implications of Metaverse technology on human society, especially concerning issues of identity, privacy, and digital interaction. The ultimate goal is to develop a theoretical framework that guides the socially responsible growth of Metaverse technologies.

2 Theoretical Background

The metaverse, a term combining "meta," meaning "beyond," and "universe," is a potential future interaction of the internet that allows users to enjoy immersive experiences through augmented reality, extended reality, and virtual realities. The metaverse project aims to develop a domain ontology (MetaOntology) that specifies the infrastructure and cutting-edge technologies relevant to the metaverse. This project adopts a four-step methodical approach to develop the ontology, but does not aim to provide a comprehensive overview of the domain.

The relevance of human-computer interaction (HCI) research in the creation of software, user interface toolkits, and the Internet was explored by Myers (1998). The research focuses on how advancements in interface technology have spurred the development of technology as a whole, and how universities and business research laboratories are then working on creating user interfaces for the next generation of computers. In recent decades the attention of researchers has shifted to the emotional aspect of user experience in mobile learning (Taharim et al., 2013; Taharim et al., 2015), website (Bidin and Lokman, 2018; Ismail and Lokman 2020; Lokman et al., 2022), e-learning (Redzuan et al., 2011), video (Abd Kadir et al., 2021) and avatar (Lokman et al. 2014).

Society is becoming more digitally, cybernetically, and informationally connected, leading to the growth of the metaverse idea. Wang et al. (2021) discuss plans for the Metaverse's development and propose a common system architecture and foundation for it. Feng et al. (2022) examined the metaverse's effects on social and economic theories and how they affect scholarly investigation.

Bibliometric analysis, a quantitative approach used to examine the significance of scientific publications, is a vital tool for identifying research trends, assessing the impact of research, and determining the future course of research for academics, institutions, and funding organizations. Donthu et al. (2021) discussed the value of bibliometric analysis for examining and analyzing scientific data, particularly in business research.

Bibliometric analysis is conducted using bibliometric software such as Gephi, Leximancer, and VOSviewer, which examines the social and structural connections among various research elements, such as authors, nations, institutions, and themes. VOSviewer is not only user-friendly but also capable of visualizing larger networks (Ahmi, 2022).

Several studies have examined the metaverse from a bibliometric perspective. Feng et al. (2022) conducted a bibliometric study of 183 metaverse-related research in the WoS core database since 2000, summarizing the characteristics, main topics, and rules of metaverse development in each stage. Verma et al. (2022) summarized the most important metaverse research using bibliometric analysis, addressing how technology develops economies, notably the industrial interest in metaverse technology.

Undale et al. (2023) stated that there is still more to learn about the metaverse, and more academics are becoming interested in it. They used bibliometric analysis to identify the top contributing authors, top nations, and top journals that publish research in the Metaverse by analyzing 504 papers from the Scopus database published between 1977 and 2022.

Shen et al. (2023) presented the findings of a 2012–2021 bibliometric investigation of academic Metaverse research, focusing on VR, AR, and other immersive technologies. The study sought to discover Metaverse research hotspots and frontiers, the most prolific nations, and their topic grouping and hotspot evolution.

According to Feng et al. (2022), university research advances more slowly than business research. Tlili et al. (2023) emphasized the need for more international collaboration to promote metaverse adoption internationally. Lukava et al. (2022) emphasized the necessity of more inclusive design principles for neurodiverse individuals when using XR technology, which can also be applied to the creation of the metaverse.

Despite numerous studies on the metaverse, there hasn't been a thorough bibliometric analysis of the field examining human interaction in the metaverse. The majority of current metaverse research is focused on computer software and applications, automation, and other scientific and technological fields.

3 The Methods

This section describes the methods used in the research implementation. As shown in Fig. 1, the research process began with an extensive search for relevant literature on January 10, 2024, resulting in 2,451 records that were included in the subsequent bibliometric study. Sample sizes were determined based on Rogers et al. (2020), who suggested that a sample size of less than 200 publications is considered a baseline for analytical robustness. Samples of 1,000 publications or more could give a reliable indication of relative citation performance at an institutional level, particularly when the research is led by high-performing individuals.

The primary data was sourced from Scopus, a comprehensive search engine that covers approximately 84% of all journal titles. The secondary data was collected through a meticulous literature review and obtained from the UiTM Online Database. Scopus was chosen as the primary data source due to its wide coverage and the availability of relevant research publications. The data collection method for this research involved keyword extraction, which was based on a prior literature review. Keywords related to human interaction and the metaverse were chosen to identify literature on this less-studied aspect and to find literature on how the metaverse is being applied in industries like education and healthcare.

The bibliometric analysis procedure proposed by Donthu et al. (2021) was used to analyze the data. The VOSviewer software was used to conduct bibliometric analysis, which examines the social and structural connections among various elements of research, such as authors, nations, institutions, and themes. This method provides a concise summary of the bibliometric and intellectual structure of a field of study. This research focused on several variables: the number of publications, subject area, countries of origin, and frequency of keywords. The number of publications indicates the amount of scholarly output and research in this area, while the subject area covers the areas of study interest. Countries of origin indicate the countries represented in the papers, while keyword frequency indicates the emphasis and prevalence of specific themes within the field.

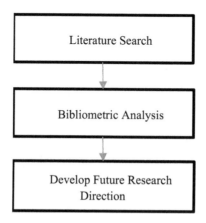

Fig. 1. The research process.

4 Results and Discussion

This section describes the research results and findings.

4.1 Number of Publications

Table 1 shown the dataset under consideration spans from 1985 to 2024, encompassing a total of 2451 scholarly papers. These papers have collectively garnered 23996 citations over a period of 39 years, resulting in an average of 615.28 citations per year. On a per-paper basis, the average number of citations stands at 9.79. The dataset exhibits an average of 8712.32 citations per author and 963.26 papers per author. Additionally, each paper, on average, has 3.65 authors. The h-index, a measure of productivity and impact, is recorded at 68, signifying that at least 68 papers in the dataset have received a minimum of 68 citations each. The g-index, another productivity measure, is 108, indicating that the top 108 papers collectively received 11664 citations. The hI,norm, a normalized h-index per year, is 36, offering a per-year measure of impact.

Table 1. Citation metrics.

Publication Year	1985–2024
Papers	2,451
Citations	23,996
Citations Years	39
Cites/Year	615.28
Cites/Paper	9.79
Cites/Author	8,712.32
Papers/Author	963.26
Authors/Paper	3.65
h-index	68
g-index	108
hI,norm	36
hI,annual	0.92
hA-index	35

The hI,annual, representing the annual increment of the h-index, is 0.92. Furthermore, the hA-index, an author-level h-index, is 35, suggesting that at least 35 papers by an author have each received a minimum of 35 citations.

The citation metrics show a reasonably high research impact and productivity over the 39-year period, with strong citation rates per paper and per author. The g-index and h-index values indicate that the top papers have had a particularly high citation impact. Overall, these metrics provide a quantitative overview of the influence and citation patterns of the scholarly output in the dataset.

Figure 2 shows the total number of publications as well as the cumulative number of publications in a particular study topic from 1985 to 2024. The first few years (1985–1999) see a slow rise in publications. But starting in 2000, there is a noticeable increase that peaks in 2004. Publications see a significant uptick between 2008 and 2011, with a peak in 2011. The number of publications is still comparatively high and consistent (although with some swings) from 2012 to 2019. There is a notable upsurge from 2020 to 2024, particularly in 2022 and 2023, when there are 551 and 972 publications, respectively. Interestingly, there is a notable spike in research production in 2023. Data for 2024 shows that there will be 41 more publications, for a total of 2451. The data indicates the research area has gone through distinct growth phases, with recent years showing accelerated expansion in scholarly publications and overall output.

4.2 Subject Area

As seen in Fig. 3, the dynamic landscape of human metaverse interaction reveals distinct roles played by various academic disciplines, with Computer Science taking the lead at 34.8%, followed by Engineering at 19.3%, emphasizing the technical foundations of

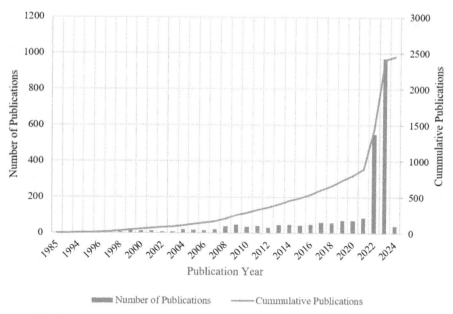

Fig. 2. The number of publications on a year-on-year basis in the field of HMI.

this field. Social Sciences (9.2%) and Decision Sciences (5.3%) contribute significantly to understanding user behavior and decision-making processes within the metaverse. Mathematics (4.5%) and Medicine (3.7%) bring quantitative and healthcare perspectives, respectively.

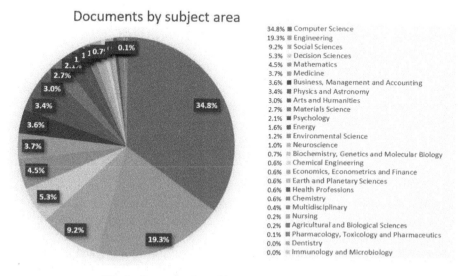

Fig. 3. Document by subject area in the field of HMI.

4.3 Countries of Origin

Table 2 presents a comprehensive overview of the top 5 countries that have made significant contributions to research on human metaverse interaction, each having published more than three documents. Topping the list is the United States, which has published 481 documents, constituting 14.32% of the total publications in the top 10. The research from the United States is notably impactful, receiving a substantial number of citations (8334) with an average of 17.33 citations per document. Collaborative efforts are evident, as indicated by a strong total link strength of 805. Following closely is China, with 368 publications (10.95% of the total) and a considerable number of citations (3195), averaging 8.68 citations per document. China also demonstrates collaborative strength with a total link strength of 619. South Korea, ranking third, has published 183 documents (5.45%), received 2626 citations, and exhibited a notable average citation per document of 14.35. The total link strength of 636 suggests a well-connected network of South Korean-authored documents. The table thus provides valuable insights into the research output, impact, and collaboration strengths of these leading countries in the specified field. Figure 4 shows network visualization of the counties with the bubble size indicating publication size from highest to lowest.

Table 2. Top 5 Countries that published more than three documents in the field of HMI.

No	Country	Quantity	Percentage (%)	Citation	Average Citation per Document	Total Link Strength
1	United States	481	14.32	8334	17.33	805
2	China	368	10.95	3195	8.68	619
3	South Korea	183	5.45	2626	14.35	636
4	United Kingdom	175	5.21	3377	19.30	616
5	Germany	173	5.15	3074	17.77	204

4.4 Frequency of Keywords

Table 3 outlines the top five keywords associated with research on human metaverse interaction, presenting their respective occurrences and total link strength. Notably, "virtual reality" stands out as the most prevalent keyword, occurring 1673 times with a substantial total link strength of 10505. Following closely are "metaverse" and "metaverses" with 930 and 901 occurrences, respectively. Other significant keywords include "augmented reality," "human computer interaction," and "virtual worlds." The "Total Link Strength" column further illuminates the interconnectedness and significance of each

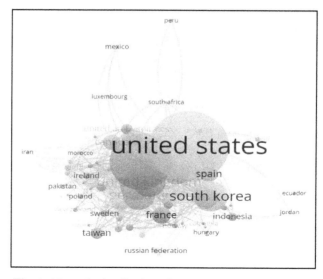

Fig. 4. Network visualization of countries in the field of HMI.

keyword within the context of human metaverse interaction research. This comprehensive overview of keyword frequencies and link strengths provides valuable insights into the prevalent themes and focal points in the literature on this subject. The network visualization can be shown in Fig. 5, with the bubble size indicating size of occurrences from highest to lowest.

Table 3. Top five keywords in the field of HMI.

No	Keyword	Occurrences	Total Link Strength
1	virtual reality	1673	10505
2	metaverse	930	5917
3	metaverses	901	6672
4	augmented reality	504	3452
5	human computer interaction	323	2221

The analysis of gaps in the literature and emerging trends in Human Metaverse Interaction (HMI) reveals several directions for future research. The lack of common terms like "virtual humans," "social virtual reality," and "human-robot interaction" highlights the need for more research on understanding and improving social interactions in virtual environments. Nevertheless, little research has been found addressing robot interaction and how it affects people who interact with robots (Aziz et al., 2015a; Aziz et al., 2015b; Ismail et al., 2016; Ismail et al., 2018). Multidisciplinary research that unites social sciences and computer science is crucial to fully examine the sociocultural dynamics of the metaverse.

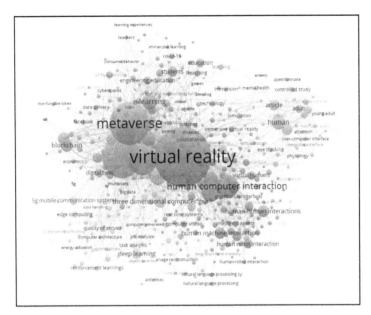

Fig. 5. Co-occurrence analysis all keywords in the field of HMI.

Research areas that require more study include Psychology, Arts and Humanities, and Neuroscience. Investigating brain correlates of immersive experiences, understanding psychological and cognitive components of people's interaction with the metaverse, and exploring artistic and creative expressions in virtual worlds can provide insights into HMI development.

The rising use of terms like "extended reality (XR)" and "digital twins" indicates a developing topic that needs further investigation. Future research may focus on the assimilation of digital twin technologies into the metaverse and the consequences of extended reality on immersive virtual environments.

Considering the metaverse's growing influence on interpersonal relationships, cross-disciplinary comparisons, investigation of cultural contexts impacting virtual environment user experiences, and examination of moral issues related to digital identity, privacy, and security in the metaverse are essential.

5 Conclusion

This study presents an in-depth bibliometric analysis that significantly contributes to the rapidly growing body of research in the field of Human Metaverse Interaction (HMI). The findings, recommendations, and future directions provided by this analysis serve as an essential roadmap for a diverse array of stakeholders, including scholars, researchers, policymakers, and practitioners. By identifying gaps in current research and highlighting areas with potential for further exploration, this study facilitates continuous evolution in HMI research to meet the diverse needs and challenges brought about by the dynamic and ever evolving metaverse (Feng et al., 2022).

This research is driven by the objective of stimulating a comprehensive, human-centric understanding of HMI. It aims to guide responsible and inclusive advancements in this dynamic and revolutionary domain. The metaverse continues its transformative journey, reshaping our digital interactions and experiences. As such, the importance of fostering a deep and inclusive understanding of human interaction within this digital universe cannot be overstated (Undale et al., 2023).

The bibliometric analysis conducted in this study reveals that research on HMI is intricate and continuously evolving. It brings to light the pressing need for researchers to engage with challenges, address the identified gaps, and commit to a thorough understanding of human interactions within the metaverse. It is this understanding that will form the foundation on which an inclusive, ethical, and human-centric metaverse can be built (Williams and Bornmann, 2016).

The study emphasizes the importance of interdisciplinary collaboration in achieving this goal. By working together across various disciplines, researchers can contribute to the creation of a metaverse that upholds ethical values, prioritizes human experiences, and promotes an inclusive digital world. The implications of this research extend beyond the academic sphere, influencing business practices, governmental decisions, and collective obligations towards creating a digital environment that enhances human well-being and connectivity (Wang et al., 2021).

This bibliometric analysis draws attention to the intricate and evolving nature of HMI research. As the metaverse continues to revolutionize our digital landscape, it is incumbent on researchers, academics, policymakers, and practitioners to embrace the challenges and opportunities presented by this dynamic domain. By doing so, they can contribute to the creation of a metaverse that is inclusive, ethical, and enhances human well-being and connectivity (Donthu et al., 2021; Myers, 1998).

Acknowledgments. The authors would like to acknowledge the Research Initiative Group for Emotion, Kansei and Design Engineering (RIG EKDE), Universiti Teknologi MARA (UiTM), as well as Malaysia Association of Kansei Engineering (MAKE), for all their assistance to the research paper.

Disclosure of Interests. The authors declare that there is no conflict of interest.

References

Abd Kadir, S., Lokman, A.M., Tsuchiya, T.: Emotional responses towards unity youtube videos: experts vs. viewers perspectives. Int. J. Affect. Eng. **20**(4), 225–235 (2021)

Ahmi, A.: Bibliometric Analysis for Beginners. UUM Press (2022)

Aziz, A.A., Moganan, F.F.M., Ismail, A., Lokman, A.M.: Autistic children's kansei responses towards humanoid-robot as teaching mediator. Procedia Computer Science **76**, 488–493 (2015)

Aziz, A.A., Moghanan, F.F.M., Mokhsin, M., Ismail, A., Lokman, A.M.: Humanoid-robot intervention for children with autism: a conceptual model on FBM. In: Soft Computing in Data Science: First International Conference, SCDS 2015, Putrajaya, Malaysia, September 2–3, 2015, Proceedings 1, pp. 231–241. Springer, Singapore (2015b). Abu-Salih, B.: MetaOntology: Toward developing an ontology for the metaverse. Frontiers in Big Data 5 (2022). https://doi.org/10.3389/fdata.2022.998648

Bidin, S.A.H., Lokman, A.M.: Enriching the comfortability emotion on website interface design using Kansei engineering approach. In: Proceedings of the 7th International Conference on Kansei Engineering and Emotion Research 2018: KEER 2018, 19–22 March 2018, Kuching, Sarawak, Malaysia, pp. 792–800. Springer, Singapore (2018)

Donthu, N., Kumar, S., Mukherjee, D., Pandey, N., Lim, W.M.: How to conduct a bibliometric analysis: an overview and guidelines. J. Bus. Res. **133**(April), 285–296 (2021). https://doi.org/10.1016/j.jbusres.2021.04.070

Feng, X., Wang, X., Su, Y.: An analysis of the current status of metaverse research based on bibliometrics. Library Hi Tech 11905042 (2022). https://doi.org/10.1108/LHT-10-2022-0467

Ismail, N.N.N.N., Lokman, A.M., Redzuan, F.: Kansei-spiritual therapeutic robot interaction design. In: Proceedings of the 7th International Conference on Kansei Engineering and Emotion Research 2018: KEER 2018, 19–22 March 2018, Kuching, Sarawak, Malaysia, pp. 580–591. Springer, Singapore (2018)

Ismail, N.N.N.N., Lokman, A.M.: Kansei engineering implementation in web-based systems: A review study. In: International Conference on Kansei Engineering & Emotion Research, pp. 66–76. Springer, Singapore (2020)

Karim, H.A., Lokman, A.M., Redzuan, F.: Older adults perspective and emotional respond on robot interaction. In: 2016 4th International Conference on User Science and Engineering (i-User), pp. 95–99. IEEE (2016)

Lokman, A.M., Isa, I.G.T., Novianti, L., Ariyanti, I., Sadariawati, R., Aziz, A.A.: Emotional user experience in web-based geographic information system: an Indonesian UX analysis. Int. J. Comp. Sci. Netw. Sec. (IJCSNS) **22**(9), 271–279 (2022)

Lukava, T., Morgado Ramirez, D.Z., Barbareschi, G.: Two sides of the same coin: accessibility practices and neurodivergent users' experience of extended reality. J. Enabling Technol. **16**(2), 75–90 (2022). https://doi.org/10.1108/JET-03-2022-0025

Myers, B.A.: A brief history of human-computer interaction technology. Interactions **5**(2), 44–54 (1998). https://doi.org/10.1145/274430.274436

Redzuan, F., Mohd. Lokman, A., Ali Othman, Z., Abdullah, S.: Kansei design model for engagement in online learning: a proposed model. In: Informatics Engineering and Information Science: International Conference, ICIEIS 2011, Kuala Lumpur, Malaysia, November 12-14, 2011. Proceedings, Part I, pp. 64–78. Springer, Berlin Heidelberg (2011)

Rogers, G., Szomszor, M., Adams, J.: Sample size in bibliometric analysis. Scientometrics **125**(1), 777–794 (2020). https://doi.org/10.1007/s11192-020-03647-7

Lokman, A.M., Mustafa, A.M., Fathir, M.F.M.: Avatar warrior: A Kansei analysis. In: Proceedings of the 3rd International Conference on User Science and Engineering (i-USEr), pp. 24–29. IEEE (2014)

Taharim, N.F., Lokman, A.M., Isa, W.A.R.W.M., Noor, N.L.M.: A relationship model of playful interaction, interaction design, kansei engineering and mobile usability in mobile learning. In: 2013 IEEE Conference on Open Systems (ICOS), pp. 22–26. IEEE (2013)

Taharim, N.F., Lokman, A.M., Isa, W.A.R.W.M., Noor, N.L.M.: Investigating feasibility of mobile learning (M-learning) for history lesson. In: International Colloquium of Art and Design Education Research (i-CADER 2014), pp. 541–550. Springer, Singapore (2015)

Shen, J., Zhou, X., Wu, W., Wang, L., Chen, Z.: Worldwide overview and country differences in metaverse research: a bibliometric analysis. Sustainability (Switzerland) **15**(4) (2023). https://doi.org/10.3390/su15043541

Tlili, A., Huang, R., Kinshuk: Metaverse for climbing the ladder toward 'Industry 5.0' and 'Society 5.0'? Service Industries Journal, 260–287 (2023). https://doi.org/10.1080/02642069.2023.2178644

Undale, S., et al.: A bibliometric analysis of citations and publications on Metaverse. In: 2023 10th International Conference on Computing for Sustainable Global Development (INDIACom), pp. 444–449 (2023)

Verma, J., Sharma, J., Sharma, A., Kaur, J.: Does Metaverse a technological revolution in artificial intelligence? a bibliometric analysis. In: PDGC 2022 2022 7th International Conference on Parallel, Distributed and Grid Computing, pp. 425–428 (2022). https://doi.org/10.1109/PDGC56933.2022.10053102

Wang, D., Yan, X., Zhou, Y.: Research on Metaverse: Concept, development and standard system. In: Proceedings 2021 2nd International Conference on Electronics, Communications and Information Technology, CECIT 2021, pp. 983–991 (2021). https://doi.org/10.1109/CECIT53797.2021.00176

Williams, R., Bornmann, L.: Sampling issues in bibliometric analysis: response to discussants. J. Informet. **10**(4), 1253–1257 (2016). https://doi.org/10.1016/j.joi.2016.09.013

Enhancement of Kansei Model for Political Security Threat Prediction Using Bi-LSTM

Liyana Safra Zaabar[1], Khairul Khalil Ishak[2], and Noor Afiza Mat Razali[1(✉)]

[1] Defence Science and Technology Faculty, National Defence University of Malaysia, Sungai Besi, Kuala Lumpur, Malaysia
noorafiza@upnm.edu.my
[2] Center for Cybersecurity and Artificial Intelligence, Management and Science University, Shah Alam, Selangor, Malaysia

Abstract. Online platforms serve as valuable sources for monitoring public sentiment related to political security threats. Thus, there is a crucial need to study the sentiment analysis and how it can be utilized to predict the threats. This study introduces a new approach to improve the political security threat prediction by proposing enhancement of Kansei model using deep learning Bidirectional Long Short-Term Memory (Bi-LSTM). Data from various sources, including social media and user comments was utilized to perform the prediction. This study discussed that utilization of manual methods to perform Kansei analysis from establishment of Kansei checklist to determine Kansei words can be improved by integrating Bi-LSTM to significantly enhanced the analysis process and predictive capabilities. Additionally, ADAM algorithm was adopted as optimizer. Experimental analysis was performed, and the results show that the enhanced Kansei model with Bi-LSTM and ADAM optimizer demonstrates improvement in accuracy and performance. This approach has the potential to assist authorities and organizations in making more informed and proactive decisions in addressing political security issues, thereby improving national stability and security.

Keywords: Kansei · Deep Learning · Bi-LSTM · ADAM · Sentiment Analysis · Security Threat · Political

1 Introduction

In today's digital era, the data generated by users through various social media platforms and other communication channels is increasing rapidly. This data contains a variety of important information including the sentiments and views of the public on current issues, especially in the political context. Sentiment analysis has become a very useful tool for detecting and understanding consumers' emotional reactions to political events, which in turn can be used to predict political security threats. The concept of Kansei, which emphasizes the influence of human emotions on perceptions and decisions, provides a solid foundation for understanding how feelings and emotional reactions can affect political security conditions [1, 2]. However, the main challenge in sentiment analysis is to derive accurate and meaningful predictions from large and complex data. Thus, to address the gap, this study investigates public sentiment towards online news related to national security through a Kansei lens. This study aims to contribute to a deeper understanding of how information dissemination influences public sentiments that can give impact to national security using hybrid Kansei method and deep learning. As reported by previous studies, the evaluation process for Kansei was divided into four phases: Instrument Preparation, Kansei Word Establishment, Kansei Evaluation, and Kansei Result Analysis [3, 4]. Kansei relied on the Kansei Checklist to manually determine Kansei Words, a process that was manual and time-consuming. For Kansei result analysis, statistical techniques of Principal Component Analysis (PCA) were utilized in the analysis process that involved manual data key-in to the tool. PCA effectively reduces the dimensionality of Kansei data while preserving essential information, facilitating the identification of underlying patterns and structures. By transforming a large set of variables into a smaller number of uncorrelated components, PCA has proven to be a method that supports discovery of latent dimensions of public sentiment and their relationships with specific news content. While PCA has been widely used, recent advancements in machine learning and deep learning offer alternative approaches for data reduction and pattern discovery and should be taken into consideration.

Autoencoders are recognized as a technique that can effectively reduce data dimensionality while preserving essential information. Additionally, clustering algorithms like K-means and Hierarchical Clustering can group similar Kansei words together swiftly with more accurate data acquisition. Meanwhile, deep learning models, including Recurrent Neural Networks (RNNs) and Long Short-Term Memory (LSTM) networks, can capture complex patterns and dependencies within data, leading to more nuanced insights for Kansei. Researchers discussed that Bidirectional Long Short-Term Memory (Bi-LSTM) networks is capable to extend the capabilities of LSTM by processing input sequences in both forward and backward directions that permits Bi-LSTM to capture contextual information from both past and future time steps and establishing a more comprehensive understanding of the sequence. As a result, in theory, Bi-LSTM can outperform traditional LSTM [5]. Furthermore, the combination of Bi-LSTM and ADAM optimizer adoption can contribute to more efficient, higher accuracy and performance in sentiment analysis for political security [6].

Thus, in this study, we introduce a new approach for Kansei analysis process automation using Bi-LSTM deep learning method optimized with ADAM optimizer for in the realm of sentiment analysis. This approach can be used to predict political security threats based on the sentiment expressed by users in social media platforms and other data sources. We believe that this approach will improve the performance and accuracy of political security threat predictions and provide deeper insight into how public sentiment evolves in a changing political context.

2 Related Works

Kansei, a Japanese term often translated as "sensitivity" or "sensitivity," refers to the subjective, emotional, and sensory aspects of the human experience [7]. Meanwhile, Kansei Engineering is a product development method that investigates human feelings and detects the quantitative relationship between affective responses and design features. It is used to improve the design of products and services, but traditional methods rely heavily on manual questionnaires, surveys, and workshops [8]. In the context of sentiment analysis, Kansei integrates these human factors into analysis to understand and predict people's emotions. It also captures and interprets more subtle emotional responses from the people. This approach goes beyond traditional sentiment analysis by focusing on the deeper layers of human emotions. Previous research discussed that, to determine Kansei Words, researchers utilized the manual process using Kansei Checklist, however, the integration of deep learning techniques such as Bi-LSTM algorithms revolutionize Kansei analysis by automating the process of extracting Kansei Words, thus, enabling faster and more accurate Kansei Words data acquisition.

2.1 Kansei Text Mining and Traditional Sentiment Analysis Method

Kansei text mining and traditional sentiment analysis have major differences in their approach and focus. Traditional sentiment analysis aims to classify texts based on emotions or opinions such as positive, negative, or neutral, using natural language processing (NLP) techniques and machine learning algorithms to determine sentiment polarity. In contrast, Kansei text mining focuses on identifying and analyzing the more subtle and profound feelings, emotions, or impressions expressed in the text, often using more specific approaches such as Kansei engineering to assess the user's perception of a particular aspect. The result of traditional sentiment analysis is usually the classification of text into sentiment categories, while Kansei text mining provides more detailed insights into how certain aspects affect user feelings, used in product design, marketing, and user experience improvement.

2.2 Kansei in Sentiment Analysis

Bibliometric Analysis. Table 1 presents main information regarding selected articles on topic Kansei in sentiment analysis from the Scopus database. This analysis employed bibliometric methods to statistically review and interpret the bibliometric data relevant to portfolio Kansei in sentiment analysis. This bibliometric study dataset includes publications from 2019 to 2024 from 13 different sources, for a total of 14 documents. There was a decrease in the annual growth rate of publications of -19.73%, yet the average citations per document was 15.29, indicating a good impact in the academic community. The average age of the document is 3.07 years. All documents were written collaboratively by 40 authors, with an average of 3.5 authors per document and 21.43% of them involved international collaboration. Although the annual growth rate of publications is declining, these documents show a good impact in terms of citations. Authors' collaboration is significant, with no single-author document and there is a significant share

Table 1. Main information regarding selected articles.

Description	Results
Timespan	2019:2024
Sources (Journals, Books, etc.)	13
Documents	14
Annual Growth Rate %	−19.73
Document Average Age	3.07
Average citations per doc	15.29
References	663
DOCUMENT CONTENTS	
Keywords Plus (ID)	104
Author's Keywords (DE)	41
AUTHORS	
Authors	40
Authors of single-authored docs	0
AUTHORS COLLABORATION	
Single-authored docs	0
Co-Authors per Doc	3.5
International co-authorships %	21.43
DOCUMENT TYPES	
article	8
conference paper	3
conference review	2
review	1

of international collaboration. The collection consists primarily of articles, with several conference papers and reviews, reflecting diverse but specialized scholarly contributions.

Annual Publication Trends. The annual publication trend in Fig. 1 shows instability with significant increases and decreases from 2019 to 2024. After peaking in 2020 and 2022, there have been repeated declines to low figures in 2021, 2023, and 2024. Factors that may influence this trend include changes in research interest, the availability of funds, and the development of the overall field of sentiment analysis.

The COVID-19 pandemic likely had a varied impact on Kansei studies in sentiment analysis [9]. While there are several factors that may increase interest in these studies, challenges such as resource constraints and disruptions to research activities can also hinder progress. The increase in publications from 2019 to 2020 may reflect continued efforts, but the full impact of the pandemic may be more clearly seen in the following years.

120   L. S. Zaabar et al.

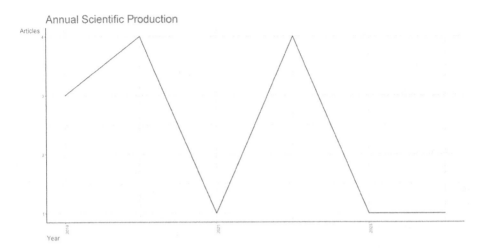

Fig. 1. Annual Publication Trends for Kansei in Sentiment Analysis

Trend Topics. Data in Fig. 2 shows that "Kansei Engineering" and "product design" both began to gain traction in 2019 and remain relevant until 2022, with a peak around 2020. Since these two terms show almost the same trend in the data, it suggests a positive correlation between "kansei engineering" and "product design." This means that an increase or decrease in the frequency of publication for one term tends to be followed by a similar increase or decrease in the other. The topic "sentiment analysis" began to gain attention a little late, starting in 2020, but also remained relevant until 2022. The frequency indicates that "sentiment analysis" is mentioned more frequently than the other two terms, suggesting that it may be a more prominent topic in this dataset.

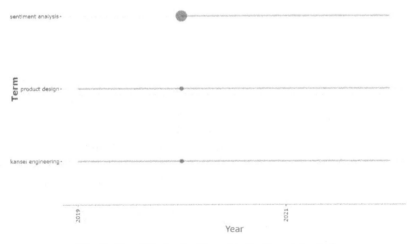

Fig. 2. Trend Topics for Kansei in Sentiment Analysis

Based on the literacy study conducted above, it is also found that Kansei is very popular in the domain of "product design" and is rarely discussed in topics related to politics and security threats. This creates a gap that needs to be studied more deeply as it has been found that there is a correlation between emotions on online platforms that leads to political security threats [2].

Emotion assessment for online news based on Kansei approach in the domain of national security was discussed in [3]. The assessment was done by employing manual Kansei evaluation methodology to measure user perceptions. The study adopted methodology that include developing Kansei checklists to administered emotional keywords or Kansei Words (KW). The research employed a manual Kansei evaluation methodology to measure user perceptions. The evaluation process was structured into four phases: Instrument Preparation, Kansei Word Establishment, Kansei Evaluation, and Kansei Result Analysis. Kansei checklists, containing emotional keywords or Kansei Words (KW), were administered to gather data.

Authors in [10] examines the implicit emotions and unity in propaganda videos posted on social networks, specifically in Malaysia. This study uses a Kansei assessment approach on 10 videos using 30 Kansei Words taken from PANAS-X emotion descriptors. Kansei's assessment also proves the existence of the concept of Kansei in propaganda videos, providing insight into the emotional response of the people to political issues through the medium of video. The results of this study are significant in understanding people's reactions to political issues and can be used as a basis for future research on political emotions and consciousness through visual images such as posters, movies, and memes.

2.3 Deep Learning

Automated extraction of Kansei words can be realised by applying advanced deep learning algorithms such as Convolutional Neural Networks (CNNs), Recurrent Neural Networks (RNNs), Generative Adversarial Networks (GANs), and Long Short-Term Memory (LSTM). CNNs and RNNs models are proficient at processing and understanding complex textual data, making them suitable for extracting Kansei words from large datasets. Meanwhile, GANs can generate new data that mirrors the emotional tone and style of the original content, enriching datasets for analysis. However, Bi-LSTM has proven to be capable in capturing contextual information from both past and future data points, enhancing the accuracy of sentiment and emotional analysis. These innovations enable researchers to uncover underlying emotional patterns more accurately and efficiently, contributing to a deeper understanding of public sentiment and informing effective strategies in various domains, including political security.

PCA is a linear dimensionality reduction technique that identifies principal components explaining the maximum variance in the data. PCA is effective for reducing data dimensions while preserving information. However, PCA assumes linear relationships between variables. Nevertheless, nonlinear patterns exist in Kansei data. Meanwhile, deep learning offers a more flexible and powerful approach to capturing complex data structures. Deep learning techniques like autoencoders can learn nonlinear representations of data, often outperforming PCA in preserving essential information. Additionally, deep learning models can extract higher-level features and patterns, leading to more accurate and informative insights. In summary, while PCA is a valuable tool, deep learning methods have demonstrated superior performance in handling the intricacies of data. Thus, deep learning analysis processes are automated and contributing greater efficiency and accuracy in identifying emotional patterns within large volumes of text compared to the PCA.

2.4 Environment and Infrastructure to Support Kansei and Bi-LSTM Model Training

The effectiveness of these approaches is contingent upon robust cloud infrastructure and stringent security measures to protect sensitive data. Thus, Kansei analysis using Bi-LSTM equipped with optimizer must be performed in high performance cloud infrastructure. However, the security of the cloud infrastructure and the underlying technology including bare-metal, firmware and software must be highly considered to ensure the overall integrity and resilience of the training environment [11]. Thus, a comprehensive understanding of cloud computing security implications is also vital to be understood to safeguard the data and systems to ensure the accuracy of the training data [12, 13]. Cloud security issues can include the access control issues and the implementation of blockchain technologies including a decentralized access control framework [14]. Latest adoption of advance technology including blockchain technology for access control in the cloud using smart contracts also require careful consideration [15, 16]. The attention also should be given to human perspective such as acceptance of cloud computing to avoid any security risks factor in delivering the training and predictive analysis [17, 18].

3 Proposed Model

In this research, we developed an enhancement of Kansei political security model to predict various threats from online platforms. Our comprehensive model comprises of several key stages illustrated in Fig. 3.

This model starts with the collection of text data from the online platform. The collected data will go through a text preprocessing process including tokenization, stop words removal and normalization. Then, the construction of the Kansei lexicon is carried out to categorize emotions such as belief, anger, fear, hope, and threat. Next, the embedding layer is used to convert words into vector forms using embeddings such as GloVe or Word2Vec [9].

This data then goes through a Bi-LSTM layer for sequential text analysis, followed by an optional dense layer for classification or regression. The main advantage of Bi-LSTM is that it can leverage both past and future context to make more accurate predictions. This is particularly useful in tasks where the context from both directions is important.

The model is trained using Adam's optimization algorithm [19]. The output layer will make a prediction as to whether there is a threat or not. This process is followed by assessment and validation, and finally data visualization and interpretation are carried out to gain insights into the analysis of political sentiment and threats.

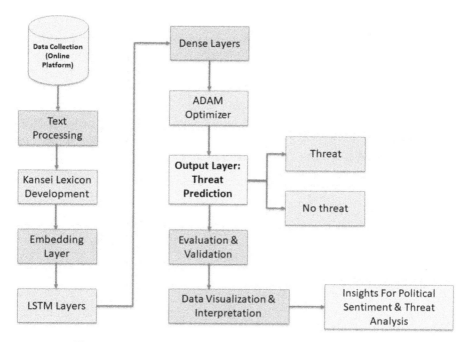

Fig. 3. Kansei Model for Political Sentiment and Threat Prediction

4 Experimental Analysis and Result Discussion

The enhanced Kansei Model that was developed and described in Fig. 3 was tested using the Python 3.12.4 environment and underwent evaluation on a PC equipped with an 13th Gen Intel(R) Core (TM) i5-1340P, 1.90 GHz, 8 GB of RAM, a 64-bit OS, and an x64-based processor. We assessed the performance of the proposed enhanced Kansei Model approach using political sentiment labelled dataset from Kaggle platform consisting of 2,084 sentences from the social media platform. The data was labelled with positive and negative sentiment. In this research, a dataset from a CSV file was loaded and deep learning techniques was used to classify text based on Kansei's emotion categories. First, it defines Kansei's emotional lexicon with related keywords. The 'tag_emotions' function is used to mark text with that emotion category. The text data is then processed using tokenization and padding. A neural network model with a Bi-LSTM and an attention layer is built, combined with emotional input. The model is trained using training data, with the use of 'early stopping' to avoid overfitting.

The final phase of this research design is to validate the analyzed data. In this phase, this study demonstrated a comparative performance evaluation to validate the proposed theoretical framework. The evaluation test compares the results of precision, recall, accuracy, and F-measure [19]. A random subset of sentences was selected to train and test the dataset that used the LSTM deep learning technique, followed by the employment of a confusion matrix, which computed the accuracy, precision, and recall of the deep learning classifiers, allowing for the comparison of algorithmic performance based on training data labels. Accuracy, defined as the proportion of correctly predicted opinions out of all input opinions to the classifier, is determined by True Positive (TP), True Negative (TN), False Positive (FP), and False Negative (FN) values. The formula is as shown in Eq. (1).

$$Accuracy = \frac{TP + TN}{TP + TN + FP + FN} \quad (1)$$

Precision is shown in Eq. (2) and is the percentage of true cases of an opinion (of an instance) among all the classified cases of the opinions (of all instances). To determine the accuracy, true positive rate (TP) was used, as shown in the formula below.

$$Precision = \frac{TP}{TP + FP} \quad (2)$$

Recall is defined as the proportion of properly categorized occurrences of a polarity over the total number of correct instances of the polarity. The formula to calculate the recall values using TP and FN is shown in Eq. (3):

$$Recall = \frac{TP}{TP + FN} \quad (3)$$

The F-score is calculated by dividing the number of true positives by the sum of true positives and false positives, as shown in Eq. (4).

$$F - score = \frac{2TP}{2TP + FP + FN} \quad (4)$$

4.1 Comparative Output

In Table 2 and Fig. 4, the results from the single LSTM model, Bi-LSTM model and existing Kansei approaches are compared. The findings indicate that the bidirectional LSTM model surpasses the existing Kansei method and the single LSTM model.

Table 2. Comparative output of the Bidirectional LSTM Kansei Model with other methods

Methods	Accuracy	Precision	Recall	F-Score
LSTM	0.6627	0.7368	0.4780	0.5799
Bidirectional LSTM	0.8527	0.8522	0.8439	0.8480
Single Kansei Approach	0.8456	0.8533	0.8244	0.8387

The results show that Bi-LSTM significantly outperforms the regular LSTM and the single Kansei approach in terms of all evaluation metrics. Bi-LSTMs achieve the highest accuracy (0.8522), accuracy (0.8439), and F1 (0.8480) scores, demonstrating their superior ability to capture contextual information from both directions in the text, resulting in better threat prediction performance. In contrast, the usual LSTM experienced a decrease in performance with an accuracy of 0.6627 and an F1 score of 0.5799, which indicates limited context understanding. Interestingly, the single Kansei approach also performs well with both precision (0.8456) and competing F1 scores (0.8387), perhaps due to its effective use of predefined categories of emotions, but it still lags in the comprehensive context handling by Bi-LSTM.

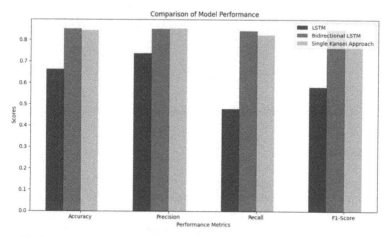

Fig. 4. Performance for LSTM model, Bi-LSTM model and existing Kansei approaches

4.2 Area Under Precision Recall (AUC-PR)

A Region Under Call-Accuracy Curve (AUC-PR) value of 1.0 indicates perfect performance in predicting positive classes (threat detection in this case) based on the predicted probability. This means that this model can maximize accuracy at every call level without any positive false errors. This performance is highly desirable in the context of model evaluation. Figure 5 below illustrates this performance.

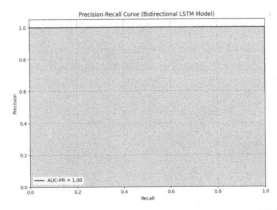

Fig. 5. Kansei Precision-Recall Curve of Bidirectional LSTM (Bi-LSTM) Model

5 Conclusion

This study proves that the Kansei model supported by Bi-LSTM provides better results than the traditional Kansei model that utilize manual Kansei analysis process in predicting political security threats. This research study also shows that combining the Adam optimizer with Bi-LSTM significantly enhances the models' abilities to predict political threats. The integration of deep learning techniques into Kansei engineering marks a significant advancement from traditional manual methods. Automating the extraction of Kansei words and utilizing advanced algorithms allow researchers to achieve more accurate and efficient analyses of emotional patterns. This not only enhances the understanding of public sentiment but also supports the development of strategic responses in various fields. Combining Kansei principles with cutting-edge deep learning techniques in the realm of natural language processing and robust infrastructure sets a new standard for analyzing and interpreting complex emotional data.

The advancements in Kansei engineering through deep learning also highlight the transformative potential of technology in understanding human emotions. By leveraging these techniques, researchers can uncover deeper insights into public sentiment, develop sophisticated early warning systems and inform strategies across diverse domains. The future of Kansei engineering lies in the continued integration of advanced technologies, robust infrastructure, and secure data practices, driving forward our ability to analyze and respond to the ever-evolving landscape of human emotions and societal trends. The integration of deep learning techniques into Kansei engineering marks a significant advancement from traditional manual methods. Automating the extraction of Kansei words and utilizing advanced algorithms allow researchers to achieve more accurate and efficient analyses of emotional patterns. This not only enhances the understanding of public sentiment but also supports the development of strategic responses in various fields. By combining Kansei principles with cutting-edge natural language processing and robust infrastructure sets a new standard for analyzing and interpreting complex emotional data.

The advancements in Kansei engineering through deep learning highlight the transformative potential of technology in understanding human emotions. By leveraging these techniques, researchers can uncover deeper insights into public sentiment, develop sophisticated monitoring and prediction systems, and inform strategies in the domain of political security domains and not only limited to product design. The future of Kansei engineering lies in the continued integration of advanced technologies, robust infrastructure, and secure data practices, driving forward our ability to analyse and respond to the ever-evolving landscape of human emotions and societal trends with scalability and high-performance characteristic. This study confirmed that our proposed enhanced Kansei methodology not only improves accuracy and efficiency in threat detection, but also offers a more autonomous, robust and adaptive solution in addressing the ever-evolving political security challenges. This model has the potential to change the landscape of predicting political security threats and support future studies in the field of national security.

Acknowledgments. The authors fully acknowledged UPNM and the Ministry of Higher Education Malaysia (MOHE) for the approved fund, which makes this research viable and effective.

Disclosure of Interests. We, the authors of this manuscript, declare that we have no significant financial, professional, or personal interests that might have influenced the performance or presentation of the work described in this manuscript.

References

1. Razali, N.A.M., Malizan, N.A., Hasbullah, N.A., Wook, M., Zainuddin, N.M., Ishak, K.K., et al.: Opinion mining for national security: techniques, domain applications, challenges and research opportunities. J. Big Data **8**(1), 150 (2021). https://doi.org/10.1186/s40537-021-00536-5
2. Razali, N.A.M., et al.: Political security threat prediction framework using hybrid lexicon-based approach and machine learning technique. IEEE Access **11**, 17151–17164 (2023). https://doi.org/10.1109/ACCESS.2023.3246162
3. Razali, N.A.M., et al.: Assessment of emotion in online news based on Kansei approach for national security. Int. J. Adv. Comp. Sci. Appl. **12**(7) (2021)
4. Mat Razali, N.A., et al.: Fear assessment in information security dialog box based on hybrid kansei engineering and KJ method. Int. J. Affect. Eng. **21**(1), 23–32 (2022)
5. Ohtomo, K., Harakawa, R., Iisaka, M., Iwahashi, M.: AM-Bi-LSTM: Adaptive multi-modal Bi-LSTM for sequential recommendation. IEEE Access (2024)
6. Touzani, Y., Douzi, K.: An LSTM and GRU based trading strategy adapted to the Moroccan market. J. big Data **8**(1), 126 (2021)
7. Lokman, A.M.: KE as affective design methodology. In: 2013 International Conference on Computer, Control, Informatics and Its Applications (IC3INA), pp. 7–13 (2013)
8. Kadir, S.A., Lokman, A.M., Muhammad, M.: Identification of positive and negative emotion towards political agenda videos posted on YouTube. In: Proceedings of the 7th International Conference on Kansei Engineering and Emotion Research 2018: KEER 2018, pp. 758–767. Kuching, Sarawak, Malaysia (2018)
9. Abdelhady, N., Soliman, T.H.A., Farghally, M.F.: Stacked-CNN-BiLSTM-COVID: An effective stacked ensemble deep learning framework for sentiment analysis of arabic COVID-19 tweets. J. Cloud Comput. **13**(1), 85 (2024)

10. Kadir, A., Lokman, A.M., Tsuchiya, T., Shuhidan, S.M.: Analysing implicit emotion and unity in propaganda videos posted in social network. J. Phys: Conf. Ser. **1529**(2), 22018 (2020)
11. Ishak, N. Rajendran, O.I. Al-Sanjary, Razali, N.A.M.: Secure biometric lock system for files and applications: a review. In: 2020 16th IEEE International Colloquium on Signal Processing & Its Applications (CSPA), pp. 23–28 (2020)
12. Noorafiza, H.M., Uda, R., Kinoshita, T., Shiratori, M.: Vulnerability analysis using network timestamps in full virtualization virtual machine. In: 2015 International Conference on Information Systems Security and Privacy (ICISSP), pp. 83–89 (2015)
13. Noorafiza, H.M., Kinoshita, T., Uda, R.: Virtual machine remote detection method using network timestamp in cloud computing. In: 8th International Conference for Internet Technology and Secured Transactions (ICITST-2013), pp. 375–380 (2013)
14. Wan Muhamad, N., et al.: Enhance multi-factor authentication model for intelligence community access to critical surveillance data. In: Lecture Notes in Computer Science, vol. 11870, pp. 560–569 (2019). https://doi.org/10.1007/978-3-030-34032-2_49
15. Noor, M., et al.: Decentralised access control framework using blockchain: smart farming case. Int. J. Adv. Comput. Sci. Appl. **14**(5) (2023)
16. Noor, N.M., Malizan, N., Ishak, K., Wook, M., Hasbullah, N.: Decentralized access control using blockchain technology for application in smart farming. Int. J. Adv. Comput. Sci. Appl. **13** (2022). https://doi.org/10.14569/IJACSA.2022.0130993
17. Bakar, R.A., Razali, N.A.M., Wook, M., Ismail, M.N., Sembok, T.M.T.: Exploring and developing an industrial automation acceptance model in the manufacturing sector towards adoption of Industry4. 0. Manuf. Technol. **21**(4), 434–446 (2021)
18. Abu Bakar, R., Mat Razali, N.A., Wook, M., Ismail, M.N., Tengku Sembok, T.M.: The mediating role of cloud computing and moderating influence of digital organizational culture towards enhancing SMEs performance. In: Advances in Visual Informatics: 7th International Visual Informatics Conference, IVIC 2021, Kajang, Malaysia, November 23–25, 2021, Proceedings 7, pp. 447–458 (2021)
19. Ali, N.Y., Sarowar, M.G., Rahman, M.L., Chaki, J., Dey, N., Tavares, J.M.R.S.: Adam deep learning with SOM for human sentiment classification. Int. J. Ambient Comput. Intell. **10**(3), 92–116 (2019)

The Integration of Kansei Engineering and Artificial Intelligence Based on Methodology and Application Perspective: A Review

Wen-Tsai Sung[1] and Indra Griha Tofik Isa[2](✉)

[1] Department of Electrical Engineering, National Chin-Yi University of Technology, Taichung 411030, Taiwan
[2] Graduate Institute, Prospective Technology of Electrical Engineering and Computer Science, National Chin-Yi University of Technology, Taichung 411030, Taiwan
indraisa89@gm.student.ncut.edu.tw

Abstract. In product development design, not only functional aspects but also human factors such as emotional, psychological, and affective elements need to be considered. Kansei engineering (KE) is a methodology that integrates human factors into product design. In line with this, artificial intelligence (AI) technology continues to develop that enables integration between KE and AI. In this study, a review of the integration of KE and AI will be carried out based on methodological and application perspectives. The stages were carried out using a systematic literature review approach consisting of (1) collecting article data; (2) article selection; (3) analysis and review of articles; (4) data documentation; and (5) conclusion. The year of publication for the articles reviewed is between 2018 and 2024. In the initial data search, 38 articles were obtained, which were then selected only for articles in journal format. Research findings show that AI can be part of KE in improving emotional relationships by extracting the hidden preferences of users deeper, carrying out automatic design creation, and personalizing application products like robots to better appeal to users' emotions.

Keywords: Kansei Engineering · Artificial Intelligence · Review · Robot

1 Introduction

User-centered design is one of the keys to product success where currently users not only look at the functional aspect but also involve the affective aspect and how the user's feelings and emotions are involved with customer needs. Kansei engineering (KE) is a methodology in product development that involves human factors such as senses, human perception, sense of appearance, and human emotions [1]. Kansei refers to Japanese terminology associated with "sensibility, impression, and emotion" and represents the methodology of affective design [2] KE implementation is often carried out in product creation, including web-based geographic information systems (GIS) [3], automotive industry [4], product information websites [5, 6], e-learning [7, 8], and other product designs. There are 8 types of KE [9] as shown in Fig. 1:

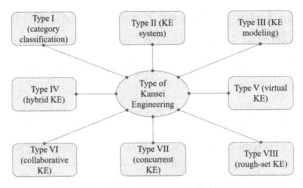

Fig. 1. Eight types of KE [9].

KE type I (KEPack) which begins with collecting Kansei words (KW) that represent user feelings or emotions. KW justification involves user perception and experts such as web designers or UX/UI experts [10] or conduct online identification to both positive and negative emotion [11]. The next stage is the development of an SD-scale questionnaire which includes KW and specimen collection which is the basis for product development. The product specimens are translated into item classifications, thereby generating basic elements that become benchmarks in product development. Next is distributing questionnaires to recommended participants where these participants have at least basic knowledge about the product to be developed. The results of the participant evaluation were analyzed using multivariate statistical analysis to produce a correlation between emotion factors and design elements. This emotion factor ultimately becomes the parameter by which design elements can be implemented into product development.

KE is an interesting thing to be discussed and continues to develop both in terms of methodology and practical application. Several researchers have discussed KE through review studies, including Ismail and Lokman [12] by reviewing how to implement KE in a web-based system which consists of three main topics: e-learning, m-learning, and website interface design. In the e-learning topic, it is described that the emotion of "harmony" has high impact value in open source-based e-learning. Principal component analysis and factor analysis were conducted in analyzing the data. The development of m-learning is carried out by combining KE with the analytical hierarchy process (AHP). Statistical analysis involves factor analysis and partial least squares. Meanwhile, in web-based systems, e-commerce websites have been developed with embedded emotion factors in the user experience. In the long term, implementing Kansei on a web-based system to increase customer loyalty, in other words, can maintain the existence of the product. The review study indicates that KE has been successful in product development by involving user emotion and psychology, which are important aspects of web-based system design. Ginting et al. [13] reviews how integration between KE and quality function deployment (QFD) is a methodological process that translates the customer needs into the technical requirements in order to develop products. The data analyzed are from international journals published from 2010 to 2019 sourced from Google Scholar and the Science Direct repository website. The integration of these two methodologies is implemented in various products, including precise cement gauge design, food snack packaging, technological-based product design, nine prototype packaging, and garbage carts. The results of the analysis of reviews in journal publications represent that KE and QFD provide customer satisfaction from both emotional and functional aspects. Lopez et al. [14] conducted a systematic literature review (SLR) in KE based on comparative study from 1995 to 2020 from several publisher repositories: ACM, MDPI, IEEE, Elsevier, and Springer. There are three phases in the SLR process, where the third phase produces 87 journal publications with relevant research questions. In the findings, the results show that methodology type III (KE modeling) is the most frequently implemented in journal article discussions with a percentage of 35%. In second and third place are type VIII (rough-set KE) and type I (category classification) with percentages of 18% and 16%, respectively. Meanwhile, collaborate KE and concurrent KE have the lowest percentage with each percentage of 4%. From the product aspect, the fields of technology/electronics and furniture/building/home products are the ones most discussed in implementation with a percentage of 29%.

KE methodology and implementation continues to develop, especially in the Industrial Revolution 4.0 era where artificial intelligence (AI) technology is a major part of it. KE can build AI-based product designs that not only focus on functional aspects but how to fulfill the user's perspective through emotion, psychological factors, and human sensing [15]. In terms of the KE methodology developed, implementing AI through deep learning technology can be done in enhancing the connection between design elements and user emotion and extract the hidden user preferences [16]. So in this study, a review will be carried out regarding the integration between KE and AI, for which there are two research questions (RQ) as follows below:

1. RQ 1: How is AI implemented in the KE methodology to produce the desired design elements?
2. RQ 2: How are AI-based products produced through KE methodology?

This article is structured as follows: Sect. 2 describes the materials and methods employed in the study, the data used, mechanisms, and stages in conducting the review study. Section 3 explains results and discussions which discusses findings related to RQ1 and RQ2. Section 4 represents the conclusion of the review study in this article.

2 Materials and Method

In this study, there are five stages in conducting the review study with a systematic literature review (SLR) approach as depicted in Fig. 2 which begins with data source and strategy of data searching as part of data collection related to constructed RQ. Article selection result which is the result of selection of reference data that has been collected in the previous stage. Article analysis and review is the synthesis stage of selected articles, where findings, results, and discussions are produced. Data documentation and evaluation is the final stage of a series of study review processes.

Fig. 2. Research stage.

2.1 Data Source and Strategy of Data Searching

Data searches were carried out on online database repositories: Scopus, IEEEXplore, ACM Digital Library, and MDPI which were limited to publication years 2018 to 2024. The data search technique was carried out by entering combination keywords that represent RQ, described in detail in Table 1:

Table 1. Keyword search strategy.

ID	Detail of Keyword
Keyword 1	"Kansei Engineering" and "Artificial Intelligence"
Keyword 2	"Kansei Engineering" and "Methodology" and "Artificial Intelligence"
Keyword 3	"Kansei Engineering" and "Application" and "Artificial Intelligence"

The initial search based on the three keywords in Table 1 obtained 38 articles in the form of journals, conferences, and book chapters, which will then be analyzed and reviewed to produce paper recommendations relevant to the review study carried out.

2.2 Article Selection Result

The next stage is the process of selecting articles that are relevant to the topic of discussion, where there are three phases. Phase 1 is initial data searching with a combination of articles in the form of journals, conferences, and book chapters with a total of 38 articles. The review study will focus on journal articles, so in phase 2 the article conference and book chapter will be excluded. Phase 3 is the final review by looking at the relevance of the content, title, and keywords to whether they represent the RQ. The final results of the article selection results are 3 journal articles related to RQ1 and 3 journal articles related to RQ2. Figure 3 represents the brief stage phase of the article selection result:

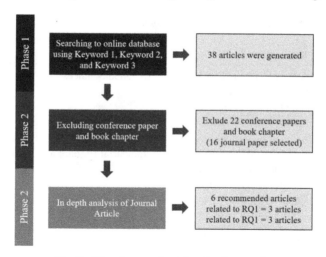

Fig. 3. The phases of article selection result.

2.3 Article Analysis and Review

The next stage is the review process for the six recommended articles that are relevant to RQ1 and RQ2. How KE methodology is integrated with AI, which in this case consists of deep learning, text mining, and robots. Meanwhile, AI applications resulting from the

Table 2. Recommended reference table.

Author	Title	Journal Name	Category
Quan, et al.	Product innovation design based on deep learning and Kansei engineering	Applied Science	RQ1
Liu, et al.	An improved Kansei engineering method based on the mining of online product reviews	Alexandria Engineering Journal	
Wang, et al.	Multiple affective attribute classification of online customer product reviews: A heuristic deep learning method for supporting Kansei engineering	Engineering Applications of Artificial Intelligence	
Lokman, et al.	Spiritual Therapeutic Robot for Elderly With Early Alzheimer's Disease: A Design Guide Based on Gender	Malaysian Journal of Medicine and Health Sciences	RQ2

(*continued*)

Table 2. (*continued*)

Author	Title	Journal Name	Category
Wu, et al.	A New Configuration Method for Glass Substrate Transfer Robot Modules Based on Kansei Engineering	Applied Sciences	
Gan, et al.	Integrating aesthetic and emotional preferences in social robot design: An affective design approach with Kansei Engineering and Deep Convolutional Generative Adversarial Network	International Journal of Industrial Ergonomics	

KE stage are dominated by robotic products. Specifically, Table 2 describes the detailed articles that will be analyzed in this review study:

2.4 Data Documentation and Conclusion

Documentation and conclusion are the final stages of this review study, which takes place throughout the study's implementation process, from data collection and searching technique to synthesis and the interpretation of the findings or results. The documentation procedure is significant because it can serve as a foundation for future research direction and as a research record demonstrating the legitimacy of the data and material employed.

3 Result and Discussion

Artificial intelligence (AI) in today's world has developed exponentially and almost all fields employ AI to make human work easier and increase productivity and efficiency. Several subsets of AI that are currently experiencing significant development are machine learning, robotics, expert systems, and computer vision [17]. Machine learning (ML) has become very popular because of the large amount of data generated from the use of the internet and computer technology. ML can carry out tasks based on structured, unstructured, or semi-structured data. In carrying out complex tasks, ML is combined with neural networks to produce accurate and precise output. The integration of ML and neural networks gave rise to a new terminology called deep learning (DL), which is a subset of ML [18]. Next is robot technology which can carry out actions or physical tasks that are widely implemented in the manufacturing, exploration, and healthcare industries. These two topics regarding DL and robots became the main discussion on the integration of KE and AI based on methodology and application perspective.

3.1 A Combined Kansei Engineering Methodology and Artificial Intelligence

Liu et al. [19] proposed the KE method combined with back propagation neural network in order to determine emotional design trends in smartphones. The strategy in

Kansei word (KW) screening employs the terms frequency-evaluation, potency, and activity (TF-EPA). The selected dataset is web-based e-commerce which has high market shares. Determining KW through TF-EPA is carried out in four stages. The first is adjective mining, where adjective words are collected that are relevant to the context of the observed study, in the initial crawling of data on 178 smartphone products, 637,979 reviews were obtained which were then carried out mining extraction and 2,878 adjective words were generated. The second stage is a grouping of similar adjectives by examining the meaning of the adjective word. The third and fourth are conducting TF statistics and word pair screening, respectively. From the TF-EPA method process in determining the KW adjective, six pairs of KW which are related to E (evaluation): good | poor and beautiful | ugly; P (potency): big | small and thin | thick; and A (activity): easy | difficult and smooth | slow. In general, the methodology stages implemented consist of objectives, collecting KW through online review, the KW screening process using the TF-EPA method; evaluation of Kansei through online review mining; and mapping model with back-propagation neural network. The stage model mapping was carried out using MATLAB software with a neural network structure consisting of input layers, hidden layers, and output layers. The input layer consists of 11 nodes which are elements of the smartphone, while the output layer is 6 pairs of KW based on TF-EPA. 28 samples were involved in the mapping model experiment where the optimal training value in parameter of mean squared error (MSE) of "good & poor" is 0.00061 at epoch of 1414, "beautiful & ugly" is 0.00053 at epoch of 1803, "big & small" is 0.00065 at epoch of 1695, "thin & thick" is 0.00038 at epoch of 962, "easy & difficult" is 0.00065 at epoch of 1192, and "smooth & slow" is 0.00093 at epoch of 2000. Multiple linear regression produces mean absolute error on KW "good & poor" is 0.0696, "beautiful & ugly" is 0.1001, "big & small" is 0.0784, "thin & thick" is 0.0815, "easy & difficult" is 0.1328, and "smooth & slow" is 0.1358. Based on the proposed KE methodology, provides a new perspective and insight into guiding the product's emotional design through the online product review data based on systematical process involving the TF-EPA method and back-propagation neural network.

Wang, et al. [20] implemented the deep learning technology in the KE method in order to classify the affective attribute in product reviews. The proposed method is an improvement from the traditional KE, which uses a questionnaire survey. Meanwhile, the proposed method employs text mining in defining KW and deep learning models in prediction and classification tasks. There are four main stages in the proposed method, which are also depicted in Fig. 4:

1. Data preparation through collecting customer product reviews sourced from web-based e-commerce. The next stage is through text processing, including segmenting the sentence, conducting the tokenization, processing the lemmatization, and defining part-of-speech (POS) tagging. From this stage, seven pairs of affective words were produced, which were sourced from 10 product samples: "like & dislike," "aesthetic & aesthetic," "soft & hard," "small & big," "useful & useless," "reliable & unreliable," and "recommended & not recommended."
2. Rule construction & model training, where there are two main processes consisting of text mining rules and classification models. Text mining rules are carried out in five stages, consisting of lemmatization, POS tagging, affective word annotating, and

extracting the candidate rules. The training model employs the deep belief network (DBF), which has three structures: the top hidden layer, middle layer, and visible layer

3. Evaluation based on four parameters consisting of F1-score, accuracy, recall, and precision. Comparative analysis was carried out with other models: MLP, CART, KNN, and LSTM.
4. Application is carried out by testing the model by providing example test data from random customers. In this stage, we will see how the model identifies and classifies affective attributes in the test data sample.

Fig. 4. Stage of the proposed method.

Model training was carried out on three review categories: all reviews, positive reviews, and negative reviews. According to the results of the experimental training model, the most optimal accuracy is produced by the combination of rule-based extraction and DBN. In the results of training with a single model (non-combination), it was found that DBN produced the most optimal accuracy and F1-score when compared with LSTM, KNN, CART, and MLP. In the positive review category, the results have better model performance compared to the negative review category. Overall, the accuracy percentage produced by the proposed method has more optimal results compared to classical machine learning and classical deep learning, where the accuracy percentage of the proposed method is 86.2%. Meanwhile, classical machine learning and classical deep learning are 75.8% and 80.8%, respectively.

Quan, et al. [21] proposed the combined KE and DL for product innovation called KENPI framework. Specifically, KE is combined with back-propagation NN (BP-NN). The case study product which is designed in the study is female coat. There are three parts to the proposed method, the first part is choosing the domain of the product which is divided into two classifications, i.e., the spanning of semantic space and properties space. Collecting KW, formulating the questionnaire to become a product style is carried out in the spanning part of semantic space. Meanwhile, in the spanning of properties space, specimens are collected to translate them into product properties. The first part produces (1) 30 KW; (2) 100 specimens of female coat design collected which were later divided by 7 element design items: pocket, opening, collar, waist, length, model, and sleeve; and (3) questionnaires with 7 SD scale points. The relevance between product style and product properties is represented in a "relationship model" which integrates the second part and the third part. In the second part is the feature extraction process via BP-NN which comes from two inputs: image content and image style, and the third part is product semantic generated, as depicted in Fig. 5.

The structure of BP-NN consists of three layers: the input layer, the hidden layer, and the output layer. The input layer consists of 7 design elements and the output layer is a style category consisting of professional-leisure, vogue-classic, simple-delicate, and

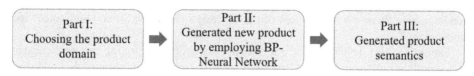

Fig. 5. The stage of proposed KENPI method.

grand-youth. In conducting KENPI, the programming language used for training data is Python, dataset sourced from MS-COCO, Adam optimizer, normalized image size of 256 × 256. The six female coats were then selected randomly as the parameter in BP-NN training as the image content. Next, neural style transfer is conducted by combining image content and image style into a result image. This resulting image was then included in the next questionnaire for 30 women customers. The representation of questionnaire results is in the form of standard deviation where a smaller value indicates a good result. The six results were then generated, where result 1 showed "professional leisure," result 2 showed "vogue classic," result 3 showed "simple delicate," result 4 showed "professional leisure," result 5 showed "grand-youth," and result 6 indicates "vogue classic."

3.2 Artificial Intelligence Related to Kansei Engineering

The application of KE methodology in designing AI products is dominated by robot design. Lokman et al. [22] developed a spiritual therapeutic robot for elderly people with early Alzheimer's disease. There are three phases in developing the robot, the first phase is determining the conceptual robot design involving spiritual elements and robot design features through in-depth interviews with the participants and thematic analysis. The second phase is implementing KE method in synthesizing the emotional UX therapeutic robot which employs the Kawakita Jiro (KJ) method. The KJ method was invented by the Japanese anthropologist named Jiro Kawakita. KJ method has four aspects: model based on problem-solving; the formulation of qualitative data and analysis tools; the method and research concept which has a new type; the concepts provided by teamwork for creativity [23]. In this case, the KJ method was implemented when the experts were conducting selecting the representative header keywords (which are called by spiritual emotion words or SEW) for the cluster emotion elements. The third phase is developing Kansei therapeutic robot based on the spiritual elements design guide which has three stages: (1) Developing the proof-of-concept using Nao Robot; (2) Developing checklist of Kansei; and (3) Kansei assessment by quantitative and qualitative approach. In the context of elements of spiritual and robot design based on five spiritual practices including praying, religious song, zikr, surah, and prayer.

There are 10 SEW concepts generated which consist of spiritual-emotional elements. Kansei assessment stage is conducted by descriptive analysis, thematic analysis, and PLS analysis. In PLS analysis results indicate both male and female gender agree most influential design elements are soft voice, smooth tone, and robotic movement while sitting. From the results of this research, a theme structure was also constructed consisting of "pious" and "devout." Meanwhile, the design element has been constructed based on gender, where there are two classifications of design elements: "sound and voice" and "movement."

Gan et al. [24] developed the social robot design by integrating the aesthetic and user emotional preferences. The proposed method employs a combined KE and deep convolutional generative adversarial network (DCGAN). There are six stages conducted in the proposed method: (1) conducting aesthetic and emotional evaluation to the participants by using the two questionnaires; (2) collect the questionnaires and analyze using regression analysis; (3) mapping the feature relationship between physical attributes and KW; (4) Training the DCGAN model; (5) Generating proposed effective new design: (6) Testing the social robot. There are nine samples of robots which are divided into four categories of anthropomorphic robots, zoomorphic robots, caricatured robots, and functional robots. 10 KW were generated and selected which is used as the user's criteria when filling in the questionnaire. There were 484 participants consisting of 180 males and 304 females in the age range of 18 to 75 years. The questionnaire analysis results generated Cronbach alpha score of 0.987 and the rank of robot based on mean value in the first to the fourth places are caricatured robot, zoomorphic robot, anthropomorphic robot, and functional robot, respectively. Next, DCGAN is conducted to generate the 64 proposed social robots based on the KE result in the previous stage. The dataset involves 428 robot images with normalized images of 256 × 256 pixels. In hyperparameter setting, the epoch is set to 600, adam optimizer, learning rate of 0.0002, LeakyReLU as discriminator step. The results of this research are physical attributes that have significant customer aesthetic features consisting of color, head, and outer shape. Meanwhile, in the emotional feature aspect, the KW of "intelligent," "pleasant," and "interesting" are the crucial KW.

Wu et al. [25] employ KE method for robot module configuration. The first stage involves collecting 400 robot specimen images which are extracted into 47 image samples which are classified into 10 classes. In order developing KW, the initial words were obtained 408 words based on online comments, questionnaires, and interviews. Next, word pre-processing, analysis, and word extraction were carried out which generated six pairs of words under shape, color, and material appearance features. The six-word pairs are "slender-stout," "hard-soft," "simple-rich," "colorful-grayish," "non-metallic-metallic," and "rough-smooth." Next, 10 specimens were integrated with the six pairs into a questionnaire, and distributed to 55 participants. From the analysis of this questionnaire, it was found that user preferences were slim and hard shape; users prefer color combinations; strong metallic; smooth and delicate surface material. Meanwhile, the resulting design element layer consists of 10 items, consisting of the "shape" category with 6 items, "color" and "material" are 2 items, respectively. The robot module was developed based on element design by developing a basic module consisting of base, turn shaft, lift shaft, arm, and fork. The robot module developed refers to the robot shape as "hard-soft" and "lender-stout;" the robot color as "colorful-grayish" and "rich-simple;" and the robot material as "smooth-rough."

4 Conclusion

In this study, a review has been carried out regarding the development of AI technology, which has had an impact on the Kansei engineering method in both methodology and application. The integration of deep learning with KE is widely used in classification tasks to determine the emotional perception of certain products. The common deep learning model combined with KE is a back-propagation neural network consisting of an input layer, hidden layer, and output layer. Generally, the input layer is in the form of design elements, and the output layer is in the form of emotion factors. Meanwhile, from an application perspective related to AI products, robots are developed by developing the functions of the robot based on emotion factors. By developing robots based on human factors, we can increase the emotion's engagement and connection between robots and humans.

KE has succeeded in integrating the AI field in both methodology and application aspects. In implementing recommended design elements into real AI prototyping, it is necessary to involve expert designers in order to address the represent the emotion into the design element of the product.

Acknowledgments. This research was supported by the Department of Electronic Engineering at the National Chin-Yi University of Technology. The authors would like to thank the National Chin-Yi University of Technology for supporting this research.

Disclosure of Interests. The authors have no competing interests to declare that are relevant to the content of this article.

References

1. Wang, W.M., Li, Z., Tian, Z.G.: Extracting and summarizing affective features and responses from online product descriptions and reviews: a Kansei text mining approach. Eng. Appl. Artif. Intell. **73**, 149–162 (2018)
2. Lokman, A.M.: KE as affective design methodology. In: 2013 International Conference on Computer, Control, Informatics and Its Applications: "Recent Challenges in Computer, Control and Informatics" (IC3INA), pp. 7–13. IEEE, Jakarta, Indonesia (2013)
3. Novianti, L., Isa, I.G.T., Ariyanti, I.: Evaluating users' emotion in web-based geographic information system. In: 5th Forum in Research. Science and Technology (FIRST), pp. 314–321. Atlantis Press, Palembang, Indonesia (2021)
4. Matsubara, T., Matsubara, Y., Ishihara, S.: Virtual prototyping with real-time rendering for Kansei engineering of leather grain patterns on car dashboard panels. Trans. Japan Soc. Kansei Eng. **9**(2), 119–128 (2010)
5. Papantonopoulos, S., Karasavova, M.: A kansei engineering evaluation of the emotional appeal of product information on E-commerce product pages. In: ACM International Conference Proceeding Series, pp. 1–8, ACM Digital Library, Athens, Greece (2021)
6. Lokman, A.M., Harun, A.F., Noor, N.L.M.: Website affective evaluation: analysis of differences in evaluations result by data population. In: First International Conference. HCD 2009, Held as Part of HCI International, pp. 643–652. Springer, Berlin Heidelberg, San Diego, CA, USA (2009)

7. Hadiana, A.: Emotional preferences towards e-learning based on analytic hierarchy process and kansei for decision making. Eur. J. Eng. Res. Sci. **5**(10), 1186–1190 (2020)
8. Hadiana, A., Lokman, A.M.: Kansei evaluation in open source e-learning system. Jurnal Teknologi **78**(12–3), 135–139 (2016). https://doi.org/10.11113/jt.v78.10031
9. Lokman, A.M.: Design and emotion: the Kansei engineering. Malaysian J. Comput. **1**(1), 1–11 (2010)
10. Lokman, A.M., Nor Laila, M.N., Nagamachi, M.: Kansei structure and visualization of clothing websites cluster. In: International Symposium on Information Technology 2008, (ITSim), pp. 1–8. IEEE, Kuala Lumpur, Malaysia (2008)
11. Kadir, S.A., Lokman, A.M., Muhammad, M.: Identification of positive and negative emotion towards political agenda videos posted on YouTube. Adv. Intell. Syst. Comput. **739**, 758–767 (2018)
12. Ismail, N.N.N.N., Lokman, A.M.: Kansei engineering implementation in web-based systems: a review study. In: 8th International Conference on Kansei Engineering and Emotion Research, pp. 66–76. Springer Nature, Tokyo, Japan (2020)
13. Ginting, R., Ishak, A., Malik, A.F.: Integration of Kansei engineering and quality function deployment (QFD) for product development : a literature review. In: 2nd International Conference on Industrial and Manufacturing Engineering, pp. 1–6. IOP Publishing, Medan, Indonesia (2020)
14. López, Ó., Murillo, C., González, A.: Systematic literature reviews in Kansei engineering for product design—a comparative study from 1995 to 2020. Sensors **21**(19), 1–19 (2021)
15. Lokman, A.M., Kadir, S.A., Hamidi, S.R.: LEIQTM as an emotion and importance model for QoL: fundamentals and case studies. J. Komun. Malaysian J. Commun. **35**(2), 412–430 (2019)
16. Chan, K.Y., Kwong, C.K., Pornpit, W.: Affective design using machine learning: a survey and its prospect of conjoining big data. Int. J. Comput. Integr. Manuf. **33**(7), 645–669 (2020)
17. Jan, Z., Farhad, A., Wolfgang, M.: Artificial intelligence for industry 4.0: systematic review of applications, challenges, and opportunities. Expert Syst. Appl. **216**(119456), 1–21 (2021)
18. Matsuo, Y., LeCun, Y., Sahani, M.: Deep learning, reinforcement learning, and world models. Neural Netw. **152**, 267–275 (2022)
19. Liu, Z., Wu, J., Chen, Q.: An improved Kansei engineering method based on the mining of online product reviews. Alexandria Eng. J **65**, 797–808 (2023)
20. Wang, W.M., Wang, J.W., Li, Z.: Multiple affective attribute classification of online customer product reviews: a heuristic deep learning method for supporting Kansei engineering. Eng. Appl. Artif. Intell. **85**, 33–45 (2019)
21. Quan, H., Li, S., Hu, J.: Product innovation design based on deep learning and Kansei engineering. Appl. Sci. **8**(12), 1–17 (2018)
22. Lokman, A.M., Ismail, N.N.N.N., Redzuan, F.: Spiritual therapeutic robot for elderly with early Alzheimer's disease: a design guide based on gender. Malaysian J. Med. Heal. Sci. **18**, 71–79 (2022)
23. Iba, T., Yoshikawa, A., Munakata, K.: Philosophy and methodology of clustering in pattern mining: Japanese anthropologist JIRO Kawakita's KJ method. In: 24th Conference on Pattern Languages of Programs., pp. 1–11. The Hillside Group, Vancouver British Columbia Canada (2017)
24. Gan, Y., Ji, Y., Jiang, S.: Integrating aesthetic and emotional preferences in social robot design: an affective design approach with Kansei Engineering and Deep Convolutional Generative Adversarial Network. Int. J. Ind. Ergon. **83**(103128), 1–17 (2021)
25. Wu, Y., Zhou, D., Cheng, H.: A new configuration method for glass substrate transfer robot modules based on kansei engineering. Appl. Sci. **12**(19), 1–18 (2022)

Competencies Required by Different Positions for Innovation

An Initial Exploration of Virtual Reality for Emotional Expression: The Case of "inescapable Grief"

Yao-Xun Chang(✉), Ki-Yan Ma, and Hei-Man Hung

Ming Chuan University, Taoyuan City 333, Taiwan
agassi@mail.mcu.edu.tw

Abstract. Virtual reality technology is not only widely used in the entertainment and military fields but has also begun to integrate with exposure therapy to assist in treating various anxiety disorders. This technology provides a controlled and safe environment, allowing for the customization of therapeutic scenarios based on individual patient needs, thereby significantly expanding its application in psychology and enhancing patients' self-understanding. This research transforms the five stages of grief into a series of virtual reality maze games, where each complex maze represents the transformation and rebirth of each stage. Players must confront challenges and negative remarks within the mazes, overcoming psychological barriers to find a way out. By metaphorically searching for the maze's exit to find psychological relief, this approach aims to help people better cope with grief and move forward to the next stage of their lives.

Keywords: Grief · Virtual Reality · Maze

1 Introduction

With the growing prevalence of Virtual Reality (VR) technology, its immersive experiences have drawn increasing attention and participation from the public. This technology is widely applied not only in the gaming industry but also in various fields including healthcare and the military. In recent years, cognitive behavioral therapy has begun utilizing VR tools to conduct exposure therapy in a controlled and safe environment by simulating real-life situations. Research indicates that VR technology can alleviate symptoms of various anxiety disorders such as obsessive-compulsive disorder, post-traumatic stress disorder, generalized anxiety disorder, social anxiety disorder, acrophobia, and fear of flying [5, 8]. As VR devices continue to improve, they not only broaden the range of symptoms that can be addressed in clinical psychology but also enable patients to interact with virtual environments in VR, enhancing self-understanding.

When individuals experience significant loss or confront death, the profound grief from losing a loved one often makes them feel as if the world has shattered. This response is a common human emotional reaction. Some quickly realize that grief is not the endpoint and move on to the next stage, while others remain trapped in their sorrow, continuously plagued by negative emotions, struggling to break free. Inspired by Elisabeth

Kübler-Ross's Five Stages of Grief theory [7], this creative project aims to develop a virtual reality game to assist people in facing and managing their grief, overcoming psychological barriers, and embracing new challenges in life.

When someone is engulfed in grief, they may feel confused and disoriented, necessitating a deep dive into personal anguish and self-exploration. In this journey, they must strive to find a psychological exit, learning to handle life's inevitable losses and challenges, ultimately transforming into a more complete self. This process can be likened to navigating a complex maze filled with potential twists and rebirths, where one must explore various paths to find an exit amidst uncertainty and setbacks. Finding this exit can feel like a sudden clearing of the skies. However, the game also reflects the reality that some may be trapped in psychological dilemmas, endlessly circling through confusion and pain, resembling a perpetual maze. Through the design concept of multiple mazes, the game portrays this seemingly endless darkness. As participants face setbacks and overcome psychological barriers, they must continue to seek a way out. This VR game presents the world of grief through a first-person perspective, allowing players to explore the maze's exit while deeply examining their inner world, accepting the reality of their grief, and seeking solutions for the next steps in their lives.

2 Literature Reviews

In this chapter, we will first use literature analysis to explain the main concept of the game—the Five Stages of Grief (Kübler-Ross model). We will then delve deeper into case studies related to the integration of virtual reality with psychotherapy.

2.1 The Five Stages of Grief

The Kübler-Ross model, also known as the Five Stages of Grief, was first introduced by Elisabeth Kübler-Ross, M.D. in her 1969 book "On Death and Dying." Originally developed to describe the psychological process that terminally ill patients experience when facing death, this model has been widely adopted to explain the psychological responses to any significant loss or life crisis, such as the loss of a loved one, divorce, or unemployment. This framework has established the foundation for discussing stages of grief and dying in Western thanatology [13]. Here are the phenomena and behaviors of the five stages:

Denial. When people suddenly face significant negative news, their immediate reaction is usually denial, a common defense mechanism that helps alleviate the immediate shock, giving them time to gradually adapt to the new reality. In this stage, individuals refuse to accept that a major change has occurred and exhibit a psychological tendency to evade reality, finding it difficult to accept unfortunate facts. Some may even appear numb or unusually indifferent, but this behavior is also a manifestation of denial and does not mean that they are without sorrow. Instead, they need this mechanism to give themselves time to explore and adjust to the changes they are experiencing at their own pace.

Anger. When denial no longer functions as a defense mechanism, deep-seated pain begins to surface, and people may become angry. According to the Kübler-Ross model, the pain stemming from loss often transforms into anger, which is a natural response to loss and a way to release emotions. This anger may manifest as irrational blame or resentment towards people and the environment around them. However, anger is not the only emotion that might be experienced in this stage; irritability, bitterness, anxiety, and impatience may also arise as different expressions of coping with loss, all part of the same grieving process.

Bargaining. In this stage, to avoid loss or alleviate pain, people might try to make compromises or deals with a higher power, such as God, hoping to change the current situation for a better outcome. This helps maintain hope and stay positive in the face of suffering. This process may also involve looking back at the past, imagining if things could have developed differently and whether the loss could have been avoided. At this point, individuals often enter a pattern of thinking characterized by "if… Then…" scenarios, attempting to find ways to escape the pain.

Depression. When bargaining no longer works, individuals begin to realize that their loss is irreversible, potentially entering a stage of depression, characterized by profound sadness and a sense of isolation, with a loss of interest in life. This emotional state manifests as a deep sorrow or despair, stemming from an insightful understanding of the loss. In dealing with depression, there is no right or wrong way to handle it, nor is there a fixed deadline to overcome these feelings.

Acceptance. Over time, individuals may gradually come to accept the reality of their loss, finding a psychological balance that allows them to coexist peacefully with their new circumstances. Entering the final stage of acceptance does not necessarily mean being content with what has happened or forgetting the pain, but rather involves acknowledging the loss experienced. At this stage, people gain a deeper understanding and acceptance of their new life reality, begin to plan for the future, and find new meaning and goals in life.

2.2 Virtual Reality with Psychotherapy

In recent years, many mental health experts have begun using virtual reality as an innovative technology-based therapy [3, 5, 11]. This technology, with its unique immersive qualities that transcend physical space limitations, allows individuals to safely confront and manage their emotions within a virtual world. When combined with exposure therapy and pharmacological treatments, virtual reality has proven effective in reducing symptoms of anxiety or depression [1]. Additionally, the adaptability of virtual reality enables therapists to design therapeutic environments tailored to individual cases, providing a safe, controllable, and repeatable treatment setting. This not only allows therapists to precisely control the virtual environment but also to record and analyze client behaviors from various angles, facilitating a deeper understanding of clients' inner worlds and assisting them in confronting and overcoming emotional traumas. Thus, virtual reality is also considered a powerful supplement and extension to traditional psychotherapy [4, 12].

Additionally, some VR games are specifically designed to explore emotional fluctuations, often aiming to provide a deep emotional experience. These games offer a new way to help players explore and understand certain psychological states, practicing coping mechanisms and emotional regulation skills. These games are not only entertainment products but also tools for self-exploration. Here is an analysis of several relevant cases.

Nevermind. This is a virtual reality psychological thriller game released in 2015, designed to guide players into exploring the dark inner world of a trauma patient [9]. The game utilizes biofeedback technology to monitor players' stress levels and adjusts the game environment based on their psychological stress levels. When a player's fear reaches its peak, the game becomes more difficult; conversely, if the player learns to remain calm and manage their stress responses effectively, the game gradually becomes easier.

DEEP VR. This is a virtual reality game focused on meditation experiences, designed to encourage mindfulness and relaxation [2]. The game incorporates a biofeedback mechanism that guides players through relaxation techniques and breathing exercises, specifically controlling the movement of the in-game character through slow, deep breaths. Research has shown that DEEP effectively alleviates players' anxiety and reduces symptoms [14].

Hellblade: Senua's Sacrifice. This game provides an in-depth portrayal of the challenges associated with schizophrenia [6]. The development team worked closely with neuroscientists and individuals suffering from mental illnesses to create a story that metaphorically represents the struggle against schizophrenia. By simulating auditory and visual hallucinations, the game aims to evoke empathy for those experiencing schizophrenia, thereby increasing public awareness of the disorder.

Phobia Exposure VR. This application offers a progressively intensifying exposure environment tailored to the user's needs, enabling them to gradually confront their fears and learn anxiety management techniques [10]. It also includes progress tracking features, which help users visualize their journey towards overcoming their fears.

3 Creative Design

The core idea of this creation is to allow participants to profoundly experience the psychological dilemmas within their inner worlds. To enhance the specificity of the experience and interactivity, we have chosen to present it in the form of a VR maze game, utilizing virtual reality technology to intensify the immersion of the game. The aim is for participants to deeply feel the emotional fluctuations and challenges faced, thereby more effectively understanding and addressing their inner conflicts.

3.1 Scene Design

In this VR maze game, the Five Stages of Grief are transformed into five different mazes, each with unique scenic visual designs and negative text on the walls. This setup provides participants with a basic understanding of the psychological aspects of grief and equips them with tools to effectively handle sorrowful emotions. The aim is to enhance people's comprehension of grief and emotional struggles. Below is an explanation of the design concept for each of the five mazes.

Dungeon. Self-doubt, why upon waking am I trapped in a dungeon? How did this happen? Where exactly is this place? How can I escape? Is this reality or an illusion?

The player character awakens in an unfamiliar dungeon, surrounded by the occasional sound of footsteps echoing nearby. The uneven, dark, and damp walls contribute to an oppressive atmosphere. Faintly visible on the walls are blurred texts and images, with dense, bloodstains and red handprints that are striking and unsettling.

The first level is a dungeon maze named "Denial" (see Fig. 1). The graphic design of this level reflects the character's psychological turmoil—specifically self-doubt and disquietude. The blurred imagery serves as a metaphor for the character's denial and disorientation towards reality, evoking a sense of psychological obscurity and instability. The bloodstains and red handprints on the walls not only visually accentuate the horror and oppressive atmosphere but also symbolically represent the character's fear and anxiety. Additionally, the background sounds, composed of ambient footsteps and laughter, are deliberately integrated to intensify the character's feelings of confusion and unease.

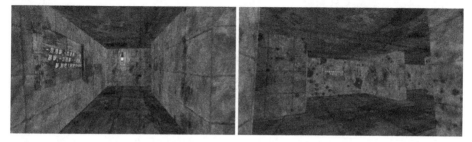

Fig. 1. The first level of the game: Dungeon.

The text on the panels reads: "Why me? It must be a dream," "This can't be happening. How can I escape?" "This can't be real." These inscriptions also symbolize the character's deep confusion and fear, as they attempt to deny the reality of being trapped in this dreadful dungeon.

The character refuses to accept the environment he is in, subconsciously choosing to deny the truth, convincing himself that everything is just a dream. As the sound of footsteps and laughter surrounds him, the character's denial intensifies, showing his inability to accept the reality of the dungeon. This design vividly conveys the character's psychological state, allowing players to deeply experience the emotions associated with the denial phase.

The Abyss. Unable to accept reality, I deceive myself. The pain is so overwhelming it drives me to rage. Why is fate so unfair to me? All my misfortunes are your fault!

The character tirelessly searches for an exit, but each effort is in vain. Anger and despair gradually erode his spirit. The writing on the walls becomes filled with hostility, expressing his inner rage and discontent.

The maze representing "Anger" is an abyss (see Fig. 2). In this level, the wall colors become deeper and more intense, primarily in shades of dark blue or green, interspersed with streaks of blood, creating a striking visual impact that symbolizes anger and despair.

These hues not only enhance the character's dissatisfaction with reality but also create an atmosphere fraught with danger.

The text on the walls becomes clearer and filled with hostility, such as "Why am I back here again? I'm so angry! Who trapped me here?" These words express the character's extreme anger and dissatisfaction. Confronted with an unchangeable reality, the character is overwhelmed with resentment and complaints. Through the design of the wall materials, changes in color, and detailed elements, these features reflect the character's rage towards reality. As the anger escalates, players will deeply experience the emotional fluctuations of this stage.

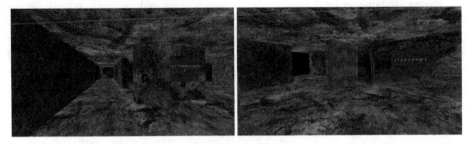

Fig. 2. The Abyss.

Ruined Hospital. Change of mind, start to beg, I'll do anything to get out, sacrifice everything... Please.

Although gradually coming to terms with his situation, he still seeks hope internally, continuously attempting to alter his circumstances through bargaining. The writing and patterns on the walls begin to show signs of pleading and prayer, conveying faint glimmers of hope and anticipation.

The third maze is a ruined hospital, designed to represent the "Bargaining" stage (see Fig. 3). The textures and colors have softened, displaying light yellow or orange hues, symbols of hope and prayer. These tones enhance the character's inner struggles and anticipation for the future. The sound effect incorporates low murmurs of prayer and occasional soft crying, allowing players to feel the character's complex emotions—both the aspirations for the future and the helplessness with the current situation. The contrast between light and shadow vividly portrays the interplay of hope and despair.

The text reflects the character beginning to converse with his own inner voice: "If I can find the exit, I will surely change... If someone can help me, I am willing to pay any price." This indicates his attempt to strike a bargain to escape his predicament. The content of these pleas and prayers shows that the character's emotions are becoming increasingly complex and contradictory, filled with hope yet also mixed with fear of the unknown as he searches for a way out. The overall design reflects the character's hope and struggle within.

Twisted N-Dimensional World. Becoming passive and negative, praying is useless. I start to become passive because there's no reason left to escape; I need to face the harsh reality.

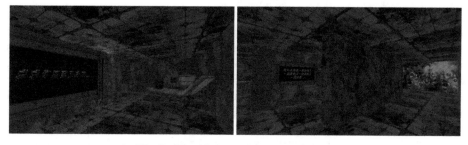

Fig. 3. The third maze: Ruined Hospital.

After experiencing the failure of bargaining, realizing that all prayers and pleas cannot change the current situation, he fell into deep depression.

The maze is designed to deliver a sense of "Depression" (see Fig. 4). It utilizes a complex, multi-dimensional twisted world as a metaphor to convey deep sorrow. The design of this stage reflects the character's passivity and negativity, enveloped in a sense of powerlessness and despair. The walls are adorned with dazzling reflective panels and shattered glass effects, creating an unstable and chaotic atmosphere. The use of vibrant, hallucinatory colors symbolizes the impotence in the character's mental world, while the broken, reflective shards suggest the fragmentation of the character's psyche. The background features soft, melancholic music that enhances the feelings of depression and despair, allowing players to experience the anguish of the character's shattered spirit. The text reads, "As life flashes before my eyes, where is paradise?" showcasing the character is filled with helplessness, with no attempts to escape, only a passive acceptance of everything.

Fig. 4. A twisted N-dimensional world.

Cage. Enshrouded in despair, I fall into an endless darkness; it's a vicious cycle from which there's no escape.

Having journeyed through denial, anger, bargaining, and depression, the character ultimately arrives at the stage of acceptance. They cease trying to escape or alter their circumstances and accept the reality of their inescapable confinement. This acceptance brings an inner calm, even though reality remains harsh.

The last maze is a sealed square cage without any exit (see Fig. 5), symbolizing a despair with no escape. The text on the wall reads, "Stop dreaming, you'll never get out," reflecting the character's acceptance of the unchangeable reality. Enveloped by an atmosphere of despair, it forms an unbreakable cycle. The color of the walls has turned to pale gray, reflecting the calm and acceptance of the character. The color of the walls has turned to pale gray, reflecting a sense of calmness and tranquility of the character, with a hint of sadness. This conveys to the player that the character has given up on the idea of escaping. The sudden shift from a surreal, psychedelic atmosphere to a confined and inescapable space intensifies the contrast and disparity with reality. Through harsh textual content and somber, slow background music, a tranquil yet resigned atmosphere is created, allowing players to deeply experience the character's ultimate acceptance of reality.

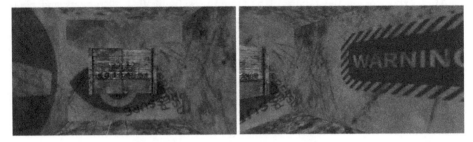

Fig. 5. The last maze of the game: Cage.

Each maze in the game is designed with a starting point and an exit, each enveloped in beams of light and distinguished by color (see Fig. 6). Green is used for the starting point of each maze, symbolizing the initial phase of dealing with loss. Despite the potential heaviness of heart, it represents the possibility of renewal and progress, helping players to start from a more peaceful mindset and gradually handle the successive emotional challenges. The end of the maze is marked by a golden yellow beam, symbolizing hope, recovery, and positive change. This represents psychological progress, offering people a positive and hopeful outlook for the future.

Fig. 6. Starting point and exit of each maze.

3.2 Interactive Design

To offer a more realistic experience of running and navigating through mazes, this VR game departs from conventional joystick-based navigation and instantaneous teleportation. Instead, it employs motion tracking of natural arm swings during simulated running to conduct the character's movement within the virtual environment. This approach uses real-time physical actions of the players – swinging their arms while holding controllers – to determine the speed of the character. The moving direction is aligned with the player's gaze direction, which is projected straight ahead through the VR headset (see Fig. 7).

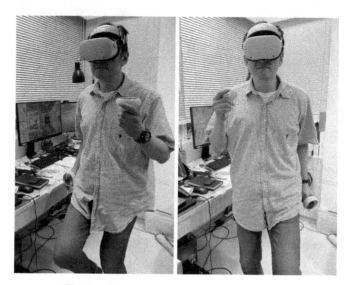

Fig. 7. The movement control system (stand).

For those susceptible to motion sickness, an alternative control scheme has been developed for seated play (see Fig. 8). Players sit in a swivel chair with controllers in hand. Movement speed is also controlled by arm swinging, and directional changes are executed by rotating the chair, thereby mimicking body turning movements without the physical strain.

This method enhances the immersive experience by leveraging natural body movements and reducing the disconnect between physical actions and virtual responses, thus potentially lowering the incidence of VR-induced motion sickness. This design not only makes the experience more intuitive but also allows for a deeper integration of physical and psychological engagement with the virtual environment.

To prevent players from getting unconsciously lost in the maze game while encouraging them to actively seek the exit, each maze has a limited playtime. Once the time is up, the experience automatically stops. However, to reduce the feeling of time pressure, the game timer is not displayed on the screen. Only when the time runs out do participants realize there was a time limit, prompting them to be more proactive in finding the exit in future sessions.

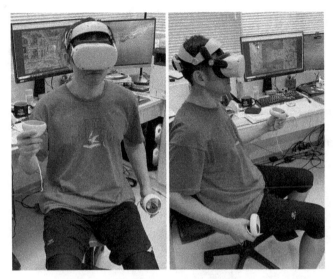

Fig. 8. The movement control system (seated).

3.3 Music and Sound Effects

Game music serves as an enhancement to the overall atmosphere of the game, providing players with varying emotional experiences. Therefore, this creation selects different background music and sound effects based on the feelings elicited by different maps in the game. The Table1 provides a summary of all music and sound effects used in this creation.

Table 1. Music and sound effects in each maze.

Maze	Music and Sound Effects	Web Link
1	Landing by Godmore 恐怖の夜の樹海 by Kamatama	https://happysoulmusic.com/audio/landing_-_godmode-mp3 https://kamatamago.com/sozai/bgm/B00006/
2	Fun House by Coyote Hearing	https://youtu.be/jjzxeRZsDRM?si=SD8TfZcBlCwUaU3n
3	Black Mass by Brian Bolger 乗り物01 by 森田交一 foot01, gun-fire, gun-ready, horror03	https://youtu.be/Z7moS4cxjOE?si=l8Aa0B37t-mauTqZ https://maou.audio/se_sound_vehicle01/ https://pixabay.com/sound-effects/
4, 5	Space Hunter by Quincas Moreira	https://youtu.be/g4yQgbuBqqg?si=qXlfu6mdCEhW6BzQ

4 Results

As an innovative auxiliary tool, virtual reality has been proven to significantly alleviate symptoms of anxiety. In this study, based on the Kübler-Ross model's five stages of grief, we have developed a virtual reality maze game. This game aims to offer players a new way to deeply experience human emotions through immersive experiences, thereby enhancing understanding of grief and helping people to cherish and empathize with the emotions of others more profoundly.

Since the five stages of grief are not a linear progression, it does not mean that a person will only go through the entire grieving process once and then permanently enter the acceptance stage. These stages are dynamic, with individuals possibly experiencing them in different orders or intensities, and may go through these stages multiple times. Thus, each person's experience varies. To better simulate this process, subsequent improvements will randomize the order of the mazes, allowing participants to have different experiences each time, thereby gaining a deeper understanding of the diversity and complexity of grief.

Acknowledgments. This research was supported by "Intelligent Manufacturing Center (iMRC)" from the Higher Education Sprout Project, Ministry of Education to the Headquarters of University Advancement at National Cheng Kung University (NCKU) and Integrated VR Research Center of Ming Chuan University.

Disclosure of Interests. The authors have no competing interests to declare that are relevant to the content of this article.

References

1. Albakri, G., et al.: Phobia exposure therapy using virtual and augmented reality: a systematic review. Appl. Sci. **12**(3), 1672 (2022)
2. Deep, V.R.: https://www.exploredeep.com/#about-deep. Last accessed 19 Nov 2023
3. Fodor, L.A., Coteț, C.D., Cuijpers, P., Szamoskozi, Ș, David, D., Cristea, I.A.: The effectiveness of virtual reality based interventions for symptoms of anxiety and depression: a meta-analysis. Scient. Reports **8**(1), 10323 (2018)
4. Geraets, C.N., Van der Stouwe, E.C., Pot-Kolder, R., Veling, W.: Advances in immersive virtual reality interventions for mental disorders: a new reality? Curr. Opin. Psychol. **41**, 40–45 (2021)
5. Hamedi, V., Hamid, N., Beshlideh, K., Marashi, S.A., Shabani, S.E.: Effectiveness of conventional cognitive-behavioral therapy and its computerized version on reduction in pain intensity, depression, anger, and anxiety in children with cancer: a randomized, controlled trial. Iranian J. Psychiatry Behav. Sci. **14**(4), e83110 (2020)
6. Hellblade: Senua's Sacrifice, https://store.steampowered.com/app/414340/Hellblade_Senuas_Sacrifice/. Last accessed 28 Oct 2023
7. Kübler-Ross, E.: On Death and Dying. Macmillan, New York (1969)
8. Mistry, D., et al.: Meditating in virtual reality: proof-of-concept intervention for posttraumatic stress. Psychol. Trauma Theory Res. Pract. Policy **12**(8), 847–858 (2020)

9. Nevermind: https://store.steampowered.com/app/342260/Nevermind/?l=tchinese. Last accessed 18 Oct 2023
10. Phobia Exposure VR: https://store.steampowered.com/app/2509570/Phobia_Exposure_VR/. Last accessed 12 Nov 2023
11. Powers, M.B., Emmelkamp, P.M.: Virtual reality exposure therapy for anxiety disorders: a meta-analysis. J. Anxiety Disord. **22**(3), 561–569 (2008)
12. Segal, R., Bhatia, M., Drapeau, M.: Therapists' perception of benefits and costs of using virtual reality treatments. Cyberpsychol. Behav. Soc. Netw. **14**(1–2), 29–34 (2011)
13. Tsai, C.H.: Interpretation of the classics in life-and-death studies – take elisabeth Kübler-ross' on death and dying as an example. General Educ. Transdisciplinary Res. **2**(1), 1–25 (2007)
14. Van Rooij, M., Lobel, A., Harris, O., Smit, N., Granic, I.: DEEP: A biofeedback virtual reality game for children at-risk for anxiety. In: Proceedings of the 2016 CHI Conference Extended Abstracts on Human Factors in Computing Systems, pp. 1989–1997. ACM, San Jose California USA (2016)

Combined Bidirectional Long Short-Term Memory and Mel-Frequency Cepstral Coefficients with Convolution Neural Network Using Triplet Loss for Speaker Recognition

Young-Long Chen[✉], Jing-Fong Ciou, Chih-Han Lin, and Shih-Sheng Lien

Department of Computer Science and Information Engineering, National Taichung University of Science and Technology, Taichung 40401, Taiwan
ylchen66@nutc.edu.tw

Abstract. In recent years, there have been significant breakthroughs in neural network technology within the field of speaker recognition, encompassing various applications such as word classification, emotion recognition, and speaker identification. This paper introduces an innovative method for speaker recognition aimed at enhancing accuracy. The method, termed bidirectional long short-term memory combined with mel-frequency cepstral coefficients and convolution neural network features using triplet loss for speaker recognition (BLSTM-MFCCCNN-TL), employs dual-factor features as the input layer of the model. Compared to traditional GMM-HMM and BLSTM-MFCC models, the new method not only provides more accurate recognition capabilities but also exhibits faster computational speeds. Consequently, these advancements significantly enhance the overall performance of speaker recognition while improving computational efficiency.

Keywords: Speaker Recognition · Convolution Neural Network · Bidirectional Long Short-Term Memory · Mel-Frequency Cepstral Coefficients · Triplet Loss

1 Introduction

Speaker identification is a biometric technology commonly used in security set-tings like access control systems, employee attendance tracking, bank vaults, robotics, and smart homes. It uses individuals' unique voice characteristics to verify their identity, enhancing both security and convenience.

Speaker recognition technology can be categorized into two types: text-dependent [1] and text-independent [2]. Text-dependent recognition requires speakers to say specific sentences or phrases, commonly used for identity verification in controlled settings. In contrast, text-independent recognition does not require specific texts and can identify speakers' voices in any conversation, making it suitable for a broader range of applications.

In wake-up word models [1], speaker recognition focuses on confirming the speaker's identity without restrictions on sentence content, unlike text-independent recognition.

Additionally, the proposes method [2] for speaker verification in noisy environments by integrating different features, suitable for text-independent recognition. Forensic speaker recognition, discussed in [3–5], involves identifying speakers by their voices when they leave audio evidence [3]. Text-dependent speaker recognition is easier to train but more constrained, requiring the same sentences during recognition. In contrast, text-independent speaker recognition is more flexible but presents challenges and often requires more data for better results. Our research aims to develop a text-independent speaker recognition method.

The speaker identification process consists of two main aspects: feature ex-traction and speaker modeling. Early research has developed various techniques for extracting features from audio signals. For example, linear frequency cepstral coefficients (LFCC) [2, 6] and mel-frequency cepstral coefficients (MFCC) [2, 6, 7] both extract features using filter bank coefficients. LFCC processes frequencies uniformly, while MFCC is designed with human ear perception in mind, emphasizing low-frequency sounds, as the human ear is more sensitive to these frequencies. The differences between LFCC and MFCC are detailed in [2], and the MFCC computation process is described in [8]. Kinnunen and Li [9] explored text-independent speaker recognition techniques and recommended feature se-lection methods, including MFCC. Sahidullah et al. [10] proposed a new windowing technique for MFCC calculation and compared it with traditional methods in terms of recognition accuracy. In dynamic time warping (DTW) [7] was used alongside MFCC-extracted features for speaker recognition. Both LFCC and MFCC techniques were applied to classify insect sounds [6], with LFCC yielding better results for high-frequency sounds. For speech recognition, gammatone cepstral coefficients are also commonly used in addition to MFCC [11]. Since our research focuses on human voices, we chose MFCC for feature extraction due to its suitability for recognizing human speech. Recently, many models have been applied in speech recognition, primarily for recognizing and processing input data. Hidden markov models (HMM) were first used for speaker-independent isolated word recognition [12]. HMM constructs implicit unknown parameters using multiple states, mainly for isolated word recognition. [13] utilized gaussian mixture models (GMM) to represent the speaker's spectral shape using multiple gaussian components. GMM-Universal background model (GMM-UBM) was developed to improve GMM performance by using a general background model [14], though it requires large datasets. The rapid development of neural network (NN) technology has led to significant advancements in AI applications, including speech recognition. A neural network architecture [15] includes an input layer, hidden layers, and an output layer, with the hidden layers being the core of the architecture. For classification tasks, the output layer provides the probability of each class. Deep denoising autoencoders, composed of neural networks, generate denoised feature vectors and use GMM-HMM for isolated word recognition [16]. [17] proposed training a single deep neural network (DNN) for both speaker and language recognition. In the Chinese text-to-speech (TTS) method [18] using re-current neural networks (RNN) was introduced. The differences between RNN and general NNs are described in [19]. RNN incorporates feedback from previous layers, allowing it to process dynamic information. However, RNN struggles with long-term memory, which long short-term memory (LSTM) addresses [20, 21].

Many applications in speech recognition use LSTM. For example, uses an attention-based LSTM for speech emotion recognition [22, 23], studied methods to improve speech emotion recognition accuracy by utilizing non-verbal sound segments, with LSTM as the model. [24] proposed bidirectional LSTM (BLSTM), consisting of two LSTMs operating in different directions, providing both for-ward and backward memory paths. BLSTM is used in applications such as gender classification in speech [24], speech emotion recognition [25], and native language recognition in short utterances [26].

This paper introduces a new speaker recognition method: bidirectional long short-term memory combined with mel-frequency cepstral coefficients and convolutional neural network features using triplet loss (BLSTM-MFCCCNN-TL). In this method, we use MFCC to extract audio features and input them into the BLSTM for training. This approach enhances the accuracy of speaker recognition. The BLSTM-MFCCCNN-TL method employs the BLSTM [27] as its model architecture.

2 Bidirectional Long and Short-Term Memory Combined Mel-Frequency Cepstral Coefficients with Convolution Neural Network Features Using Triplet Loss Method (BLSTM-MFCCCNN-TL)

In previous audio processing methods, a large volume of signals imposed a significant computational burden on computers. Extracting features from many signals posed challenges due to the abundance of noise and irrelevant signals, necessitating extensive audio processing. The development of feature extraction techniques like MFCC [2] has made significant contributions to audio signal processing. In this paper, we investigate text-independent speaker recognition [2], which does not require preserving the original audio signal. Therefore, we adopt the MFCC technique to extract features from the audio sequences. This study constructs a speaker recognition model using neural network [15] technology. The results of MFCC feature extraction are used as input to the model. After passing through the neural network, a set of feature vectors representing speaker identities is obtained. The accuracy of the identification results is then deter-mined through similarity distance calculations. The paper proposes a new speaker recognition method. For deep learning models, the input data size must comply with the requirements of CNN. The encoded CNN features are concatenated with MFCC features to serve as the input for the deep learning model. Because CNN features result from speaker encoding, their combination can effectively enhance the model's learning performance.

In the method by Chen et al. [28], dual-factor features are used as the input layer of the model. It mentions the use of an autoencoder for feature extraction, with the hidden layer being an NN layer. In this method, we modified the neural network architecture in the autoencoder. We used CNN for feature extraction in the second factor. Therefore, we proposed the new method, called BLSTM-MFCCCNN-TL.

The architecture of BLSTM-MFCCCNN-TL is shown in Fig. 1. First, the audio starts feature extraction from the lower left corner, using MFCC technology for all feature extraction. After the extracted features are preprocessed, all MFCC features are constructed into a uniform size (N, 39). On the other hand, the audio is fed into the input layer of the convolutional neural network (CNN) on the right side to train the

CNN model. After training, the previous layer will output 1014 features, which are then redefined as a two-dimensional feature of size (26, 39). Next, the mel-frequency cepstral coefficients (MFCC) features of size (N, 39) are merged with the features of size (26, 39) extracted by the neural network model, constructing the final feature matrix of size (N + 26, 39). The final features are then fed into the main bidirectional bong short-term memory (BLSTM) model as input, and the model is trained using triplet loss as the loss function.

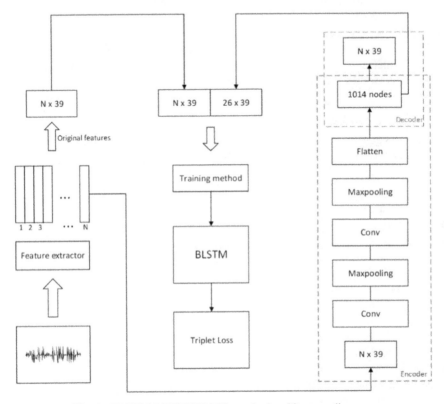

Fig. 1. BLSTM-MFCCCNN-TL method architecture diagram.

In the first step, after the MFCC feature extraction process, our signal is trans-formed from a one-dimensional signal to a two-dimensional spectrogram. How-ever, due to the varying lengths of each audio clip, the size of the MFCC features will differ. MFCC feature extraction effectively denoises the audio and captures its characteristics, which is why we use MFCC for feature extraction. Next, the second step is feature preprocessing, which focuses on standardizing the size of the MFCC features. We redefine all MFCC sizes to be (N, 39) because the input to a neural network typically needs to have a uniform size to effectively train the model. However, in the third step, before feeding the features into the main mod-el, we first construct an additional CNN model. We then train this CNN model, and ultimately, we extract the feature vectors from the output layer of this pre-trained model as additional features beyond the MFCC. According to research [16], utilizing additional features obtained from the pre-trained neural network model can significantly improve the performance of the final results. For the pre-trained model, we constructed a CNN-based Autoencoder for feature extraction. In building this network, we used two convolutional layers and max-pooling layers, followed by a neural network layer with 1014 neurons. The final output is a matrix of the same size as the input layer. To ensure that the additional features extracted from the middle layer match the size of the MFCC features, we configured the neural network layer to construct a two-dimensional matrix of size (26, 39). For training this model, we used mean squared error (MSE) as the loss function to train the Autoencoder model. After training, we extracted the aforementioned feature vectors, which have a size of (26, 39). The BLSTM model, as shown in Fig. 2, includes the model architecture detailed in Table 1.

Fig. 2. BLSTM model architecture diagram in the BLSTM-MFCCCNN-TL method.

Table 1. BLSTM model parameter table in BLSTM-MFCCCNN-TL method.

Layers	size	information
input	(512, 39, 1)	Units: 32
Conv2D	(256, 20, 32)	
MaxPooling	(128, 20, 32)	
Conv2D	(64, 20, 16)	Units: 16
MaxPooling	(32, 20, 16)	
flatten	(1, 5120)	
Dence	(1, 1014)	
Dence	(1, 19968)	
Output	(1, 128)	

From Table 1, we can see that our pre-trained model is constructed using a CNN. It consists of a convolutional layer with 32 neurons, a max-pooling layer, another convolutional layer with 16 neurons, another max-pooling layer, and a flatten layer. Finally, it has two neural network layers for output, with the layer containing 1014 neurons being the one from which we extract additional features. This layer constructs feature vectors of size (26, 39).

3 Experiment Results

To evaluate the performance of the method proposed in this paper, we conducted speaker recognition experiments using the AISHELL-1 dataset [28]. The AISHELL-1 dataset contains Mandarin speech recorded with a high-fidelity microphone (44.1 kHz, 16-bit).

The audio in the AISHELL-1 dataset is down sampled to a sampling rate of 16 kHz. This dataset is a public speech corpus recorded by 400 participants from different regions of China, each with various accents. In this paper, we used the AISHELL-1 dataset for speaker recognition. In our experiments, the entire dataset of 400 speakers was divided into a training set and a validation set, with a split ratio of 90% for training and 10% for validation. Table 1 shows the input and output feature shapes, representing the size of the features fed into and outputted by the main model, respectively. Table 2 lists the experimental parameters for the proposed BLSTM-MFCCCNN-TL method.

Table 2. Experimental parameters of our proposed method.

Method Name	BLSTM-MFCCCNN-TL
Input features	MFCC&CNN
Main model	BLSTM
Batch size	64
Learning rate	10^{-5}
Loss function	Triplet loss
Optimizer	Adam
Input feature shape	(538, 39)
Output feature shape	(1, 128)

Table 2 presents the architectures of the main models in the proposed BLSTM-MFCCCNN-TL method. In this method, the input section utilizes a combination of MFCC and CNN features, resulting in a feature vector size of (538, 39). The main model adopts a BLSTM architecture. During training, a learning rate of 0.00001 was employed with the Adam optimizer, and the batch size was set to 64. The triplet loss function was utilized as the loss function, and training was con-ducted for 20 epochs, resulting in an output feature vector size of (1, 128). Throughout the training process, the loss values and accuracy were recorded for each epoch to evaluate the training performance. Initially, a comparison of the loss values between our proposed BLSTM-MFCCCNN-TL method and other approaches was conducted, as illustrated in Fig. 3 and Table 3.

Figure 3 shows the changes in training and validation data loss values for the BLSTM-MFCCCNN-TL method across each epoch. In the second epoch, the loss values for each method decrease significantly. Subsequently, the gap between the training and validation loss values becomes smaller, and the training loss values gradually stabilize. Table 3 compares the loss values of the proposed method at the 20th epoch, showing a training loss value of 0.66% and a validation loss value of 0.62%. The comparison indicates that there is not a significant difference between the training and validation loss values, and our method achieved lower loss values. Therefore, this proposed method has been effectively trained. Next, we compared the accuracy of the proposed BLSTM-MFCCCNN-TL method, as shown in Fig. 4 and Table 4.

Figure 4 shows the accuracy trends of the training and validation data for the proposed BLSTM-MFCCCNN-TL method over the first 20 epochs. Initially, each method

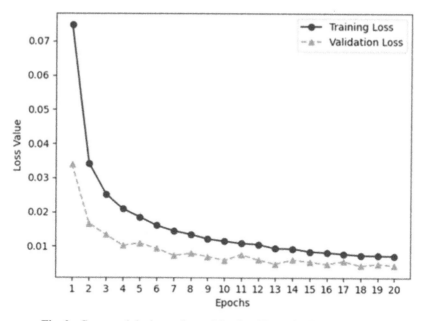

Fig. 3. Compared the loss values of the first 20 epochs for the method.

Table 3. We provide a comparison table showing the loss values at the 20 epochs.

Method Name	Train Loss Value	Validation Loss Value
BLSTM-MFCC-TL [28]	1.19%	0.42%
BLSTM-MFCCNN-TL [28]	0.75%	0.49%
BLSTM-MFCCCNN-TL (our)	0.66%	0.62%

experiences a significant change in accuracy amplitude in the second epoch, followed by a gradual increase. The results at the first 20 epochs are listed in Table 4. In this speaker recognition experiment, accuracy is defined as the ratio of correctly identified speakers to the total number of predictions. At the 20th epoch, the proposed BLSTM-MFCCCNN-TL method achieved a validation accuracy of 93.69%, while the GMM-HMM-MFCCAE method only reached 64.48%, indicating the poorest performance (as shown in Table 4). Although the GMM-HMM-MFCCAE method was initially designed for word recognition, we referenced its CNN feature architecture and applied it to speaker recognition in our experiments. This ensured that the proposed BLSTM-MFCCCNN-TL method could utilize CNN features and achieve better accuracy in speaker recognition. The accuracy improvement seen in the BLSTM-MFCCCNN-TL method indicates that adding additional CNN features can effectively increase validation accuracy by 0.61%.

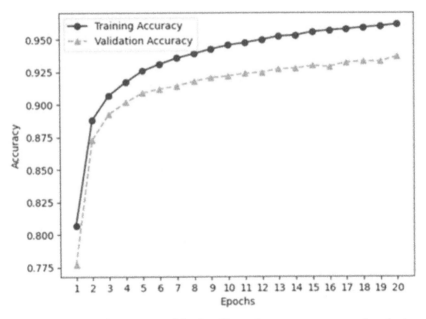

Fig. 4. Compare the accuracy of the first 20 epochs among our proposed method.

Table 4. We propose a comparison of the accuracy within the first 20 epochs among our method.

Method Name	Train Accuracy	Val Accuracy
GMM-HMM-MFCCAE	65.60%	64.48%
BLSTM-MFCC-TL	93.68%	91.18%
BLSTM-MFCCAE-TL	96.01%	93.08%
BLSTM-MFCCCNN-TL (our)	96.16%	93.69%

Table 5 lists the computation times for each method: GMM-HMM-MFCCAE, BLSTM-MFCCC-TL, BLSTM-MFCCAE-TL, and BLSTM-MFCCCNN-TL. By comparing these computation times, we can understand the operational efficiency of each method in actual prediction scenarios.

Table 5. Computation time comparison table of each method.

Method Name	First Computation Time (s)	Average Computation Time (s)
GMM-HMM-MFCCAE	1.5195	1.5286
BLSTM-MFCC-TL	0.5844	0.3856
BLSTM-MFCCAE-TL	0.9677	0.4052
BLSTM-MFCCCNN-TL (our)	0.5236	0.2456

From the perspective of initial computation time and average computation time, the new method proposed in this paper is faster than the traditional GMM-HMM-MFCCAE method and the BLSTM-MFCCAE-TL method. The GMM-HMM-MFCCAE method involves the GMM-HMM model, which requires higher computation time, in addition to the time needed to process AE features. Therefore, our proposed method is superior in computation time compared to both the GMM-HMM-MFCCAE method and the BLSTM-MFCCAE-TL method.

4 Conclusions

This paper proposes a novel speaker recognition method: Bidirectional long and short-term memory combined mel-frequency cepstral coefficients with convolution neural network features using triplet loss for speaker recognition, abbreviated as BLSTM-MFCCCNN-TL. This proposed method integrates pre-trained CNN features for extraction, forming the BLSTM-MFCCCNN-TL approach.

After training the proposed method for 20 epochs, the validation accuracy for the BLSTM-MFCC-TL method reaches 93.69%. The BLSTM architecture used in this study is primarily based on the literature [28], which we adopted for experimental comparison. As the research in [16] focuses on denoised speech recognition, we adapted the architecture described in [16] to the GMM-HMM-MFCCAE method for speaker recognition to compare training outcomes. The validation accuracy for the GMM-HMM-MFCCAE method is 64.48%. The average computation times for the BLSTM-MFCC-TL, BLSTM-MFCCAE-TL, and BLSTM-MFCCCNN-TL methods are 0.3856 s, 0.4052 s, and 0.2456 s, respectively. Our proposed method has a higher average computation time compared to other methods. From the results, the BLSTM-MFCCCNN-TL meth-od, utilizing the BLSTM model with CNN features, improves speaker recognition accuracy by 2.51% compared to the BLSTM-MFCC-TL method. Furthermore, the proposed method outperforms both the reference framework's GMM-HMM-MFCCAE method and the BLSTM-MFCC-TL method in terms of speaker recognition accuracy. In conclusion, our proposed method outperforms the traditional GMM-HMM-MFCCAE method, BLSTM-MFCC-TL method and BLSTM-MFCCAE-TL [28] method in computation time and speaker recognition accuracy.

Acknowledgments. This study was partly supported by the MOE Teaching Practice Research Program of the Republic of China under Grant No. Psk1123096.

Disclosure of Interests. The authors have no competing interests to declare that are relevant to the content of this article. The authors declare no conflict of interest.

References

1. Tsai, T.H., Hao, P.C., Wang, C.L.: Self-defined text-dependent wake-up-words speaker recognition system. IEEE Access **9**, 138668–138676 (2021)

2. Mohammadi, M., Mohammadi, H.R.S.: Robust features fusion for text independent speaker verification enhancement in noisy environments. In: Proceedings of the Iranian Conference on Electrical Engineering (ICEE), pp. 1863–1868. IEEE (2017)
3. Wang, Z., Hansen, J.H.: Multi-source domain adaptation for text-independent forensic speaker recognition. IEEE/ACM Trans. Audio, Speech Lang. Process. **30**, 60–75 (2021)
4. Campbell, J.P., Shen, W., Campbell, W.M., Schwartz, R., Bonastre, J.F., Matrouf, D.: Forensic speaker recognition. IEEE Signal Process. Mag. **26**(2), 95–103 (2009)
5. Hansen, J.H., Hasan, T.: Speaker recognition by machines and humans: a tutorial review. IEEE Signal Process. Mag. **32**(6), 74–99 (2015)
6. Noda, J.J., Travieso-González, C.M., Sánchez-Rodríguez, D., Alonso-Hernández, J.B.: Acoustic classification of singing insects based on MFCC/LFCC fusion. Appl. Sci. **9**(19), 4097 (2019)
7. Muda, L., Begam, M., Elamvazuthi, I.: Voice recognition algorithms using mel frequency cepstral coefficient (MFCC) and dynamic time warping (DTW) techniques. arXiv preprint arXiv:1003.4083. (2010)
8. Dighore, V.D., Thool, V.R.: Analysis of asthma by using mel frequency cepstral coefficient. In Proceedings of IEEE International Conference on Recent Trends in Electronics, Information and Communication Technology (RTEICT), pp. 976–980. IEEE (2016)
9. Kinnunen, T., Li, H.: An overview of text-independent speaker recognition: from features to supervectors. Speech Commun. **52**(1), 12–40 (2010)
10. Sahidullah, M., Saha, G.: A novel windowing technique for efficient computation of MFCC for speaker recognition. IEEE Signal Process. Lett. **20**(2), 149–152 (2012)
11. Alashban, A.A., Qamhan, M.A., Meftah, A.H., Alotaibi, Y.A.: Spoken language identification system using convolutional recurrent neural network. Appl. Sci. **12**(18), 9181 (2022)
12. Lee, K.F., Hon, H.W.: Speaker-independent phone recognition using hidden Markov models. IEEE Trans. Acoust. Speech Signal Process. **37**(11), 1641–1648 (1989)
13. Reynolds, D.A., Rose, R.C.: Robust text-independent speaker identification using Gaussian mixture speaker models. IEEE Trans. Speech Audio Process. **3**(1), 72–83 (1995)
14. Kumar, V.R., Vydana, H.K., Vuppala, A.K.: Significance of GMM-UBM based modelling for Indian language identification. Procedia Comput. Sci. **54**, 231–236 (2015)
15. Sze, V., Chen, Y.H., Yang, T.J., Emer, J.S.: Efficient processing of deep neural networks: a tutorial and survey. Proc. IEEE **105**(12), 2295–2329 (2017)
16. Grozdić, ÐT., Jovičić, S.T., Subotić, M.: Whispered speech recognition using deep denoising autoencoder. Eng. Appl. Artif. Intell. **59**, 15–22 (2017)
17. Richardson, F., Reynolds, D., Dehak, N.: Deep neural network approaches to speaker and language recognition. IEEE Signal Process. Lett. **22**(10), 1671–1675 (2015)
18. Chen, S.H., Hwang, S.H., Wang, Y.R.: An RNN-based prosodic information synthesizer for Mandarin text-to-speech. IEEE Trans Speech Audio Process. **6**(3), 226–239 (1998)
19. Malhi, A., Yan, R., Gao, R.X.: Prognosis of defect propagation based on recurrent neural networks. IEEE Trans. Instrum. Meas. **60**(3), 703–711 (2011)
20. Adam, K., Smagulova, K., James, A.P.: Memristive LSTM network hardware architecture for time-series predictive modeling problems. In: Proceedings of IEEE Asia Pacific Conference on Circuits and Systems (APCCAS), pp. 459–462. IEEE (2018)
21. Xie, Y., Liang, R., Liang, Z., Huang, C., Zou, C., Schuller, B.: Speech emotion classification using attention-based LSTM. IEEE/ACM Trans. Audio, Speech, Lang. Process. **27**(11), 1675–1685 (2019)
22. Hsu, J.H., Su, M.H., Wu, C.H., Chen, Y.H.: Speech emotion recognition considering nonverbal vocalization in affective conversations. IEEE/ACM Trans. Audio Speech Lang. Processing **29**, 1675–1686 (2021)
23. Graves, A., Schmidhuber, J.: Framewise phoneme classification with bidirectional LSTM and other neural network architectures. Neural Netw. **18**(5–6), 602–610 (2005)

24. Alamsyah, R.D., Suyanto, S.: Speech gender classification using bidirectional long short term memory. In: Proceedings of 3rd International Seminar on Research of Information Technology and Intelligent Systems (ISRITI), pp. 646–649. IEEE (2020)
25. Sajjad, M., Kwon, S.: Clustering-based speech emotion recognition by incorporating learned features and deep BiLSTM. IEEE Access **8**, 79861–79875 (2020)
26. Adeeba, F., Hussain, S.: Native language identification in very short utterances using bidirectional long short-term memory network. IEEE Access **7**, 17098–17110 (2019)
27. Chen, Y.L., Wang, N.C., Ciou, J.F., Lin, R.Q.: Combined bidirectional long short-term memory with mel-frequency cepstral coefficients using autoencoder for speaker recognition. Appl. Sci. **13**(12), 7008 (2023)
28. Bu, H., Du, J., Na, X., Wu, B., Zheng, H.: Aishell-1: An open-source mandarin speech corpus and a speech recognition baseline. In: Proceedings of 20th conference Oriental Chapter of the International Coordinating Committee on Speech Databases and Speech I/O Systems and Assessment (O-COCOSDA), pp. 1–5. IEEE (2017)

An Improved YOLOv5 Model with FRB Method for Product Surface Defect Detection

Young-Long Chen(✉), Chuan-Cheng Chung, and Li-Hong Qin

Department of Computer Science and Information Engineering,
National Taichung University of Science and Technology, Taichung 40401, Taiwan
ylchen66@nutc.edu.tw, s1710932015@ad1.nutc.edu.tw

Abstract. In factory production, defect inspection is a crucial section in quality assurance. However, manually is inefficient, expensive, and unreliable. As Industry 4.0 becomes mainstream, the enhancement of production processes through Industrial Artificial Intelligence is an important topic. The deployment of defect inspection systems in automated production lines can reduce employee training, save labor costs, and enhance product quality. This study proposes a new approach for defect detection tasks based on the YOLOv5 object detection method. In this paper, this proposed method combines the YOLOv5 model with a Feature Retaining Block (FRB) to preserve fine features and is referred to as FRB-YOLO. It aims to tackle the issue of declining product quality caused by minor defects during the manufacturing process for most products. Minor defects which may resemble the natural patterns of the product can result in defective products being incorrectly classified as normal. The study conducts experiments using the publicly available NEU-DET datasets. The FRB-YOLO method proposed outperforms YOLOv5 in terms of mean average precision (mAP) across the NEU-DET datasets.

Keywords: Deep Learning · Convolutional Neural Networks · Object Detection · YOLOv5 · Defect Inspection · Data Augmentation · AutoEncoder · Real-Time Processing

1 Introduction

In recent years, with the continuous advancement of industrial product manufacturing, the detection of surface defects in products has become an indispensable part of the process and a key factor in ensuring product quality. In the past, it was manually less efficient for inspecting product defects. Manual screening is not only highly susceptible to issues such as the inspector's current emotions, fatigue, and subjective judgments, but it also leads to unreliable conditions such as missing inspections and slowing down the inspection speed. Additionally, the additional costs of hiring and training employees make manual defect detection very expensive. In recent years, Industry 4.0 [1] has gradually become a mainstream trend, improving the existing industrial production environment through Industrial Artificial Intelligence (IAI), which is a hot topic. Deploying defect detection systems on automated production lines can significantly reduce the time required to train quality inspectors, save more manpower costs, and enhance the reliability of products. This direction is very promising for both the present and the future.

The estimation methods related to automated defect detection can be categorized into two types: traditional computer vision methods and detection methods based on Convolutional Neural Networks (CNNs) [2]. Traditional computer vision methods are typically classified into approaches based on structural statistics and filters to achieve surface defect detection. For products with repetitive texture features such as steel or tree trunks, structural-based methods like modeling through Gaussian functions can be utilized [3]. For products with handmade texture features, statistical-based methods can be employed to calculate the random distribution of handmade texture features, such as Local Binary Patterns (LBP) [4] and Histogram of Oriented Gradient (HOG) [5]. The filter-based method will utilize Gabor filters [6] to distinguish different texture features, as well as curvature filters [7] and other techniques.

In recent years, due to the increasing popularity of CNN technology, some automated defect detection methods have started to use Convolutional Neural Networks to solve related problems [9]. However, due to the complexity of the feature extraction process in these methods and the possibility of losing feature information during extraction, detection performance may be poorer when defects are small [10]. Overall, there is still plenty of room for improvement in surface defect detection methods based on Convolutional Neural Networks.

Inspired by SDDNet [11], efforts are being made to enhance the performance of surface defect detection. This paper introduces a novel method named FRB-YOLOv5, which is based on YOLOv5 and integrates the Feature Retaining Block (FRB) from SDDNet. This method aims to achieve both speed and high accuracy in surface defect detection. The paper conducted experiments on public datasets, comparing them with other object detection methods, showcasing the effectiveness of the proposed method in surface defect detection tasks. The main contributions of the proposed method are as follows: First, the FRB-YOLOv5 method is proposed to improve the performance of surface defect detection by combining YOLOv5 with feature retaining blocks. Second, through cross-comparison with other object detection methods on the public datasets NEU-DET, it has been demonstrated that the proposed method can effectively improve the accuracy of surface defect detection while maintaining fast and efficient execution speeds.

2 The FRB-YOLOv5 Method

The network architecture of the FRB-YOLOv5 method is based on the YOLOv5 architecture with a Feature Retaining Block (FRB) added after each BCSP block in its backbone. The architecture of FRB-YOLOv5 is shown in Fig. 1.

FRB-YOLOv5 mainly consists of three components. Upon inputting an image into FRB-YOLOv5, it first extracts features through the Backbone network, then enhances the feature information through the Neck, and finally outputs feature maps of large, medium, and small sizes through the Prediction head to complete the prediction.

The Backbone first undergoes downsampling convolution through two CBS components, followed by feature extraction through the BCSP block. After feature extraction,

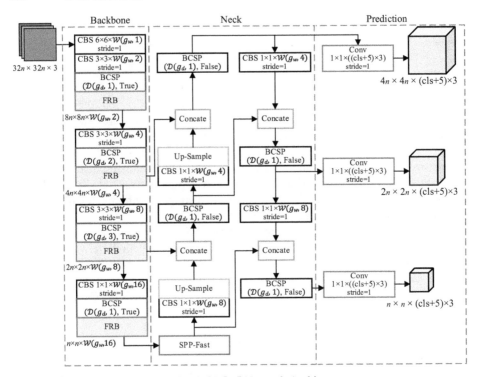

Fig. 1. FRB-YOLOv5 Network Architecture

there is an adjacent downsampling convolution to further reduce the size of the feature map. The CBS component consists of a convolutional layer, a batch normalization layer [12] and a SiLU [13] activation function. Apart from the prediction head, all convolutional computations in FRB-YOLOv5 are performed by CBS components. The structure of the CBS component is shown in Fig. 2.

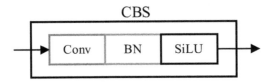

Fig. 2. Architecture diagram of the CBS component

The BCSP (Bottleneck with Cross Stage Partial) block is a feature extraction block composed of several CBS components or residual units, connected with Cross Stage Partial Connections (CSP). This block first uses a 1×1 convolutional CBS component to reduce the number of input channels by half.

Subsequently, depending on the second argument of the BCSP block, the block will perform feature extraction with using residual units or CBS components with 1×1 and 3×3 convolutional kernels for times. Residual units perform feature extraction using

1 × 1 and 3 × 3 convolutional CBS components. The results are then merged through element-wise summation and undergo identity mapping with the original feature map. The structure of the residual unit block is shown in Fig. 3.

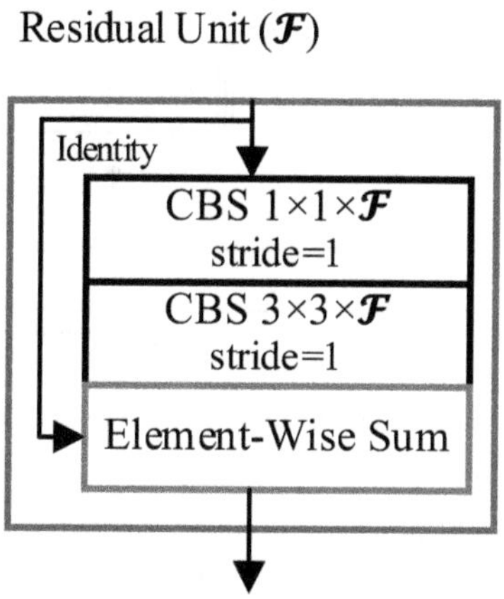

Fig. 3. Architecture diagram of the residual unit block

Finally, the BCSP block concatenates the extracted features with the cross-stage connection. The cross-stage connection records the variation information between input and output features directly, allowing more gradient information to flow during backpropagation training, thereby reducing computational burden and enhancing feature extraction capability. Finally, the concatenated feature information undergoes another convolution with a convolution kernel size of 1 in the CBS convolution to complete the merging of cross-stage connections. The architecture of the BCSP block is shown in Figs. 4 and 5.

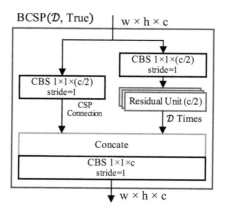

Fig. 4. BCSP(, True) Block Architecture Diagram.

The Feature Retaining Block (FRB) used in FRB-YOLOv5 also consists of two branches, as shown in the diagram in Fig. 6. These are the Patches branch which attempts

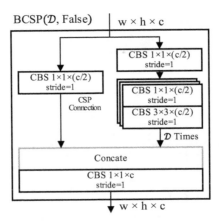

Fig. 5. BCSP(, False) Block Architecture Diagram.

to recover lost features due to downsampling and the Extraction branchwhich continues feature extraction. Let the input feature map size be W × H × C, where W、H and C represent the width, height and number of channels of the feature map, respectively. For the left branch, the network first reduces the number of channels of the input feature map with a 1 × 1 × (C ÷ 2) reducing the computational burden. Secondly, it uses a 2 × 2 transposed convolution with a stride of 2 to upsample the feature map to 2W × 2H × (C ÷ 2) This transposed convolution layer reconstructs the feature map to its size before pooling for attempting to recover lost feature information. Thirdly, a 3 × 3 convolution with a stride of 2 is used to scale the feature map back to W × H × (C ÷ 2). Finally, a 1 × 1 convolutional layer restores the number of feature channels to C.

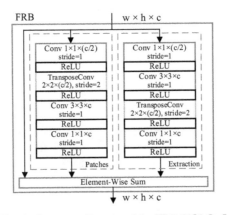

Fig. 6. Structure diagram of the FRB-YOLOv5.

For the right branch, it is similar to the left branch structure, except the order of the 2 × 2 transposed convolution and the 3 × 3convolution is reversed. Firstly, the network reduces the number of channels of the input feature map to(C ÷ 2) with a 1 × 1 convolutional layer used to reduce the computational burden. Secondly, through a 3x3 convolution with a stride of 2, an output feature map with size (W ÷ 2) × (H ÷ 2) × (C ÷ 2) is obtained. This convolutional layer extracts features from the input feature map to acquire higher-level feature information. Thirdly, a 2 × 2 transposed convolution with a stride of 2 restores the size of the feature map to W × H × (C ÷ 2). Finally, a 1 × 1 convolutional layer restores the number of feature channels to C. All convolution and transposed convolution calculations in both branches are followed by a Rectified Linear Unit (ReLU) [13] activation function. The branches are then merged using element-wise summation and undergo an identity mapping with the original feature map.

The Neck module first takes the output features from the Backbone and processes them through a Fast Spatial Pyramid Pooling (SPP-Fast) [14] block. This block segments and reorganizes the feature map from coarse to fine scales, enabling the model to handle images of different resolutions.

Subsequently, through a Path Aggregation Network (PANet) [14] based on the Feature Pyramid Network (FPN) [15], the feature information at different scales from various layers of the Backbone is integrated from shallow to deep to provide multi-scale detection capability. Then, the high-level and low-level feature information is combined from deep to shallow to enhance the hierarchical features of the network Firstly, the Neck processes the output feature maps of the Backbone through a specially designed parallel pooling structure based on Spatial Pyramid Pooling [16], allowing feature information to be segmented based on size from small to large with different receptive fields. Then, these features of different scales, which simultaneously possess global and local information, are recombined to enable the network to recognize objects of the same shape but different sizes or proportions. This design enables the network to adapt and accurately detect objects of various sizes and proportions. Due to the special design of the Spatial Pyramid Pooling (SPP) the network is able to adapt and correctly detect objects of the same shape but different sizes or proportions. SPP-Fast achieves this by iteratively performing pooling operations with a size of 5 where the second iteration is equivalent to a pooling size of 9 and the third iteration is equivalent to a pooling size of 13. Subsequently, these iterative results are concatenated with the original feature map and a 1 × 1 convolutional layer is used to adjust the channel numbers of the feature map. The architecture of the SPP-Fast block is shown in Fig. 7.

Next, the feature pyramid network integrates features from different scales to enhance the detection capability for objects of various sizes. In a convolutional neural network, low-level feature [17] in the shallow layers reflect texture, edges, and other object details, which are helpful for detecting small objects. As feature extraction progresses, high-level feature [17] in the deeper layers can comprehensively reflect abstract information about the objects based on the earlier low-level features for aiding in the detection of large objects. In the past, the single prediction head structure, which only predicted on the final feature map of the convolutional neural network, could only utilize the high-level feature information in the model without considering the low-level features provided by intermediate feature maps of different sizes. This resulted in the loss of information from different scales. In the feature pyramid segment of the Neck, the network first performs upsampling on the deep features of the Backbone through nearest-neighbor interpolation. Then, it concatenates these upsampled features with the shallower features and applies convolution to the concatenated feature map through the BCSP block. This

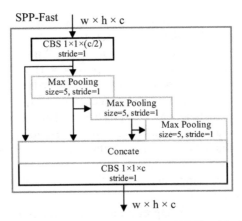

Fig. 7. Architecture diagram of the SPP-Fast block.

process enables deep high-level features to obtain multi-scale feature information from shallow low-level features, thereby enhancing the object detection capability at different scales.

3 Experiment Results

3.1 Evaluation Metrics

This paper evaluates the object detection performance of relevant methods using Mean Average Precision (mAP) across all classes. mAP is a commonly used metric to assess the effectiveness of an object detection method. It considers both precision and recall providing a comprehensive measure of detection accuracy and the likelihood of missing objects.

Precision is used to assess whether a model can make correct predictions defined as the ratio of correct predictions to all predictions. Recall is used to evaluate the model's ability to find objects defined as the ratio of predictions that contain samples to all relevant samples, The Precision and Recall can be defined as follows:

$$\text{Precision} = \frac{TP}{TP+FP}, \text{Recall} = \frac{TP}{TP+FN} \quad (1)$$

The paper uses the calculation method defined by Pascal VOC2010.which involves flattening the precision-recall curve and then calculating the area under the curve. The mean average precision (mAP) can be defined as follows:

$$AP_{VOC2010} = \sum (r_{n+1} - r_n) \max_{\hat{r} \geq r_{n+1}} p(\hat{r}), \text{mAP} = \frac{1}{C} \sum_{c=1}^{C} AP(c) \quad (2)$$

Here, p(r) represents the precision-recall curve which shows the model's precision performance at different recall levels. It is created by sequentially calculating the precision and recall for each prediction result sorted by confidence scores.

3.2 NEU-DET

The Northeastern University Surface Defect Database (NEU) [18] is a dataset used for detecting defects in hot-rolled steel manufacturing. It includes six typical defects found in hot-rolled steel: Rolled-in Scale, Pitted Surface, Crazing, Patches, Inclusion and Scratches. Each type of defect in the NEU dataset consists of 300 samples with each sample image having a resolution of 200 × 200 pixels. The NEU dataset has two versions: NEU-CLS labeled with text for image classification tasks and NEU-DET labeled with bounding boxes for object detection tasks. This study uses NEU-DET for training and evaluation with a training-validation split ratio of 8:2.

NEU-DET Experiment Results. The mAP accuracy for YOLOv5 after 300 epochs of training is 70.59%, while for our method, it is 76.18%, as shown in Fig. 8.

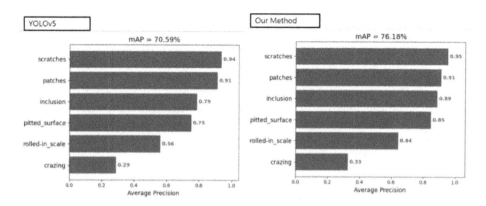

Fig. 8. Comparison of mAP for NEU-DET Dataset after 300 rounds of training.

The mAP accuracy for YOLOv5 after 400 epochs of training is 76.31%, while for our method, it is 77.35%, as shown in Fig. 9.

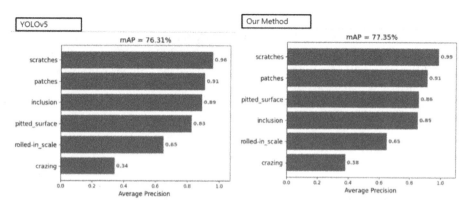

Fig. 9. Comparison of mAP for NEU-DET Dataset after 400 rounds of training.

The mAP accuracy for YOLOv5 after 500 epochs of training is 76.18%, while for our method, it is 78.75%, as shown in Fig. 10.

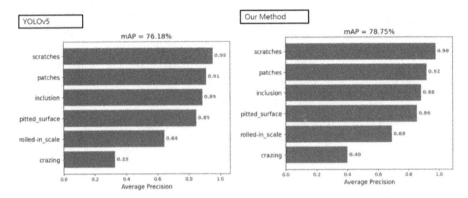

Fig. 10. Comparison of mAP for NEU-DET Dataset after 500 rounds of training.

In Table 2, we compare the mAP of our method using the NEU-DET dataset with YOLOv5. Significant improvements in mAP are observed after 300, 400 and 500 epochs of training. At 300 epochs, our method shows an increase in precision compared to YOLOv5 by 5.59%. This improvement decreases slightly to 1.04% at 400 epochs but increases again to 2.57% at 500 epochs.

Table 2. Compares the mAP on NEU-DET Dataset.

Method	The mAP for after training 300 epochs	The mAP for after training 400 epochs	The mAP for after training 500 epochs
yolov5	70.59%	76.31%	76.18%
Our Method	76.18%	77.35%	78.75%

In Table 3, we can observe the execution time comparison between our method and YOLOv5 for 300, 400, and 500 epochs of training. At 300 epochs, our method shows a speed difference of 10.41 ms compared to YOLOv5. This difference decreases to 7 ms at 400 epochs and 7.86 ms at 500 epochs. Overall, the execution speed difference is less than 11.2 ms. However, despite this slight difference in execution speed, our method achieves a detection improvement of 1% to 2.57%.

Table 3. Compares the execution speeds on NEU-DET Dataset.

Method	The execution speeds for after training 300 epochs (ms)	The execution speeds for after training 400 epochs (ms)	The execution speeds for after training 500 epochs (ms)
yolov5	25.3	25.4	23.0
Our Method	35.71	33.26	34.2

4 Conclusion

The paper proposes a new object detection method called FRB-YOLOv5 and applies it to the surface defect detection of industrial products. In order to improve the difficulty of identifying small defects by convolutional neural networks, this paper combines feature retention blocks after the BCSP block in the Backbone network of YOLOv5 to propose the FRB-YOLOv5 method and verifies the effectiveness of the proposed method by comparing the detection results of three datasets. The experiments demonstrate an improvement in the performance of surface defect detection using the proposed method. From the experimental data in this paper, it is evident that under the same environment and conditions, the detection results of various datasets show significant enhancement while maintaining competitive execution efficiency. The proposed method achieves an increase in mAP from 76.18% to 78.75% on NEU-DET. Improvements in detecting minor defects can be observed, particularly in the Blowhole category of the Magnetic-Tile dataset. Using the original YOLOv5, the AP value for detecting this category is 0.80, whereas with the FRB-YOLO method, the AP value increases to 0.99. Overall, although our method enhances detection accuracy, there is a slight decrease in execution speed. In the future, we will continue to optimize our network to improve precision while maintaining or even enhancing execution speed.

Acknowledgments. This study was partly supported by the MOE Teaching Practice Research Program of the Republic of China under GRANT NO.PSK1123096.

Disclosure of Interests. The authors have no competing interests to declare that are relevant to the content of this article. The authors declare no conflict of interest.

References

1. Jan, Z., Ahamed, F., Mayer, W., Patel, N., Grossmann, G., Stumptner, M., et al.: Artificial intelligence for industry 4.0: systematic review of applications challenges and opportunities. Expert Syst. Appl. **216**, 119456 (2023)
2. He, K., Gkioxari, G., Dollar, P., Girshick, R.: Mask R-CNN. In: Proceedings of the IEEE International Conference on Computer Vision (ICCV), Venice, Italy, 22–29 October 2017
3. Wang, H., Zhang, J., Tian, Y., Chen, H., Sun, H., Liu, K.: A simple guidance template-based defect detection method for strip steel surfaces. IEEE Trans. Industr. Inf. **15**(5), 2798–2809 (2019)

4. Aghdam, S.R., Amid, E., Imani, M.F.: A fast method of steel surface defect detection using decision trees applied to LBP based features. In: Proceedings of the 7th IEEE Conference on Industrial Electronics and Applications (ICIEA), Singapore, 18–20 July 2012
5. Kim, C.-W., Koivo, A.J.: Hierarchical classification of surface defects on dusty wood boards. Pattern Recogn. Lett. **15**(7), 713–721 (1994)
6. Kamarainen, J.-K., Kyrki, V., Kalviainen, H.: Invariance properties of Gabor filter-based features—Overview and applications. IEEE Trans. Image Process. **15**(5), 1088–1099 (2006)
7. Zhang, H., Jin, X., Wu, Q.M.J., Wang, Y., He, Z., Yang, Y.: Automatic visual detection system of railway surface defects with curvature filter and improved gaussian mixture model. IEEE Trans. Instrum. Meas. **67**(7), 1593–1608 (2018)
8. Li, C., Gao, G., Liu, Z., Huang, D., Liu, S., Yu, M.: Defect detection for patterned fabric images based on GHOG and low-rank decomposition. arXiv preprint arXiv:1702.05555 (2017)
9. Tu, Y., Ling, Z., Guo, S., Wen, H.: An accurate and real-time surface defects detection method for sawn lumber. IEEE Trans. Instrum. Meas. **70**, 1–11 (2021)
10. Su, H., Liu, F., Xie, Y., Xing, F., Meyyappan, S., Yang, L.: Region segmentation in histopathological breast cancer images using deep convolutional neural network. In: 2015 IEEE 12th International Symposium on Biomedical Imaging (ISBI), pp. 55–58 (2015)
11. Luo, Z., Yang, S., Li, Y., Wu, Y.: FPCB surface defect detection: a decoupled two-stage object detection framework. IEEE Trans. Instrum. Meas. **70**, 1–11 (2021)
12. Santurkar, S., Tsipras, D., Ilyas, A., Madry, A.: How does batch normalization help optimization? In: Advances in Neural Information Processing Systems 31, pp. 2483–2493. Curran Associates, Inc., Red Hook, NY, USA (2018)
13. Elfwing, S., Uchibe, E., Doya, K.: Sigmoid-weighted linear units for neural network function approximation in reinforcement learning. Neural Netw. **107**, 3–11 (2018)
14. Morrow, H.: Identifying and mapping community vulnerability. Disasters **23**(1), 1–18 (1999)
15. Agarap, A.F.: Deep Learning using Rectified Linear Units (ReLU). arXiv preprint arXiv:1803.08375 (2018)
16. He, K., Zhang, X., Ren, S., Sun, J.: Spatial pyramid pooling in deep convolutional networks for visual recognition. IEEE Trans. Pattern Anal. Mach. Intell. **37**(9), 1904–1916 (2015)
17. Lee, G., Tai, Y.-W., Kim, J.: Deep saliency with encoded low level distance map and high level features. In: Proceedings of the IEEE Conference on Computer Vision and Pattern Recognition (CVPR), Las Vegas, NV, USA, 27–30 June 2016
18. Song, K., Yan, Y.: A noise robust method based on completed local binary patterns for hot-rolled steel strip surface defects. Appl. Surf. Sci. **285**, 858–864 (2013)

Application of Digital Gamification Systems in Intelligent Automated Learning

Chein-Hui Lee[1,2], Evelyn Saputri[3], and Min-Chi Chiu[4(✉)]

[1] Department of Information Technology, Ling Tung University, Taichung, Taiwan
[2] Doctoral Program of Intelligent Engineering, National Taichung University of Science and Technology, Taichung, Taiwan
[3] Department of Intelligent Manufacturing Technology, Ling Tung University, Taichung, Taiwan
[4] Department of Multimedia Design, National Taichung University of Science and Technology, Taichung, Taiwan
minky413@gmail.com

Abstract. This study aims to develop an innovative digital game-based teaching platform designed to facilitate students' learning of intelligent automation-related knowledge through immersive game scenarios. This method offers significant advantages, including the provision of an engaging and challenging learning environment that encourages active student participation while fostering critical thinking and problem-solving skills. Students can apply the knowledge they acquire in practical game contexts, thereby enhancing their learning outcomes, aligning with the contemporary educational demand for diversified teaching approaches. Furthermore, this pedagogical strategy effectively stimulates students' motivation to learn by engaging them in problem-solving and task completion activities, consequently improving their learning efficiency. More importantly, it emphasizes not just the transmission of knowledge but also the cultivation of students' cognitive abilities and problem-solving skills, which are crucial for their future academic and professional development. By integrating educational content with interactive game elements, the platform can adapt to various learning paces and styles, making education more personalized and accessible. Overall, this study aims to provide students with a highly engaging and efficient learning method that meets the modern educational requirements for innovative teaching models, thereby significantly improving students' learning outcomes and addressing their educational needs.

Keywords: Digital game-based teaching platform · Intelligent automation · Immersive learning

1 Introduction

In contemporary industries, the widespread application of intelligent automation technology is evident [1–3]. However, in the traditional learning process for the intelligent automation engineer certification exams, candidates often encounter issues such as the overwhelming amount of knowledge points and the monotonous nature of the question banks. This leads to a focus on rote memorization rather than a genuine understanding of the knowledge required for the certification. The study of intelligent automation is crucial for modern industrial development, as it enhances production efficiency, reduces costs, and fosters innovation. As industries undergo digital transformation, there is a growing demand for professionals skilled in intelligent automation [4].

Digital gamification learning has gradually emerged as an innovative teaching method [5]. This approach leverages the engaging and challenging aspects of games, integrating game narratives with real-world educational contexts. It guides learners to complete tasks within the game, thereby cultivating correct thinking patterns and problem-solving abilities [6]. This enriches the educational content, making it more diverse and encouraging active participation in learning [7]. Not only does this improve learning efficiency, but it also allows students to apply the acquired knowledge in practical scenarios, providing industries with higher-quality talent [8].

To increase candidates' willingness to take the exam and their interest in studying the question banks, digital gamification learning offers a promising and effective solution [9, 10]. This method not only enhances the appeal and interactivity of learning but also promotes a comprehensive understanding and mastery of intelligent automation knowledge. Consequently, it lays a solid foundation for students' future career development and supports the growth of the industry. By incorporating elements such as rewards, progress tracking, and interactive challenges, gamified learning platforms can create a more stimulating and immersive learning experience. This approach also encourages collaboration among learners, fostering a sense of community and shared goal achievement.

Moreover, digital gamification learning platforms can be tailored to individual learning paces and styles, providing a personalized educational experience that traditional methods often lack. The integration of adaptive learning technologies allows the platform to adjust the difficulty level and content based on the learner's performance, ensuring that each student is appropriately challenged and engaged. This personalization helps maintain high levels of motivation and reduces the likelihood of disengagement or dropout.

Overall, digital gamification learning transforms the educational landscape by making the acquisition of complex knowledge more engaging and accessible. It supports the preparation of proficient professionals who can meet the evolving demands of modern industries, ultimately contributing to the advancement of intelligent automation technology. By addressing the limitations of traditional learning methods and leveraging the benefits of gamification, this approach represents a significant step forward in educational innovation and workforce development.

2 Related Works

In the realm of contemporary intelligent automation education, gamified learning methods have emerged as highly effective and advantageous. Traditional educational approaches often grapple with challenges such as an overwhelming volume of knowledge points and monotonous content, leading students to focus more on rote memorization rather than genuinely understanding and applying the acquired knowledge. As intelligent automation technology continues to advance rapidly, the need for professionals equipped with practical skills and innovative thinking becomes increasingly crucial.

The concept of gamification has introduced a revolutionary teaching approach in the educational field. By incorporating the engaging and challenging elements of games into the learning process, gamification allows students to develop critical thinking patterns and problem-solving skills while completing game-based tasks. This method not only enhances the appeal and attractiveness of learning but also significantly improves student engagement and educational outcomes. As educators, we can leverage gamification to create more interactive and motivating learning experiences that foster deeper understanding and application of knowledge [11].

Khaldi et al. [12] conducted a study on gamification in e-learning, identifying 17 methods suitable for higher education e-learning systems. These methods aim to increase students' motivation through interactivity and competition. Cavus et al. [13] further explored gamification in both high school and higher education settings, finding only 24 studies specifically focused on higher education from 2012 to 2022. However, these studies indicated that gamification holds significant potential in boosting learning motivation and participation.

Ratinho and Martins [14] demonstrated that gamification has a substantial impact on learning motivation. Their review of 35 articles related to higher education, published between 2013 and 2022, revealed that gamification enhances students' interest and motivation in learning. Montenegro-Rueda et al. [15] concluded that while gamification remains an underexplored area in higher education, its application potential is considerable.

Wiggins [16] conducted an in-depth study on the use of games, simulations, and gamification in higher education, emphasizing that successful gamification requires careful planning, design, and implementation. Barna and Fodor [17] tailored their gamification approach to the characteristics of Generation Y and Z students, who are accustomed to using the internet and social media but tend to have shorter attention spans. Their study showed that gamified courses delivered through the Moodle platform significantly increased student engagement and received higher evaluations.

Moreover, gamification in intelligent automation education offers practical benefits beyond increased motivation and engagement. By incorporating elements such as rewards, progress tracking, and interactive challenges, gamified learning platforms create a more stimulating and immersive learning experience. This approach also encourages collaboration among learners, fostering a sense of community and shared goal achievement. The integration of adaptive learning technologies allows the platform to adjust the difficulty level and content based on the learner's performance, ensuring that each student is appropriately challenged and engaged. This personalization helps maintain high levels of motivation and reduces the likelihood of disengagement or dropout [18, 19].

Furthermore, digital gamification learning can transform the educational landscape by making the acquisition of complex knowledge more engaging and accessible. It supports the preparation of proficient professionals who can meet the evolving demands of modern industries, ultimately contributing to the advancement of intelligent automation technology. By addressing the limitations of traditional learning methods and leveraging the benefits of gamification, this approach represents a significant step forward in educational innovation and workforce development.

In conclusion, gamification methods show great promise in the field of intelligent automation education. They not only enhance students' interest and motivation but also facilitate the practical application of knowledge, thereby cultivating high-quality talents with innovative thinking and practical skills in intelligent automation. This research direction merits further exploration and implementation to fully realize the potential of gamified learning in education. As the demand for skilled professionals in intelligent automation continues to grow, integrating gamification into educational frameworks can play a pivotal role in meeting this need, ensuring that students are well-equipped for the challenges and opportunities of the future. Through continued research and development, gamified learning has the potential to revolutionize education and significantly impact the preparation of the next generation of intelligent automation professionals.

3 System Architecture Design

The system architecture design of this study aims to develop an innovative digital gamified teaching platform to enhance students' learning outcomes in intelligent automation. The system architecture is divided into three primary modules: the gamified learning system, supplemental professional knowledge materials, and the learning behavior recording system, as shown in Fig. 1.

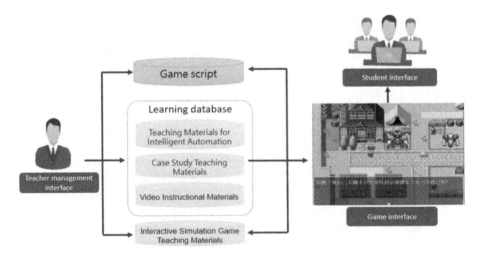

Fig. 1. System architecture diagram.

The gamified learning system is the core module that incorporates game design elements into the educational content. This module is designed to create an engaging and interactive learning environment. It includes various game-based tasks and challenges that align with the curriculum of intelligent automation. These tasks are designed to foster critical thinking and problem-solving skills by simulating real-world scenarios that students might encounter in their professional careers. The system uses elements such as point scoring, leaderboards, and achievement badges to motivate students and keep them engaged. By making the learning process more enjoyable, this module aims to improve students' retention of knowledge and their ability to apply what they have learned in practical settings.

The supplemental professional knowledge materials module provides additional resources to support the core learning activities. This module includes a comprehensive library of reference materials, case studies, video tutorials, and interactive simulations related to intelligent automation. These resources are curated to provide deeper insights into the subject matter and to help students understand complex concepts more thoroughly. The materials are accessible at any time, allowing students to study at their own pace and revisit topics as needed. This module ensures that students have access to the latest information and industry practices, which is crucial for their professional development.

The learning behavior recording system is a critical component that tracks and analyzes students' learning activities and progress. This module collects data on various aspects of the learning process, such as the time spent on different tasks, the number of attempts made to complete challenges, and the overall performance in quizzes and assignments. By analyzing this data, educators can gain valuable insights into each student's learning patterns and identify areas where additional support may be needed. The system also provides personalized feedback to students, helping them understand their strengths and areas for improvement. Additionally, the data collected can be used to refine and optimize the gamified learning system, ensuring that it remains effective and relevant.

Overall, the system architecture design of this innovative digital gamified teaching platform aims to create a holistic and dynamic learning experience for students. By integrating gamified elements with supplemental professional knowledge materials and a robust learning behavior recording system, the platform seeks to enhance students' engagement, motivation, and learning outcomes in intelligent automation. This comprehensive approach not only prepares students for their certification exams but also equips them with the practical skills and knowledge needed to excel in their future careers. Through continuous improvement and adaptation, the platform aims to remain at the forefront of educational innovation in the field of intelligent automation.

3.1 Gamified Learning System

The gamified learning system is the core module of the platform, designed to stimulate students' learning motivation through engaging and challenging game scenarios. This module includes various game elements such as mission challenges, level designs, and real-time feedback mechanisms. The game content is based on intelligent automation knowledge, presented in the form of an RPG (role-playing game). During the design process, we considered learners' knowledge levels and learning habits, gradually increasing the difficulty and diversifying tasks to allow students to continuously challenge themselves and consolidate their learning achievements.

3.2 Gamified Learning System

To ensure students acquire in-depth professional knowledge during the game process, the system provides extensive supplemental materials. These materials include textual content, instructional videos, and interactive multimedia resources, covering the core concepts and applications of intelligent automation. The material content is closely integrated with game tasks, allowing students to refer to relevant knowledge points in real-time while completing game tasks, achieving seamless integration of learning and gaming.

3.3 Learning Behavior Recording System

This system records students' behavior data during the learning process, including learning progress, task completion status, and question accuracy. Through this data, the system generates detailed learning reports to help students understand their learning performance and identify areas for improvement. This module also provides personalized learning recommendations, dynamically adjusting learning content and difficulty based on students' performance, ensuring that each student receives an optimal learning experience.

Overall, the system integrates gamified learning strategies with supplemental professional knowledge materials, allowing learners to gradually build their understanding of professional knowledge while enjoying the fun of the game. This approach enhances learners' motivation and engagement in the learning process. The digital game designs developed in this study are illustrated in Figs. 2, 3, 4, 5 and 6.

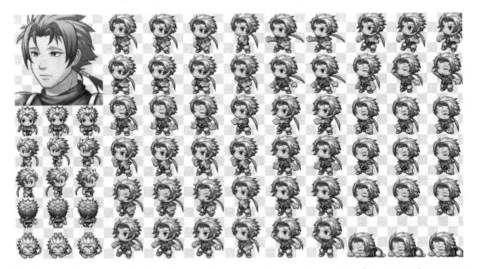

Fig. 2. The character design for the game's protagonist allows players to choose any character as their main character.

Application of Digital Gamification Systems   183

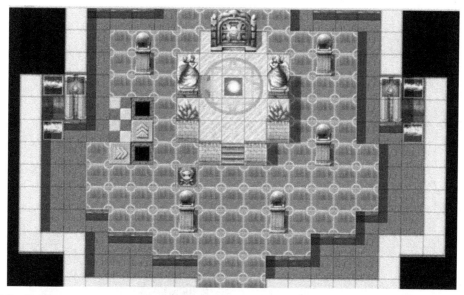

Fig. 3. The characters in the game search for quest hints and treasures in the dungeon to pro-ceed to the next level

Fig. 4. Introduction to the mission leading to the next level.

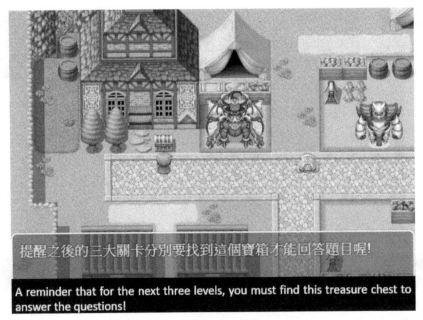

Fig. 5. The characters in the game need to find treasure chests and complete the specified tasks within them to advance to the next level.

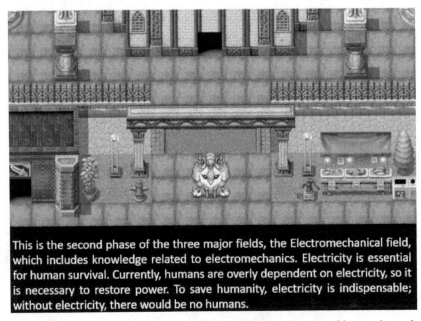

Fig. 6. The game's protagonist encounters NPCs who provide quest hints and supplement professional knowledge.

4 System Design and Development

The system design aims to enhance the learning outcomes of intelligent automation knowledge through a meticulously designed architecture and interactive game scenarios. The front end utilizes the latest web technologies to create a clean and engaging user interface, while the back end employs Node.js and MongoDB to manage data and handle user requests. In game development, we used the Unity engine to ensure that the games are educational and enjoyable. After multiple rounds of testing and optimization, the system was officially launched and continues to receive updates to maintain its long-term effectiveness and advancement.

4.1 System Design and Development

During the system design phase, we first developed the overall architecture, defining the functions and interrelations of each module to ensure seamless integration of system components. We meticulously designed the game scenarios and levels, writing detailed scripts to seamlessly integrate learning objectives into the gameplay. Character designs underwent multiple iterations to maintain high engagement levels. Front-end development utilized HTML, CSS, and JavaScript to create a clean and appealing interface, while back-end development used Node.js and MongoDB to manage data and user requests, providing a robust foundation.

The system architecture is divided into three primary modules: the gamified learning system, supplemental professional knowledge materials, and the learning behavior recording system. The gamified learning system incorporates game design elements into the educational content, creating an engaging and interactive environment. This module includes various game-based tasks and challenges that align with the curriculum of intelligent automation. The supplemental professional knowledge materials module provides additional resources, such as reference materials, case studies, video tutorials, and interactive simulations, to support core learning activities. The learning behavior recording system tracks and analyzes students' learning activities and progress, collecting data on various aspects of the learning process.

4.2 Supplemental Professional Knowledge Materials

Game development involved writing game logic, building scenes, and implementing interactive features using the Unity engine to create highly interactive and immersive learning games. After multiple rounds of testing and optimization, the game levels were fine-tuned to meet learners' needs. Following comprehensive testing, the system was officially launched. We continuously monitor performance, collect feedback, and perform updates to ensure the system's long-term effectiveness and advancement, adapting to user needs and technological advancements.

Throughout the development process, we focused on creating an engaging and educational experience. By integrating detailed game mechanics and interactive scenarios, we aimed to enhance the learning outcomes of intelligent automation concepts. Continuous user feedback is crucial for refining game levels and features, ensuring that the platform remains relevant and effective. Regular updates and optimizations are performed to address any issues and incorporate new advancements in technology and education methodologies. This ongoing commitment to improvement ensures that our gamified learning platform provides a dynamic and effective educational tool for students.

5 Conclusions

Through systematic design and development, we successfully established an innovative digital gamified teaching platform. This platform enhances the learning outcomes of intelligent automation-related knowledge while significantly increasing students' motivation and engagement. The system integrates a user-friendly interface created with the latest web technologies on the front end and a robust back end using Node.js and MongoDB to manage data and handle user requests. Utilizing the Unity engine, we ensured the games are both educational and enjoyable. After thorough testing and optimization, the system was launched and continues to receive updates, maintaining its long-term effectiveness and adapting to user needs and technological advancements.

The system architecture design of this study aims to develop an innovative digital gamified teaching platform to enhance students' learning outcomes in intelligent automation. The system architecture is divided into three primary modules: the gamified learning system, supplemental professional knowledge materials, and the learning behavior recording system.

The gamified learning system is the core module that incorporates game design elements into the educational content. This module is designed to create an engaging and interactive learning environment. It includes various game-based tasks and challenges that align with the curriculum of intelligent automation. These tasks are designed to foster critical thinking and problem-solving skills by simulating real-world scenarios that students might encounter in their professional careers. The system uses elements such as point scoring, leaderboards, and achievement badges to motivate students and keep them engaged. By making the learning process more enjoyable, this module aims to improve students' retention of knowledge and their ability to apply what they have learned in practical settings.

The supplemental professional knowledge materials module provides additional resources to support the core learning activities. This module includes a comprehensive library of reference materials, case studies, video tutorials, and interactive simulations related to intelligent automation. These resources are curated to provide deeper insights into the subject matter and to help students understand complex concepts more thoroughly. The materials are accessible at any time, allowing students to study at their own pace and revisit topics as needed. This module ensures that students have access to the latest information and industry practices, which is crucial for their professional development.

The learning behavior recording system is a critical component that tracks and analyzes students' learning activities and progress. This module collects data on various aspects of the learning process, such as the time spent on different tasks, the number of attempts made to complete challenges, and the overall performance in quizzes and assignments. By analyzing this data, educators can gain valuable insights into each student's learning patterns and identify areas where additional support may be needed. The system also provides personalized feedback to students, helping them understand their strengths and areas for improvement. Additionally, the data collected can be used to refine and optimize the gamified learning system, ensuring that it remains effective and relevant.

In the future, we will continue to optimize system functions and explore more applications of gamified learning. We plan to expand the question bank, introduce NPC interaction features, provide answer explanations, and use intelligent technology for personalized learning recommendations. Additionally, we will implement a user feedback mechanism to facilitate continuous system improvement. This comprehensive approach not only prepares students for their certification exams but also equips them with the practical skills and knowledge needed to excel in their future careers. Through continuous improvement and adaptation, the platform aims to remain at the forefront of educational innovation in the field of intelligent automation.

Acknowledgments. This work was financially supported by the Ministry of Science and Technology of Taiwan, with project numbers: NSTC 113-2410-H-025 -042.

Disclosure of Interests. The authors have no competing interests to declare that are relevant to the content of this article.

References

1. Alenizi, F.A., Abbasi, S., Mohammed, A.H., Rahmani, A.M.: The artificial intelligence technologies in Industry 4.0: a taxonomy, approaches, and future directions. Comput. Ind. Eng. **185**, 109662 (2023)
2. Zhang, W., Ye, S., Mangla, S.K., Emrouznejad, A., Song, M.: Smart platforming in automotive manufacturing for Netzero: intelligentization, green technology, and innovation dynamics. Int. J. Product. Econ. **274**, 109289 (2024). https://doi.org/10.1016/j.ijpe.2024.109289
3. Zhong, R.Y., Xu, X., Klotz, E., Newman, S.T.: Intelligent manufacturing in the context of industry 4.0: a review. Engineering **3**(5), 616–630 (2017)
4. Wang, L., Zhao, H., Cao, Z., Dong, Z.: Artificial intelligence and intergenerational occupational mobility. J. Asian Econ. **90**, 101675 (2024)
5. Hong, Y., Saab, N., Admiraal, W.: Approaches and game elements used to tailor digital gamification for learning: a systematic literature review. Comput. Educ. **212**, 105000 (2024)
6. Hughes-Roberts, T., Brown, D., Boulton, H., Burton, A., Shopland, N., Martinovs, D.: Examining the potential impact of digital game making in curricula based teaching: Initial observations. Comput. Educ. **158**, 103988 (2020)
7. Aldalur, I., Perez, A.: Gamification and discovery learning: motivating and involving students in the learning process. Heliyon **9**(1), e13135 (2023). https://doi.org/10.1016/j.heliyon.2023.e13135
8. Laakso, N.L., Korhonen, T.S., Hakkarainen, K.P.: Developing students' digital competences through collaborative game design. Comput. Educ. **174**, 104308 (2021)
9. Suárez-López, M.J., Blanco-Marigorta, A.M., Gutiérrez-Trashorras, A.J.: Gamification in thermal engineering: does it encourage motivation and learning? Educ. Chem. Eng. **45**, 41–51 (2023)

10. Ma, Z., Li, W.: Design of online teaching interaction mode for vocational education based on gamified-learning. Entertainment Comput. **50**, 100647 (2024). https://doi.org/10.1016/j.entcom.2024.100647
11. Deterding, S., Dixon, D., Khaled, R., Nacke, L.: From game design elements to gamefulness: defining" gamification". In: Proceedings of the 15th international academic MindTrek conference: Envisioning future media environments, pp. 9–15 (2011)
12. Khaldi, A., Bouzidi, R., Nader, F.: Gamification of e-learning in higher education: a systematic literature review. Smart Learn. Environ. **10**(1), 10 (2023)
13. Cavus, N., Ibrahim, I., Okonkwo, M.O., Ayansina, N.B., Modupeola, T.: The effects of gamification in education: a systematic literature review. Broad Res. Artific. Intell. Neurosci. **14**(2), 211–241 (2023)
14. Ratinho, E., Martins, C.: The role of gamified learning strategies in student's motivation in high school and higher education: a systematic review. Heliyon **9**(8), e19033 (2023). https://doi.org/10.1016/j.heliyon.2023.e19033
15. Montenegro-Rueda, M., Fernández-Cerero, J., Mena-Guacas, A.F., Reyes-Rebollo, M.M.: Impact of gamified teaching on university student learning. Educ. Sci. **13**(5), 470 (2023)
16. Wiggins, B.E.: An overview and study on the use of games, simulations, and gamification in higher education. Int. J. Game-Based Learn. **6**(1), 18–29 (2016)
17. Barna, B., Fodor, S.: An empirical study on the use of gamification on IT courses at higher education. In: Auer, M.E., Guralnick, D., Simonics, I. (eds.) ICL 2017. AISC, vol. 715, pp. 684–692. Springer, Cham (2017). https://doi.org/10.1007/978-3-319-73210-7_80
18. Polito, G., Temperini, M.: A gamified web based system for computer programming learning. Comput. Educ.: Artific. Intell. **2**, 100029 (2021)
19. Kao, M.-C., Yuan, Y.-H., Wang, Y.-X.: The study on designed gamified mobile learning model to assess students' learning outcome of accounting education. Heliyon **9**(2), e13409 (2023). https://doi.org/10.1016/j.heliyon.2023.e13409

Exploring the Intersection of Kansei Engineering and Affective Computing in Digital Media Design Research

Constructing a Curatorial Awareness-Based Sustainable Development Model—A Kansei Engineering Perspective

Huang-Yin Chen[1] and Teng-Wen Chang[2(✉)]

[1] National Yunlin University of Science and Technology, Yunlin County, Taiwan
[2] 123 University Road Section 3, Douliou, Yunlin 640, ROC
tengwen@yuntech.edu.tw

Abstract. This study innovatively integrates curatorial awareness design and Kansei engineering, applying them to sustainable markets to enhance participants' emotional experiences and promote sustainable development. Through a literature review, the definitions and characteristics of sustainable markets were established, emphasizing their environmental, social, and economic impacts, and exploring the application of curatorial awareness design and Kansei engineering in this context. The research methodology included a survey of sustainable market participants and multiple regression analysis. The results indicate that curatorial awareness design effectively communicates sustainability concepts, while Kansei engineering plays a crucial role in enhancing participants' emotional engagement and depth of experience. The contribution of this study lies in providing theoretical support and practical recommendations for optimizing the design and management of sustainable markets, thereby advancing the achievement of sustainable development goals.

Keywords: Kansei Engineering · Sustainable Market · Curatorial Awareness · Emotional Design · Innovative Multimodal Model

1 Introduction

1.1 Research Background and Motivation

With the increasing severity of global environmental issues, sustainable development has become an important concern for countries worldwide. In 2015, the United Nations adopted 17 Sustainable Development Goals (SDGs) aimed at achieving a more sustainable and equitable world by 2030. In this context, sustainable markets have emerged as a crucial means of promoting sustainable development. These markets emphasize not only economic benefits but also environmental protection and social equity, striving to achieve balanced development across economic, social, and environmental dimensions [1, 2]. Additionally, the development of sustainable markets contributes to building green, inclusive, and resilient cities and communities, making significant progress towards achieving multiple SDGs [3, 4].

In the design of sustainable markets, curatorial awareness and Kansei engineering design are two key theoretical foundations. Curatorial awareness is an integrated and specific curatorial practice method that encompasses the curator's vision, planning, practice, and reflection, emphasizing the interactive experience between the exhibition content and the audience [5]. Kansei engineering, on the other hand, emphasizes that design should consider users' emotional responses, aiming to enhance users' emotional experiences, which is particularly important in planning sustainable markets [1].

The background and motivation of this study lie in exploring how to achieve sustainable development goals in market design by integrating sustainable markets, curatorial awareness, and Kansei engineering design. The aim is to enhance participants' emotional experiences and understanding, thereby promoting broader public participation and support. Specifically, this study will use a questionnaire survey method, inviting participants with market and curatorial experience to experience domestic sustainable markets and complete a retrospective subjective experience questionnaire to analyze and evaluate market characteristics, SDG linkages, and their impact assessments [6, 7].

This study aims to provide an integrated curatorial awareness market model and propose specific design strategies and practical recommendations to promote the development and impact of sustainable markets. The research results will contribute to theoretical development in the relevant field and provide valuable references for practice.

1.2 Research Objectives and Questions

In response to the existing shortcomings of sustainable markets in concept dissemination and public engagement [1, 8], this study aims to explore the subjective feelings and diverse perceptions of participants after experiencing sustainable markets from the perspective of Kansei engineering. By analyzing participants' feedback, this research will examine the interrelationships among the dimensions affecting market experience and perception, providing scientific evidence and practical guidance for the emotional design strategies and practical pathways of the "curatorial awareness market." Therefore, this study focuses on the following questions:

1. What are the strengths and weaknesses of sustainable markets in conveying sustainable concepts and shaping emotional experiences? What are the key influencing factors behind these differences?
2. How do participants' emotional experiences in sustainable markets correlate with the four main constructs identified in the survey, and what are the interrelationships among these constructs in terms of influencing perception?
3. How can subjective perceptions and participation feedback be used to construct emotional design strategies and practical pathways for "curatorial awareness markets" to resonate with sustainable values more effectively?

1.3 Research Significance and Innovation

The significance of this study lies in exploring how to optimize the design of sustainable markets using curatorial awareness and Kansei Engineering to enhance public awareness and acceptance of sustainable development goals. The innovations include Proposing a multimodal sustainable market model combining curatorial awareness and Kansei Engineering; Validating the interrelationships and contributions of various dimensions through questionnaire surveys and multivariate regression analysis; Providing specific Kansei Engineering design strategies and practical paths, offering theoretical and practical foundations for the design and management of sustainable markets.

2 Literature Review

2.1 Definition and Characteristics of Sustainable Markets

Sustainable markets are trading environments that integrate the core goals of the Sustainable Development Goals (SDGs) into their operations, aiming to promote community resilience and sustainability [2]. These markets not only focus on economic benefits but also emphasize environmental protection and social equity, striving to achieve balanced development in economic, social, and environmental dimensions [9]. The main characteristics of sustainable markets include promoting eco-friendly products and services, supporting local economic development, fostering community interaction, and learning, and encouraging innovation and diversity [3, 10]. Through these characteristics, sustainable markets not only provide consumers with opportunities to access sustainable products but also create platforms for producers and community members to exchange, learn, and innovate, thereby promoting the sustainable development of the region [1]. With the deepening of sustainable development concepts, sustainable markets have experienced significant growth and increasing support in recent years [10]. These markets not only provide important pathways for building green, inclusive, and resilient cities and communities but also make positive contributions to achieving multiple SDGs [2].

2.2 Curatorial Awareness Design

"Curatorial awareness" is an integrated and specific curatorial practice method encompassing the curator's vision, planning, practice, and reflection. This concept is considered an essential part of the curator's process in designing and implementing exhibitions. Curatorial awareness focuses not only on the selection and arrangement of exhibits but also on the audience's interactive experience, the narrative of the exhibition, and the cultural and social implications it conveys. According to [5], curatorial awareness is a dynamic and evolving process that requires curators to possess keen observation, innovative thinking, and a profound understanding of historical, cultural, and social contexts. [5] also notes that curatorial awareness includes the design of exhibition spaces, the use of lighting, and the careful consideration of audience flow, all of which collectively create a comprehensive curatorial experience.

The core of curatorial awareness lies in the strategic thinking and creativity of the curator. [4] suggests that curators should be able to translate abstract concepts into tangible exhibition forms, establishing deep emotional connections with the audience through exhibition content. [4] believes that curatorial awareness also involves how curators use new media technologies to enrich exhibition forms and interactivity, thereby enhancing audience engagement and experience. Another important aspect is its social responsibility. [11] emphasizes that curators should consider the social impact of exhibits and exhibition themes, including cultural diversity, historical justice, and social equity.

2.3 Application of Kansei Engineering in Sustainable Markets

Kansei engineering emphasizes that design should consider users' emotional responses, which is particularly important in planning sustainable markets. [12] "emotional design" theory suggests that design should shape users' emotional experiences on sensory, behavioral, and reflective levels. This approach can be applied to curatorial awareness in sustainable markets to enhance participants' understanding and emotional connection to sustainability. [7] found that integrating slow design with curatorial awareness can create more mindful forms of interaction, enhancing users' engagement and emotional experience. Chang, Wu, Datta, and pointed out that this combination can explore individuals' life trajectories, reveal unnoticed issues, and strengthen users' emotional connections with interactive behaviors. Additionally, curators should consider using sustainable materials to construct booths and set up highly interactive displays and workshops to enhance the audience's emotional experience and resonance. Combining Kansei space technology and interactive design can provide visitors with richer emotional experiences [13].

3 Research Methodology

3.1 Research Design

This study employs a questionnaire survey method, inviting 40 participants with market and curatorial experience to complete a retrospective subjective experience questionnaire after experiencing domestic sustainable markets (such as SDG markets, green markets, and environmental markets). The questionnaire design is based on a literature review, including dimensions of market characteristics, SDG linkages, and impact assessment. The design of the questionnaire is informed by [12] emotional design theory and the research findings of [7]. To ensure the reliability and validity of the questionnaire, a pilot test was conducted, and reliability analysis showed that Cronbach's α values exceeded 0.7, indicating good internal consistency.

3.2 Research Framework

The research framework integrates three core elements: sustainable markets, curatorial awareness, and Kansei engineering design, forming a comprehensive curatorial awareness market model as shown in (see Fig. 1). This model aims to optimize market experiences through emotional design and promote the realization of sustainable development goals. Specifically, Sustainable Markets: Analyzes market design and operations, emphasizing environmental protection and sustainability, and explores their connections with SDGs; Curatorial Awareness: Studies how curatorial concepts enhance market attractiveness and participation, and improve overall participant experiences; Kansei Engineering Design: Applies Kansei engineering theory to analyze the emotional design factors influencing market experiences and proposes optimization strategies.

Fig. 1. Research Framework.

3.3 Questionnaire Constructs

To thoroughly investigate participants' experiences and perceptions of sustainable markets, this study designed an expert questionnaire survey and conducted interviews, covering four main dimensions as shown in (see Fig. 2):

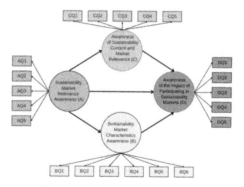

Fig. 2. Questionnaire Constructs.

1. Perception of Sustainable Market Relevance (A): Participants' awareness of the relevance of markets to sustainability.
2. Perception of Sustainable Market Characteristics (B): Participants' understanding of the specific characteristics of the market.
3. Perception of the Relationship Between Sustainability and Market (C): Participants' awareness of the correlation between the market and the SDGs (Sustainable Development Goals).
4. Perception of the Impact of Participation in Sustainable Markets (D): Participants' recognition of the environmental, social, and economic impacts of their participation.

These dimensions aim to comprehensively understand participants' multi-dimensional perceptions and evaluations of sustainable markets, extracting specific emotional design factors and optimization strategies to provide a practical basis for future sustainable market design.

3.4 Data Analysis Methods

The data analysis methods include correlation analysis, multivariate regression analysis, and Kansei engineering analysis to examine the interrelationships between different constructs and their contributions to participants' perception of impact. The detailed analysis methods are as follows:

Correlation Analysis
Correlation analysis is used to calculate the correlation coefficients between the constructs to understand their mutual influences. The specific method includes calculating the average scores of each construct and generating a correlation coefficient matrix.

Multivariate Regression Analysis
Multivariate regression analysis uses the average scores of each construct as independent variables to conduct regression model analysis. The specific steps include setting hypotheses, testing hypotheses, and analyzing results. These analyses help to understand the interrelationships between the constructs and their contributions to participants' perception of impact.

Kansei Engineering Analysis
Kansei engineering analysis aims to analyze the impact of each construct on participants from the perspective of emotional design. This analysis method includes using bar charts, pie charts, and radar charts to visually present the average scores and relative proportions of each construct, understanding participants' emotional experiences and perceptions in market design.

3.5 Research Limitations and Future Directions

This study has the following limitations:

1. Sample Size: The sample size of this study is limited. Future research can expand the sample range to improve the generalizability of the results.
2. Cross-Sectional Data: This study uses cross-sectional data for analysis. Future research can use longitudinal studies to observe changes in constructs of SDGs markets over time.
3. Regional Limitation: The data in this study is from a specific region. Future research can expand to other regions for comparative studies.
4. Future research can build on these results to further explore the relationships and impacts of SDGs market constructs in different regions and cultural contexts, providing more comprehensive management recommendations and strategies.

4 Research Analysis

4.1 Overview

This study uses a questionnaire survey method, inviting 40 participants with market and curatorial experience to complete a retrospective subjective experience questionnaire after experiencing domestic sustainable markets (such as SDG markets, green markets, and environmental markets). Through literature review and quantitative analysis, the study aims to summarize participants' experiences and perceptions of sustainable markets and explore the emotional design factors that influence these experiences. The questionnaire covers four main constructs: market characteristics, SDG linkage, and impact assessment, combining participants' subjective feelings and evaluations to propose human-centered market optimization strategies and practical recommendations.

4.2 Reliability and Validity Analysis

Cronbach's α was used as the reliability index, and the results show that the Cronbach's α values of all constructs exceeded 0.7, indicating good internal consistency of the questionnaire.

Reliability Analysis
As shown in Table 1 and (see Fig. 3), Cronbach's α was used as the reliability index, and the results show that the Cronbach's α values of all constructs exceeded 0.7, indicating good internal consistency of the questionnaire.

Validity Analysis
As shown in (see Fig. 4), Validity analysis includes content validity and construct validity. Content validity was evaluated by multiple experts who assessed the relevance and representativeness of the questionnaire items, while construct validity was examined using factor analysis to verify whether the questionnaire effectively measured each construct.

Table 1. Reliability Analysis.

Construct	Cronbach's α
A	0.75
B	0.78
C	0.80
D	0.82

- Content Validity: Participants were invited to assess the relevance and representativeness of the questionnaire items. The results show that the content validity index (CVI) of all items exceeded 0.8.

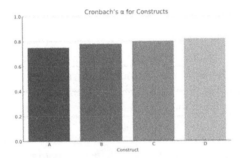

Fig. 3. Cronbach's α for Constructs.

- Construct Validity: Through factor analysis, the factor loadings of each item were examined. The results show that the factor loadings of the items on their respective constructs all exceeded 0.6, confirming good construct validity of the questionnaire.

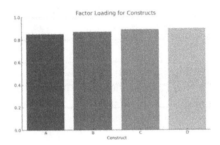

Fig. 4. Factor Loading for Constructs.

4.3 Correlation Analysis Between Constructs

First, the correlation coefficients between the constructs are calculated to understand their mutual influences. The specific method includes calculating the average scores of each construct and generating a correlation coefficient matrix.

Correlation Matrix

As shown in Table 2 and (see Fig. 5). The analysis results show that:

- The perception of sustainable market relevance (A) has a moderate positive correlation with other dimensions.
- The perception of sustainable market characteristics (B) has a relatively weaker but still positive correlation with other dimensions.
- The perception of the relationship between sustainability and markets (C) has a stronger correlation with the perception of the impact of participation in sustainable markets (D).

Constructing a Curatorial Awareness   199

Table 2. Correlation Matrix.

Dimension	A	B	C	D
A		0.472717	0.445746	0.472467
B	0.472717		0.161195	0.339995
C	0.445746	0.161195		0.399064
D	0.472467	0.339995	0.399064	

These results indicate that enhancing public perception of the relationship between the market and sustainability can improve their positive evaluation of market characteristics and impact.

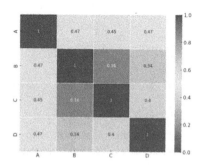

Fig. 5. Correlation Matrix.

4.4 Multivariate Regression Analysis

As shown in (see Fig. 6), this study conducts multivariate regression analysis to examine the contributions of each dimension to the overall impact. The average scores of each dimension are used as independent variables, and the regression model is used for analysis.

4.5 Kansei Engineering Analysis

As shown in Table 3, Based on Kansei Engineering theory, the subjective feelings and evaluations of respondents on each dimension were analyzed. The emotional evaluations of dimensions A, B, C, and D indicate that respondents rated the perception of sustainable market characteristics (B) the highest, showing that they consider the characteristics of sustainable markets to be the most prominent and important. This is followed by the perception of the impact of participation in sustainable markets (D), indicating that participants believe sustainable markets have great potential in terms of impact.

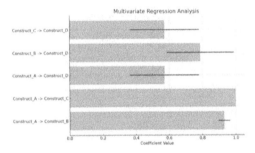

Fig. 6. Multivariate Regression Analysis.

Table 3. Kansei Evaluations of Constructs A, B, C, and D.

Dimension	Average Score
A	3.865
B	4.196
C	3.675
D	.905

Based on the results of multivariate regression analysis, we further analyze the impact of each construct on participants from the perspective of Kansei engineering. The following charts help to visually present these results:

Bar Chart Analysis
The bar chart(see Fig. 7) shows the average scores of each construct, as follows:

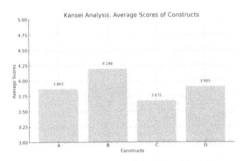

Fig. 7. Kansei Analysis: Average Scores of Constructs.

1. Sustainable Market Relevance Perception (A): The average score is 3.865, indicating that participants have a high perception of the relationship between the market and sustainability.

2. Sustainable Market Characteristics Perception (B): The average score is 4.196, the highest among all constructs, indicating that participants have the strongest perception of market characteristics.
3. Perception of Relationship Between Sustainability and Market (C): The average score is 3.675, indicating that participants have a high perception of the specific relationship between the market and sustainability.
4. Perception of Impact of Participating in Sustainable Markets (D): The average score is 3.905, indicating that participants believe that participating in sustainable market activities has a significant impact.

Pie Chart Analysis
The pie chart (see Fig. 8) shows the relative proportions of the average scores of each construct, making it easier to see the share of each construct in the overall perception. From the pie chart, the perception of sustainable market characteristics (B) accounts for the largest proportion, reaching 27.3%, indicating that participants consider the specific characteristics of the market to be the most important in their emotional evaluation.

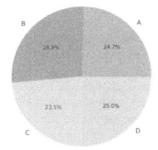

Fig. 8. Kansei Analysis: Average Scores Distribution.

Radar Chart Analysis
The radar chart (see Fig. 9) provides a multi-dimensional perspective, showing the comprehensive evaluation of each construct, which helps to compare the differences between different constructs. From the radar chart, the perception of sustainable market characteristics (B) is significantly higher than other constructs, further reinforcing the findings of the bar chart and pie chart.

Comprehensive Analysis
These charts show that participants have the highest perception of sustainable market characteristics, which means that the specific characteristics of market design (such as the environmental friendliness of products and the social nature of activities) are crucial for enhancing participants' emotional experiences and perceptions. Therefore, planners should focus on showcasing and reinforcing these characteristics when designing sustainable markets, and design highly interactive displays and activities based on Kansei engineering principles to enhance participants' emotional experiences and sense

Fig. 9. Kansei Analysis: Average Scores Radar Chart.

of identity. Additionally, the perception of the relationship between the market and sustainability and its impact also shows high importance, further supporting the necessity of emphasizing sustainability connections and educational outreach in market design. Through these strategies, sustainable markets can more effectively convey the concept of sustainability, enhance participants' emotional experiences and sense of identity, thereby promoting broader public participation and support.

4.6 Research Findings and Discussion

Based on the regression analysis and correlation matrix results, we found that: the perception of sustainable market relevance significantly positively influences the perception of market characteristics and the perception of the relationship between sustainability and markets; the perception of sustainable market relevance significantly influences the perception of the impact of participation in sustainable markets, though the influence is lower; the perception of sustainable market characteristics and the perception of the relationship between sustainability and markets both significantly positively influence the perception of the impact of participation in sustainable markets. These findings suggest that promoting sustainable markets should focus on enhancing public overall perception of market relevance, characteristics, and SDG content to encourage broader participation and awareness.

5 Conclusion

This study explored the interrelationships between various dimensions of sustainable markets and their contributions to the overall impact. These relationships were validated through correlation analysis and multivariate regression analysis. The results indicate significant correlations between the perception of sustainable market relevance, the perception of sustainable market characteristics, the perception of the relationship between sustainability and markets, and the perception of the impact of participation in sustainable markets.

References

1. Cheng, Y., Randall-Parker, A.: The role of marketplaces in sustainable development. J. Sustain. Dev. **10**, 45–60 (2017)

2. Morales, A.: Public marketplaces: resilience and sustainability. Sustainability **13**, 6025 (2021)
3. Hinrichs, C., Gillespie, G., Feenstra, G.: Social learning and innovation at retail farmers' markets. Rural. Sociol. **69**, 31–58 (2004)
4. Johnson, L.: Transforming abstract concepts into tangible exhibits. J. Museum Educ. **43**, 123–138 (2018)
5. Smith, M.: The evolving nature of curatorial practice: strategies for contemporary curators. Int. J. Arts Manage. **18**, 56–70 (2015)
6. Chang, T.-W., Wu, Y.-S., Datta, S.: Developing a slow sensor for elderly. Sensors Mater. **32**, 2425–2432 (2020)
7. Wu, Y.-S., Chen, C.-Y., Chang, T.-W.: Carbon footprint interaction through slow design computing and visual design. In: Ahram, A., Falcão, C. (eds.) Human-Centered Design and User Experience. AHFE (2023) International Conference., vol. 114, pp. 446–455. AHFE International, Hawaii, USA (2023)
8. Liao, S., Chen, W.: Emotional design in eco-friendly markets. Int. J. Des. **13**, 78–90 (2019)
9. Cheng, J., Randall-Parker, T.: Conceptualizing sustainable consumption at farmers markets. In: Cheng, J. (ed.) Consumption, Identity and Style, pp. 119–132. Routledge (2017)
10. Li, W., Wang, F., Liu, G., Gao, H.: A bibliometric analysis and systematic review on e-marketplaces, open innovation, and sustainability. Sustainability **14**, 5456 (2023)
11. White, R.: Curatorial responsibility: addressing social issues through exhibition design. Curator: The Museum J. **63**, 345–361 (2020)
12. Norman, D.A.: Emotional design: why we love (or hate) everyday things (2005)
13. Wu, Y.-S., Chang, T.-W., Datta, S.: Developing the interaction for family reacting with care to elderly. In: Rau, P.-L.P. (ed.) HCII 2020: Cross-Cultural Design. Applications in Health, Learning, Communication, and Creativity, vol. 12193, pp. 200–210. Springer International Publishing, Cham, Cham (2020)

Design For/From Pray-The Building Process of Virtual Pilgrimage Site

Chia Hui Nico Lo(✉)

Department of Art and Design, Yuan Ze University, Taoyuan, Taiwan
nicoisworking@gmail.com

Abstract. This study explores the process of constructing a virtual pilgrimage site, aimed at providing an immersive pilgrimage experience for believers who cannot personally visit sacred sites. Virtual pilgrimage not only offers believers the opportunity to participate in religious rituals but also provides young people from different religious or cultural backgrounds with a way to understand and appreciate religious culture. By using the Minecraft platform and Virtual Reality (VR) devices, this study has constructed a virtual environment of the Boudhanath Stupa in Nepal, offering a space for pilgrims to perform rituals. The study employs "Kansei Design" theory to explore the correspondence between human feelings and specific objects. Under the collective prayers of a group of people, the researcher conducted a literature review and on-site observations. The study has designed sensory experiences for the virtual space and provided a space for ritual practice for pray. The preliminary results from Tibetan Buddhist practitioners' experiences indicate that the design can serve as a support for daily practice for believers, helping people understand the meanings conveyed by Buddhism. Additionally, the practitioners believe that the Minecraft interface is user-friendly for the younger generation, which is beneficial for the dissemination of Dharma. Future research will continue to explore the experiences of Buddhists, using observation, interviews, and questionnaires to validate the possibility of collective virtual pilgrimage.

Keywords: Virtual Pilgrimage · Kansei Design · Minecraft · Virtual Reality

1 Introduction

1.1 The Need for Virtual Pilgrimage

Pilgrimage is a process in religious or spiritual life that seeks spiritual meaning, usually involving a journey to a sacred place or a location of significant importance to one's faith. For religious individuals, pilgrimage to a holy site is not just about physically reaching a sacred space, but more importantly, it is about experiencing an inner journey of exploration and purification. However, due to issues such as geographical location, health conditions, and economic capacity, not all believers can personally make the journey. The virtual pilgrimage game Muslim 3D, released by Germany's Bigitec Studio, has exceeded one million downloads. This situation highlights the necessity of creating virtual pilgrimages for Buddhists as well. The virtual pilgrimage space in this study is built upon the collective prayers and wishes of many people. Through virtual pilgrimage, not only can believers participate, but younger generations and individuals from other religious or cultural backgrounds can also understand and appreciate religious culture through virtual experiences.

1.2 Research Objectives

Since virtual pilgrimage is not a true "immersive" experience but rather a "conceptually immersive" one, the focus of this research is to explore how to enhance believers' religious faith and emotional connection through an immersive experience, providing a sense of spiritual fulfillment similar to that of an actual pilgrimage. This research employs "Kansei Design," which discusses the correspondence between human feelings and specific "objects."

The study aims to design virtual pilgrimage spaces based on user needs, using Kansei research and sensory correspondence theory. Through literature review and on site observation, the research seeks to design the sensory experiences of virtual spaces. Additionally, by utilizing user-friendly interactive technology and Virtual Reality (VR) equipment, the study aims to provide a pilgrimage experience that feels as if one is "present" at the sacred site. This allows believers who cannot personally visit holy sites to perform religious rituals in a virtual sacred space, creating pilgrimage experiences and memories. The objectives are as follows:

1. To collect and analyze spatial resources of pilgrimage sites, constructing a digitized cultural heritage space. This research adopts a user-centered perspective to shape a unique pilgrimage experience by integrating holy site tourism with Buddhist rituals and spatial Kansei engineering, exploring the possibilities of virtual religious tourism.
2. To deepen the participants' understanding of Buddhist culture and artistic heritage throughout the process, from constructing digital replicas to experiencing virtual pilgrimage.
3. To construct digital cultural heritage through an open-world platform, allowing users to create their own pilgrimage memories, which can also become a part of others' pilgrimage experiences.

2 Literature Review

2.1 Pilgrimage Site

A stupa serves as a place for preserving the relics of the Buddha, regarded by Buddhists worldwide as a "sacred space." According to Eliade's theory of space, he posits that the space experienced by religious individuals contains interruptions or breaks [1]. It is a space of reality and existence, in opposition to the secular experiential space. Compared to sacred space, the sanctification of space involves the transformation of secular space, even in the construction of a house. Pilgrimage to a holy site is challenging, but each visit to any church or Buddhist temple is a form of pilgrimage. Moreover, sacredness is not confined to the space of caves; it is mobile. As long as it is attached to sacred objects, sacredness can be omnipresent. According to the Dirgha Agama Sutra [2], Volume Three, Sakyamuni Buddha instructed that after his cremation, his relics should be enshrined in stupas for public veneration, to inspire devotion in people. The earliest discovered stupas are in the shape of an inverted bowl, known as stupas, which are hemispherical mounds, such as the Sanchi Stupa from the 7th century CE. Pilgrimage to holy places is difficult, but each visit to any church or Buddhist temple is itself a pilgrimage. Moreover, sacredness is not confined to the space of caves; it can be mobile. As long as it is attached to sacred objects, sacredness can be omnipresent [3].

The Boudhanath Stupa in Kathmandu, Nepal, is a well know pilgrimage site for Buddhists, attracting approximately 800,000 visitors annually. The stupa is characterized by its white dome and the four pairs of Buddha's eyes. It enshrines the relics of Kashyapa Buddha. In 1975, it was designated a UNESCO World Heritage Site and is commonly referred to as the "Wish-Fulfilling Stupa." There are various legends about its name, one of which is attributed to the Tibetan Buddhist master Guru Rinpoche, who said, "This great stupa is like a wish-fulfilling jewel, making all prayers and wishes come true" [4] Due to its significant architectural space and cultural meaning, as well as its relatively remote geographical location, it is well-suited as a site for virtual pilgrimage.

In Keith's book, Power Places of Kathmandu, he highlights the architectural features of Boudhanath Stupa [5]. Dowman describes the stupa as having three terraced plinths, upon which a drum and dome are constructed. The first plinth is approximately seven feet in height, while the upper plinths are each around six feet. All three plinths are 'twenty-corner' and 'gated' to symbolize doorways into the mandala. The drum, about four feet high, supports the dome and is unadorned except for an added terrace with one hundred and eight niches filled with stone deities. The stupa's dome is one hundred and twenty feet in diameter, topped by the harmika box and the thirteen steps of the spire, which are covered with gilt copper sheets. A textile apron encircles the top of the harmika, 'framing the powerful eyes of the stupa'. A large lotus surmounts the spire, capped by a jewel pinnacle.

Additionally, a stupa represents the outward expression of the Buddha's enlightened body, speech, and mind. Sylvia Somerville states that every architectural feature of a stupa represents a facet of the Buddhist spiritual path. The five geometric shapes of a stupa correspond to the five elements that in turn correspond to attributes of a fully awakened being. The sputa's create a mandala that represents a pure distillation of the universe [6].

2.2 Pilgrimage Ritual

Rituals, as a form of cultural and social practice, have always been regarded as an important means of community cohesion and cultural inheritance. Rituals often occur through actions performed by humans in real environments. In the Encyclopedia of Religion, Evan Zuesse defines rituals as "conscious, voluntary, repetitive, and stylized symbolic bodily actions centered on cosmic structures and/or sacred beings" [7]. The emphasis of rituals is on bodily actions and actual practice.

In the book Pilgrimage, Tibetan Buddhist master Dzongsar Khyentse Rinpoche [8] mentions that the main practice during the pilgrimage is to accumulate wisdom and merit and to purify defilements. Visiting sacred sites can accumulate great merit through practices such as prostrations, offerings of water, flowers, incense, and food, reciting prayers, circumambulating stupas, and dedicating merits. Since stupas symbolize the Buddha and are seen as a continuation of the Buddha, all Buddhist sects agree that offering to stupas by laypeople can result in rebirth in the heavens [9].

Circumambulation, or walking around a stupa or statue with the right side facing it, is a ritual practice. According to the Right Circumambulation of the Stupa Sutra [10], the Buddha taught that circumambulating a stupa can accumulate great merit and benefits, including avoiding eight difficulties, obtaining a beautiful and dignified body, being born into noble families, gaining wealth, ascending to heaven, and obtaining a good voice. Since stupas symbolize auspiciousness and purity, Indians show respect through right circumambulation. The ritual of circumambulation has always been an important practice for Buddhists. On the carvings of the Bharhut Stupa from the 2nd century BCE, followers can be seen with hands folded in prayer, devoutly circumambulating in a clockwise direction.

Therefore, based on the above literature and the observations made by the researcher during the study, the ritual of circumambulating a stupa can be summarized as follows:

Circumambulation involves walking clockwise around the stupa, aiming to circle the stupa at least once from the starting point. The scriptures recommend circling three times for greater auspiciousness, with the ideal number being 108 circles. Buddhists generally believe that circumambulating the stupa accumulates merit, and they participate with great devotion and joy. Non-Buddhists, although they may not understand the meaning of circumambulation, often join in the ritual, believing it brings good fortune.

During circumambulation, it is not just about physically walking around the stupa; Buddhists recite mantras while circumambulating. Some even progress slowly with full-body prostrations. The area around the stupa often provides spaces for offering water, flowers, incense, food, and for prostrations, allowing these rituals to be performed simultaneously. Circumambulation is a familiar and continuously practiced method for Buddhists. This study aims to facilitate repetitive use of this practice through a game mode, serving as an interface for Buddhists to pray by the ritual. In addition to physical movement, circumambulation involves chanting, listening to mantras, and performing different offerings depending on the scene. It synchronizes the practice of body, speech, and mind, completing the entire process and achieving the meaning of the practice. These aspects will be the essential elements in the design of the virtual pilgrimage ritual.

2.3 Tools for Virtual Pilgrimage

Building a virtual space is not difficult with current 3D technology, such as 360-degree photography or drone image capture. However, a pilgrimage space involves more than just spatial form; it must also evoke emotional responses in pilgrims. Shinya Nagasawa.

[11] proposed that emotional satisfaction arises from sensory and perceptual stimulation through different sensory organs. For pilgrims, the visual and auditory elements in the environment are the most direct ways to evoke the sacred site's atmosphere. Rituals can create a sense of "conceptual immersion." Therefore, interactivity in the experience is essential; merely watching videos cannot fulfill this need.

To customize experiences and create more content in the virtual sacred site, as well as to enable the possibility of collective online pilgrimages, this study uses the Minecraft game platform and VR headsets as tools for spatial experience. In recent years, game interfaces have often been used as tools for digitized cultural heritage spaces. Fernandez and Medeiros [12] pointed out that the video game Minecraft is one of the best solutions for digital cultural assets, and is an effective tool to convey architectural heritage. The project "Tate Worlds" 5) is the digital interpretation of landscapes based on works in the museum's collection and is an inspiration for a successful collaboration between educators and heritage professionals in Scotland called Crafting the Past, which saw communities undertake digital archaeological excavations and reconstruct historic buildings in Minecraft, attracting audiences of over 100,000 people. Indeed, the potential benefits of Minecraft and video gaming for heritage visualization more generally have been recognized by UNESCO6.

It is important to consider that Minecraft's unit is $1 \times 1 \times 1$ cube not an accurate reproduction of proportions of architectural elements of particular importance. Méndez, et.al., outline Minecraft can be a good tool for making basic elements about circulation in buildings, its types, functionality, and accessibility [13].

The gamefic interface allows to perceive different aspects, nuances and details similar to the people who will live in real life buildings. It is a good teaching strategy when we realize that playing, creating, experimenting and making mistakes improve the building process.

In this paper, the Minecraft program was used to work on the stupa architectural heritage. The virtual building, they were asked to create could have represented the architecture's characteristics and interactivity with users, so the activity, in addition to the objective of getting to know the pilgrimage space, also aimed to empower the imagination of the participants.

Virtual Reality" (VR) is a medium which can represent reality or create an alternate scene to experience the real site. It could potentially be used for triggering memory recollections by connecting users with their past or provide an interactive environment to accumulating new memories. It can be a great tool for people to visit a cultural site without time and location limitations. It is in order to give the impression of being immersed in that environment" offering the possibility of interacting with additional contextual cultural data. VR can provide the tourism industry with another useful tool to experience. Woodard and Sukittanon said to the present by using virtual walkthroughs in a three-dimensional environment is the most efficient way to present two-dimensional sketches [14]. Eugene Ch'hg et.al., revealed that VR is a powerful medium for communicating cultural sites [15]. Using immersive technology/VR for the visualization of concepts as it enhances learning and immersive experience.

3 The Design Process

The virtual Boudhanath Stupa in Nepal serves as a space for virtual pilgrimage, considering that many people are unable to visit the sacred site for various reasons. Additionally, those who have visited the site can use digital tools to perform rituals and evoke the sense of pilgrimage. This study began with the collective prayers and needs of a group of people, with a total of 42 sponsors supporting the project. The goal is to create a virtual pilgrimage space where people can make their wishes.

Based on Tony Wang presents Nagamachi's relationship between sensation and emotion [16], the design process of this study aims not just to construct a model but to create an emotionally resonant virtual pilgrimage space. The construction process involves the following relationships (Fig. 1):

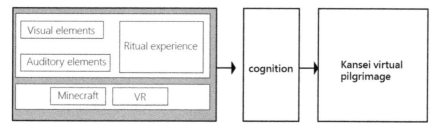

Fig. 1. The relationships of the design virtual pilgrimage process.

3.1 Collection of Visual Elements

With the assistance of Samantabhadra Society of Translation & Compassion and the lamas from the Sakya of Tibetan Buddhism, the project team conducted on-site inspections (Fig. 2).

Fig. 2. On-site observation and measurement.

Through on-site observations and the collection of online photos, several specific visual elements can instantly help those wishing to make a pilgrimage recognize the location as the Boudhanath Stupa in Nepal. These key visual characteristics are essential elements for constructing the virtual scene, as shown in Table 1.

Table 1. The key visual elements of on-site and virtual site.

key visual elements	On-site photos	Virtual site
The eyes on the top of stupa		
The shape of an inverted bowl		
Tibet Buddha prayer Colorful Flags		
The smaller stupa on the east side		
Many pigeons gather in the northeast corner		
Guru Rinpoche Temple at the main entrance with bell.		

3.2 The Auditory Elements for Pray

At the Boudhanath Stupa site, the area is filled with various sounds due to the presence of pilgrims, restaurants, and shops catering to their needs. However, the predominant sound is the "mantra chanting." Whether recited by people or played through speakers, you are always surrounded by mantra chanting no matter where you go around it. The most common mantras are the six-character great bright mantra and the seven line prayer to Guru Rinpoche. "Om Mani Padme Hum" is a mantra associated with the bodhisattva of compassion, Avalokiteshvara (Chenrezig in Tibetan Buddhism). The Seven Line Prayer to Guru Rinpoche is a revered and powerful prayer in Tibetan Buddhism, invoking the blessings of Guru Rinpoche, also known as Padmasambhava, is considered the "Second Buddha" who brought Buddhism to Tibet in the 8th century. The prayer is believed to summon his presence and blessings, offering protection and spiritual growth. The auditory elements detailed in Table 2.

Table 2. The sound and meaning of on-site auditory elements

auditory elements	The sound on-site	The meaning
six-character great bright mantra	om mani padme hum	"the jewel is in the lotus," indicating the transformative journey of rising above suffering and achieving enlightenment through the union of wisdom and compassion. The mantra is often chanted for its protective and purifying effects, bringing inner peace and fostering a compassionate heart
seven line prayer to Guru Rinpoche	Hung örgyen yul gyi nubjang tsam pema gesar dongpo la yamtsen choggi ngödrub nyé pema jungné shyesu drak khordu khadro mangpö kor khyé kyi jesu dag drub kyi jingyi lab chir shek su sol guru pema siddhi hung	On the northwest border of the country of Uḍḍiyāna. On the anthers of a lotus flower. You attained the marvelous supreme siddhi, renowned as the Lotus Born. Surrounded by a retinue of many ḍākinīs, following you, I will reach accomplishment. Please come and bestow your blessings GURU PADMA SIDDHI HŪṂ

3.3 The Ritual Experience

To experience virtual pilgrimage, it is crucial to record the behaviors on-site. Beyond the visual and auditory aspects of the environment, pilgrimage must be "performed" through actions such as circumambulating the stupa, making smoke offerings, lighting lamps, and turning prayer wheels (Table 3).

Table 3. The meaning of on-site pilgrimage ritual

The pilgrimage ritual	The on-site photo	The meaning
Circumambulating a stupa		Circumambulating a stupa, also known as "kora" in Tibetan Buddhism, is a significant practice in many Buddhist traditions. This ritual involves walking clockwise around a stupa or other sacred object, typically while reciting prayers or mantras. The practice is believed to offer numerous spiritual benefits.
Smoke offering		Smoke offering, also known as "sang" in Tibetan Buddhism, is a ritual practice that involves burning substances such as herbs, incense, and aromatic woods to produce fragrant smoke. This practice is believed to have several spiritual benefits.
Lamp offering		a common and significant practice in various Buddhist traditions. It involves lighting and offering lamps, candles, or butter lamps as an act of devotion. The light represents the wisdom of the Buddha, dispelling the darkness of ignorance. It is a symbol of enlightenment and the illumination of the mind.
Turning prayer wheel		A prayer wheel is a cylindrical wheel mounted on a spindle, which contains scrolls of paper or other materials inscribed with "Om Mani Padme Hum." Turning a prayer wheel helps accumulate merit, purify negative karma, and spread blessings. It enhances meditation, mindfulness, and spiritual connection while promoting peace, compassion, and healing.

These rituals, in the understanding of Buddhists, represent offerings to the Three Jewels, purification of negative karma, accumulation of merit, and dedication of good deeds to all beings. Only by unifying space and actions can a pilgrimage experience that encompasses body, speech, and mind be constructed.

4 The Virtual Experience

Based on on-site observations, data collection, and literature review, this study uses Minecraft to construct a virtual environment of the Boudhanath Stupa in Nepal, providing pilgrims with a space for ritual practice (see Fig. 3).

Fig. 3. The virtual pilgrimage site for pray: digital Boudhanath Stupa.

The aim is to design an emotionally resonant virtual pilgrimage experience.Pilgrimage sites differ from other spaces in that they carry symbolic sacredness. In addition to simulating the physical environment, it is essential to evoke the sensory aspects of pilgrimage. This means that beyond visual and auditory aids, the key is to trigger internal feelings through actions. To verify the state of the virtual Boudhanath Stupa, before it was opened to Buddhists for use, two Tibetan Buddhist Rinpoches were invited to experience it (see Fig. 4).

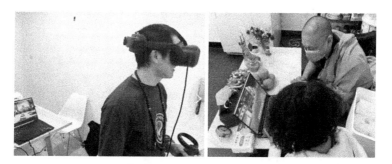

Fig. 4. Rinpoches' virtual experiences in digital Boudhanath Stupa.

Both Rinpoches are Tibetans who frequently lead followers on pilgrimages to Nepal and are very familiar with the local conditions. After using the virtual pilgrimage, the Rinpoches stated that its design could serve as a support for daily practice for believers. They also believed this design can help people to understand the meanings conveyed by Buddhism, and that the Minecraft interface is user-friendly for the younger generation, which is beneficial for the transmission of Dharma.

5 Conclusion and Future Research

This research tries to make contributions to the understanding of religious culture and utilizes technology to simulate virtual pilgrimage. The conclusions are:

1. The gamified style of Minecraft does not affect the pilgrimage experience. The key is how to reflect the local context and create an emotional and cognitive connection to the pilgrimage.
2. Designing a virtual pilgrimage space must recreate important visual elements so that people can immediately recognize the location. Once sensory recognition is triggered, it creates a feeling of "being" in the sacred site.
3. Auditory elements can enhance the sense of sacredness in the virtual holy site. From the Rinpoches' experience, it is evident that when mantras are played, they naturally follow along, experiencing the virtual holy site through "speech."
4. Providing a space for performing rituals, without the limitations of geography, time, or economics, allows for repeated entry into the virtual holy site. Therefore, the Rinpoches believe it can serve as a means for the public to engage in regular practice, achieving the significance of "conceptually" experiencing the holy site.

In the future, this study will continue to explore the experiences of Buddhists, using observation, interviews, and questionnaires to investigate the possibilities of collective virtual pilgrimage.

Future research should further validate the general public's acceptance of virtual pilgrimage through additional observations, interviews, and surveys to comprehensively assess its effectiveness and impact.

Acknowledgments. This study expresses gratitude to the Samantabhadra Society of Translation & Compassion, Jamchen Lhakhang Monastery, the sponsors, the Rinpoche who participated in the experience, and the project assistants for their assistance.

References

1. Eliade, M.: Traité d'histoire des religions. Editions Payot & Rivages, Paris (1949). (Yan, K., Yao, B. trans.: 神聖的顯現：比較宗教、聖俗辯證，與人類永恆的企盼.. PsyGarden Publishing Company, Taiwan (2022))
2. Dirgha Agama Sutra. CBETA 2023.Q3, T01, no. 1, p. 20b4-11
3. Eliade, M.: Histoire des croyances et des idées religieuses I. (1989). (Wu, J., Chen, J. trans.: 世界宗教理念史:卷一.. Business Weekly Publications, Taipei (2015))
4. Guo, R.: 朝聖尼泊爾:走入蓮師密境努日.. Oak Tree Publishing Co, Taipei (2020)

5. Dowman, K.: Power Places of Kathmandu: Hindu and Buddhist Holy Sites in the Sacred Valley of Nepal. Thames and Hudson, London (1995)
6. Young, T.: Boudhanath Stupa: Reflections on Living Architecture. Deakin University (2014)
7. Zuesse, E.: Ritual. In: Jones, L. (ed.) Encyclopedia of Religion, vol. 11, 2nd edn., pp. 7833–7848. Macmillan Reference USA, Detroit (2005)
8. Dzongsar Khyentse Rinpoche: 朝聖:到印度佛教聖地該做的事.. (Yao, K. trans.) Crone Publishing, Hong Kong (2016)
9. Zhanru: 印度早期佛教史研究.. Zhonghua Book, Beijing (2006)
10. Right Circumambulation of the Stupa Sutra. CBETA 2023.Q3, T16, no. 700, p. 801c4-18
11. Nagasawa, S.: Present state of Kansei engineering in Japan. In: IEEE International Conference on Systems. Man and Cybernetics, vol. 1, pp. 333–338. IEEE, The Hague (2004)
12. Garcia-Fernandez, J., Medeiros, L.: Cultural heritage and communication through simulation videogames—a validation of minecraft. Heritage **2**(3), 2262–2274 (2019)
13. López Méndez, M.D.C., González Arrieta, A., Queiruga Dios, M., Hernández Encinas, A., Queiruga-Dios, A.: Minecraft as a Tool in the Teaching-Learning Process of the Fundamental Elements of Circulation in Architecture. In: Graña, M., López-Guede, J.M., Etxaniz, O., Herrero, Á., Quintián, H., Corchado, E. (eds.) SOCO/CISIS/ICEUTE -2016. AISC, vol. 527, pp. 728–735. Springer, Cham (2017). https://doi.org/10.1007/978-3-319-47364-2_71
14. Woodard, W., Sukittanon, S.: Interactive virtual building walkthrough using oculus rift and microsoft kinect. In: SoutheastCon 2015. IEEE (2015)
15. Ch'ng, E., Li, Y., Cai, S., Leow, F.T.: The effects of VR Environments on the acceptance, experience, and expectations of cultural heritage learning. J. Comput. Cult. Herit. **13**, 1–21 (2020)
16. Wang, T.: From Kansei Engineering to Kansei Design: the Basic and Application for Researching of Kansei Engineering. 感性工學研究的基礎與應用.. Chuan Hwa Book, Taipei (2023)

A Research of Kansei Engineering and Affective Computing for Generating Game Controller

Yinghsiu Huang[✉] [iD] and Syue-Ting Lung

Department of Industrial Design, National Kaohsiung Normal University, Kaohsiung, Taiwan
yinghsiu@mail.nknu.edu.tw

Abstract. In product design, computer-aided design (CAD) is used for the efficient output and accurate presentation of products. Early CAD elements were employed in the later stages of the design process, such as for product simulation and manufacturing. With the advancement of computer hardware and software technology, CAD and computer-aided manufacturing has been widely used in the early development of computer-aided concept design. In addition, a generative design system has been applied to the stage of concept development, which can quickly present diverse and systematic product ideas, allowing the designer to stimulate more possibilities in initial sketches. However, during the design process, designers might not know which forms users like. Thus, the research problem of this study is how to use heavy gamers' emotional preferences for a controller's appearance and integrated them with a generative concept-based CAD system in the initial design stage to generate a design concept preferred by users. Thus, this study focused on the design of game controllers, emphasized the emotional needs of heavy gamers, summarized the correlation of game controller appearance and emotional needs, and integrated a generative design system in the process to assist with conceptual design.

Keywords: Generative Design System · Game Controller Design · Kansei Engineering · Hayashi's Quantitative Theory Type I

1 Introduction

With the advancement and vigorous development of industrial technology, socioeconomic structure and consumption patterns have gradually shifted from a production orientation to a focus on user needs. In addition to meeting the basic needs of consumers, product design must give products a unique image through the design process to meet higher-level consumer demand. This change has transformed product appearances from "form follows function" to "form follows fun" and "form follows emotion" [1]. However, the current design orientation is a user-centered design concept, and the needs and feedback of users have also been emphasized.

In product design, computer-aided design (CAD) is used for the efficient output and accurate presentation of products. Early CAD elements were employed in the later stages of the design process, such as for product simulation and manufacturing. With the advancement of computer hardware and software technology, CAD and computer-aided manufacturing has been widely used in the early development of computer-aided concept design. Therefore, studies have focused on computer use in early design behavior to assist in one dimensional and three-dimensional sketching. In addition, a generative design system has been applied to the stage of concept development. On the basis of parameters set by a designer, a generative design system can quickly present diverse and systematic product ideas, allowing the designer to stimulate more possibilities in initial sketches. As such, the design process becomes clearer, and these generative models also provide a conceptional space for designers [2].

The appearance of a product is symbolic, and they may affect consumer preferences. Moreover, consumers prefer products that fit their own image [3]. In addition to allowing the user to quickly engage with the product, product appearance determines a user's first impression of a product. Game controller design is crucial to heavy gamers because they value both function and appearance. Currently, studies related to game peripheral devices have mainly focused on the computer mouse, discussing appearance, materials, and tactile impression. In addition, most studies related to game control have discussed the human–machine interface in terms of human factors [4] but not from an emotional perspective. Therefore, the research problem of this study is how to use heavy gamers' emotional preferences for a controller's appearance and integrated them with a generative concept-based CAD system in the initial design stage to generate a design concept preferred by users. Therefore, this study focused on the design of game controllers, emphasized the emotional needs of heavy gamers, summarized the correlation of game controller appearance and emotional needs, and integrated a generative design system in the process to assist with conceptual design.

Research objectives, methods, and procedures are outlined in the following three stages: Stage 1: Understand the expectations and appearance preferences of heavy gamers toward game controllers through interviews and determine related emotional vocabulary. After sorting and screening, 10 groups of emotional adjectives were obtained, and a questionnaire survey was conducted to reveal emotional adjectives related to game controllers. Stage 2: Propose a correlation between emotional level of heavy gamers and appearance of game controller through a morphological analysis and Hayashi's quantitative theory type I analysis. Stage 3: Five control variables of the morphological analysis (i.e., upper edge, lower edge, grip length, chamfering level, and chamfering point) were used to construct a game controller model using the Grasshopper coding application. Moreover, emotional adjectives such as "unoriginal," "creative," "stable," and "lightweight" are used to produce the game controller model preferred by heavy gamers.

2 Related Works

This study mainly investigated emotional evaluations of game controllers and the application of a generative design system. The game controller designed in this study is discussed in Sect. 1. Related theories and studies of Kansei engineering and appearance perceptions are discussed in Sect. 2. The parametric design, generative model, and currently used generative design tools are discussed in Sect. 3.

2.1 Game Controller

Game pads are common game controllers (Fig. 1, left). They are isometric joystick analog controls that can judge the direction of force exerted by the user when operating the game controller [5]. A game pad is a handheld game controller that uses electronic signals as inputs. In most interface designs, the left side contains direction control keys and the right side contains action control keys. In terms of key functions, game controllers mainly enable direction operations and matches them with action keys to control actions in a game [4].

Fig. 1. Left: Game Pad; Right: Nintendo N64 [6].

From 1972 to 1985, most game controller designs were rectangular. Gradually, design shifted to a curve shape, (e.g., the first model: Sega Genesis/Mega Drive in 1988). The makers of Sony PlayStation, released in 1994, considered that users would feel tired after using the console's game controller for a long time based on human factor considerations; thus, the hand grip of the controller was lengthened, rendering it more comfortable to use. The Nintendo N64, released in 1996, has a grip, allowing users to choose their preferred grip (Fig. 1, right). The appearances of most controllers by 2005 were fine-tuned for grip length, number of buttons, and functions. In 2006, The Nintendo Wii was launched with a detached controller, with its having new uses and operations. However, people seem to be more accustomed to the previous controller form. Until 2017, the appearance of grip was still in the mainstream of the game pad (Fig. 1, left).

2.2 Kansei Engineering and Appearance Perceptions

Kansei engineering (KE), originates from the emotional engineering proposed by Mitsuo Nagamachi in 1970 [7], predicting that emotional demand will be prioritized by human beings after material civilization. President Kenichi Yamamoto of Mazda Motor Corporation, Japan, summarized a systematic organization and structure in 1986 [8]. In 1988, Nagamachi renamed the concept KE [7]. KE translates the customer's perceptions and needs on a product into the concrete domain of product design; it is a novel consumer-oriented product development technology. It is generally uses users' emotional responses and cognition as the basis of analysis and research, and products that meet the sensory or psychological needs of consumers are constructed through statistical analysis and the application of computer technology [7].

In terms of effectively assisting designers in clarifying the relationship between key appearance features and emotional attributes, Chen [9] noted that KE is a method of rationalizing consumers' emotional needs; as such, it can analyze the relationship between components of a game controller's appearance components and emotional attributes.

Therefore, effective product attribute prediction models and procedures are key to the success of an emotive product design. The current forecasting methods for the evolution of product attributes generally employ linear analysis methods, such as early analysis techniques used in linear model regression analysis and Hayashi's quantitative theory type I analysis. Hayashi's quantitative theory type I analysis is similar to qualitative regression analysis, which is based on the assumption that each appearance component has a linear combination. The advantage is that the partial correlation coefficient be used to evaluate the importance of each component and the priority of selected items according to their evaluation scores [10].

2.3 CAD and Generative Design

In recent years, Grasshopper has been rapidly developed and popularized in the field of digital architecture, with diverse applications. In particular, freeform surface modeling and digital architecture are achieved using this parametric modeling tool. A computer's ability to perform monotonously large calculations is markedly helpful in dealing with numerous components with small changes in digital architecture, and rapid calculations that show changes in real time also allow designers to repeatedly use dynamic and nonlinear design techniques to determine optimal appearance and structural display. A generative model is built on a parametric model through the input of functional requirements (e.g., size, performance, and cost). Generative models are standardized through product architectures (e.g., CAD model files, design procedures, rules, feature restrictions, and formulas) and produce results that meet designer requirements.

Generative design is performed using a CAD system and software component with a programming interface. A system designer must convert possibilities and abstract ideas that make up a product into a digital geometric model structure and calculation commands before a system is constructed. Krish [2] proposed the generative design method, which is a generative system based on parametric CAD. The parameters of the system include a certain range from the maximum to minimum values, and this scope is the search space of the program. The search range of each parameter can form a design space and limit the search range of a computer. The generative design system randomly generates new design outputs according to the values of these parameters, and these parameters are regarded as design genes. Numerous parameters can be synthesized using the Genoform plug-in, allowing the system to generate hundreds of design ideas in a short time; this saves time required for idea generation. Krish also introduced a filtering mechanism in Genoform; a performance filter selects useful ideas that meet design requirements and conditions. Another proximity filter removes highly similarity concepts to ensure that unique concepts are produced. These data are then stored for further modification [2]. This filtering mechanism can be adjusted by the designer to improve the efficiency of screening numerous unfitting variations or searching for the optimal solution in the solution space.

Following Krish's research in 2011, Khan, Gunpinar, and Dogan [11] proposes a design framework for the parametric design and shape modification of a yacht hull. In this framework, designers can create hulls from classical to modern yachts with different chine line(s) and bow shapes. Finally, a 3D surface model of a yacht hull is obtained by generating Coons patches using feature curves.

3 Experiment on the Relationship Between the Emotional Vocabulary and Appearance of a Game Controller

The primary research methods and procedures of this study are as follows. Vocabulary related to the appearance of game controllers from relevant studies and interviews with heavy gamers was summarized. Subsequently, a representative sample was selected by experts in industrial design. Finally, an online questionnaire was used to obtain users' adjectives to their preferences for the appearance of game controllers.

3.1 Emotional Adjective Selection

Personal involvement inventory (PII) is a tool proposed by Zaichkowsky [12] to measure the degree of product involvement of participants. This scale was developed on the basis of a person's perceptions of how relevant a product is to their internal needs, interests, and values. Ten polarized adjectives obtained through a semantic differential technique are used to measure participants' views of a product. The score range of this scale is 10–70 points. Higher scores indicate a higher degree of product involvement. In this study, those with PII scores of 10–29, 30–50, and 51–70 were defined as having low, medium, and high involvement, respectively.

The questionnaire design consisted of three parts. The first part involved using the revised PII by Zaichkowsky [12] to identify users with high involvement. According to the research objectives, items tested in the questionnaire and rated using a 7-point Likert scale included "a game controller is important to me," "a game controller can bring joy to my life," "a game controller is closely related to my life," "the use of a game controller excite me," "a game controller is meaningful to me," "a game controller is attractive to me," "a game controller is fascinating to me," "a game controller is valuable to me," "I am passionate about using a game controller," and "game controllers are indispensable in my life." The second part of the questionnaire examined how frequent the respondents used a game controller; and the third part obtained participants' demographic variables and contact information.

To comply with the definition of heavy gamers in this study, the participants had to (1) have high product involvement with PII scores of 51–70, and (2) be a heavy user, that is, one who uses a game controller at least three times a week. Finally, 12 responses were received (online questionnaire settings require each item to be answered for submission). The sum of the revised PII scores was between 10 and 70 points, and a score of 51 points or more was regarded as indicating high involvement. Six heavy gamers (4 male and 2 female participants) who had used game controllers more three times per week were invited and agreed to participate in in-depth interviews. The participants' professions included students, engineers, video game artists, librarians, and multimedia marketers; their involvement scores were all above 51 points. Basic information of the participants is shown in Table 1.

Relevant vocabulary was obtained through participant interviews with the participants, and the collected semantics were converted into adjectives of psychological perceptions and style imagery through analyzing parts of speech. The adjectives were matched with antonyms to obtain 10 sets of emotional adjectives, namely "common-unique," "complex-simple," "obstructed-smooth," "strenuous-effortless," "curve-linear," "stable-lightweight," "unoriginal-creative," "handmade-technological," "cool-warm," and "nostalgic-futuristic".

Table 1. Basic information of the heavy gamer participants for in-depth interview.

NO	gentle	ages	occupation	Frequency (per week)	PII scores
1	male	25	Engineer	3	52
2	male	28	Game art	4	58
3	female	25	Engineer	7	69
4	male	24	Student	7	55
5	male	25	Librarian	3	55
6	female	24	Multimedia marketing	4	60

3.2 Game Controller Sample Selection

At this stage, samples were selected by first collecting various product images from manufacturers, the Internet and catalogs. Subsequently, considered the image resolution and similarity of product appearance, three professionals with industrial design experience for 5 years and more selected 12 representative samples from 41 samples (Fig. 2). This study aimed to explore the appearances of home-console game controllers. Therefore, at the beginning of this study, the brand characteristics of the samples were removed and their colors were displayed in grayscale to eliminate participants' preferences for brands and color combinations that may affect the accuracy of this experiment.

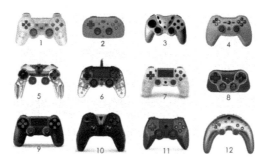

Fig. 2. Selected 12 representative samples.

3.3 Questionnaire Design and Experiment

The adjectives in Subsect. 3.1 were integrated with game controller samples of Subsect. 3.2 to form the main questionnaire. The experiment was then conducted online. A

5-point Likert scale was used to measure the adjectives, where 5 points indicate that the respondent strongly agrees with the adjective on the right, 3 points means the respondent is indifferent to the provided adjectives, and 1 point implies a strong agreement with the adjective on the left side.

The questionnaire was distributed on Internet, randomly and anonymous. Before formal questionnaire, there is a PII test, which mentioned in Subsect. 3.1, to group heavy or mid/low gamers. Therefore, 86 participants were collected, including 33 participants of heavy gamers, and 53 participants of mid/low gamers. Only 33 participants of heavy gamers will be analyzed, and the age range of heavy gamers was 18–38 years old (mean: 26.9 years old).

4 Morphological Analysis and Hayashi's Quantitative Theory Type I Analysis

Differences in the preferences of high and middle-low users for adjectives related to the appearance of game controllers were obtained from previous online questionnaires. However, this study only focused on morphological and quantitative analyses for heavy gamers. Key appearances of game controllers in the morphological analysis were analyzed and classified, and quantitative analysis was performed on 12 game controller samples with different classifications to determine the correlation between heavy gamers' emotional adjectives and the appearance of game controllers.

4.1 Morphological Analysis

Morphological analysis, first used in mechanical engineering, was an idea collection method developed by Zwicky and Wilson [13]. It mainly involves listing the independent elements of a particular object; variable parameters for each independent element are listed to compose a morphological chart for studying the creative ideas of all combinations in detail. The aim is to conceive all possible combinations. Most ideas can be covered, and powerful association combinations can also be used to test concepts that are usually not considered in ideas. This enables a rapid, cost-effective, and systematic organizational analysis. However, occasionally, the idea cannot converge in time after divergence, and whether the quality and quantity of the selected solution can meet specific requirements depends on assistance of those with professional knowledge [14]. Table 2 presents a morphological analysis of the appearance features of samples 1–12, and Table 3 details the component classification that records the morphological attributes of each sample.

4.2 Hayashi's Quantitative Theory Type I

Hayashi's quantitative theory was developed by Chikio Hayashi of the Institute of Statistical Mathematics of Japan. In Hayashi's original papers for predicting of phenomena from qualitative data and the quantification of qualitative data in terms of the mathematico-statistical point of view, there is a serial of papers [15], in which the methods of quantification of qualitative statistical data obtained by measurements and observations. In Japan, Hayashi's methods of quantification are well known and widely used in various fields, such as social and marketing surveys, psychological research, and medical research.

Table 2. Game controllers of the morphological analysis.

items	category		
A grip length	A1 shorter grip		A2 longer grip
B chamfering level	B1 lower		B2 higher
C upper edge	C1 straight	C2 arced	C3 concave
D lower edge	D1 straight		D2 convex
E panel style	E1 simple		E2 complex
F transparent case	F1 transparent		F2 non-transparent

Tanaka [16] reviewed Hayashi's first-fourth methods of quantification. Based on cases with or without external criterion, methods of quantification are divided into two main classes. In first class, case with an external criterion is for prediction or analyzing the effects of factors, and there are First method and Second method. The external criterion of first method is observed quantitatively, and is to maximize the correlation coefficient [9]; on the other hand, the external criterion of second method is observed qualitatively, and is to maximize the correlation ratio [17]. Moreover, in second class, case with no external criterion is for classification or constructing a spatial configuration. The second class contains Third method and Fourth method. Third method is to maximize the correlation coefficient between subjects and categories [18]; whereas, Fourth method is to maximize the objective function, and Similarities between pairs of subjects are observed quantitatively.

It is mostly used for appearance feature investigation, emotional discussions, and style identification [9]. Feature data such as degree, state, yes–no, and true–false are assigned numerical values for multivariate analyses including multiple regression analysis, principal component analysis, or discriminant analysis. To identify the approximate functional relationship between a certain variable (target variable) and other qualitative item groups (used 0 or 1 as dummy variables), multiple regression analysis is adopted to

Table 3. The morphological attributes of each sample.

items	sample category	1	2	3	4	5	6	7	8	9	10	11	12
A grip length	A1 shorter grip	o	o	o	o		o	o	o	o		o	
	A2 longer grip					o					o		o
B chamfering level	B1 lower		o		o					o			
	B2 higher	o		o		o	o	o	o			o	o
C upper edge	C1 straight		o			o	o		o				
	C2 arced												o
	C3 concave	o		o	o		o			o	o	o	
D lower edge	D1 straight		o						o				
	D2 convex	o		o	o	o	o			o	o	o	o
E panel style	E1 simple	o	o	o	o			o		o		o	o
	E2 complex					o	o		o		o		
F transparent case	F1 transparent	o				o	o						
	F2 non-transparent		o	o	o				o	o	o	o	o

determine the effect of each qualitative item on the target variable. Each item is composed of several categories. It is assumed that all samples have to be selected for each item and participants could only select one; such method can be used to establish regression formulas to predict the variance of data and events. The statistical relationship between imagery response and morphological attributes were obtained from Hayashi's quantitative theory type I. The morphological effect can be identified from relevant factors obtained from imagery that is conducive to design. The partial correlation coefficient is the weight of each item; a larger value indicates that the imagery has a more profound effect, and category scores can show the effect of a category on an item. Values are either positive or negative, representing the positive and negative effects of an item. A higher value indicates a greater effect.

In previous Sect. 4.1, in Table 3, 12 samples have been categorized by the morphological attributes. In order to processing the calculation of Hayashi's Quantitative Theory Type I [15], we reformatted the data from Table 3 into Table 4. For example, the morphological attribute of type A (grip length) is including 2 sub-categories, A1 (shorter grip), and A2 (longer grip) in Table 4; thus, in Table 5, type A is transfer to X1, and sub-category A2 is recorded as 2 in sample 5, 10, and 12. The rest of samples are recorded as 1. Another example, type C (upper edge) in table 4 is including 3 sub-categories, C1 (straight), C2 (arced), and C3 (concave); thus, type C is transfer to X3, and sub-category C1 is recorded as 1, C2 is recorded as 2, and C3 is recorded as 3, respectively.

Table 4. The morphological attributes of each sample.

	X1 grip length	X2 chamfering level	X3 upper edge	X4 lower edge	X5 panel style	X6 transparent case
Sample 1	1	2	3	2	1	1
Sample 2	1	1	1	1	1	2
Sample 3	1	2	3	2	1	2
Sample 4	1	1	3	2	1	2
Sample 5	2	2	1	2	2	2
Sample 6	1	2	1	2	2	1
Sample 7	1	2	3	2	1	1
Sample 8	1	2	1	1	2	2
Sample 9	1	2	3	2	1	2
Sample 10	2	1	3	2	2	2
Sample 11	1	2	3	2	1	2
Sample 12	2	2	2	2	1	2

4.3 Analysis Results of Emotional Attributions

After prepared Independent variables (X1 to X6) in Table 5 and Dependent variables (Y1 to Y10) in Table 5, we calculated the results based on Hayashi's Quantitative Theory Type I [15], and explained correlation coefficient of 10 emotional adjectives in following section.

Table 5. The morphological attributes of each sample.

Sample	Y1 Common-Unique	Y2 Complex-Simple	Y3 Obstructed-Smooth	Y4 Strenuous-Effortless	Y5 Curved-Linear	Y6 Stable-Lightweight	Y7 Unoriginal-Creative	Y8 Handmade-Technological	Y9 Cool-Warm	Y10 Nostalgic-Futuristic
1	1.9	4.6	4.3	4.2	2.7	3.5	2.0	3.8	2.4	2.6
2	2.8	4.1	3.5	3.5	3.2	3.3	2.8	3.2	2.9	2.4
3	4.2	2.3	3.1	2.8	1.9	2.5	3.6	4.3	2.6	4.0
4	3.0	3.2	3.5	3.3	2.8	2.5	2.8	3.6	2.9	3.1
5	4.9	1.8	3.0	3.0	3.0	3.7	4.8	4.8	1.8	4.8
6	3.9	2.8	3.4	3.0	2.2	3.1	3.6	4.2	2.5	3.8
7	2.5	4.2	3.9	3.9	2.4	3.5	2.6	3.6	3.0	2.8
8	3.4	3.8	3.7	3.6	3.3	3.2	3.1	3.2	2.8	2.9
9	2.2	4.1	4.2	4.0	2.6	3.3	2.4	3.5	2.8	2.8
10	4.3	2.9	3.2	3.0	3.4	2.8	3.9	4.4	2.8	4.1
11	3.2	4.2	3.8	3.8	3.2	3.1	3.1	3.5	2.9	3.3
12	4.7	3.5	3.7	3.4	1.4	3.6	4.5	4.5	2.0	4.5

2. Analysis Results of "Complex-Simple" Vocabulary: The analysis results in Table 6 reveal that the multiple correlation coefficient of "complex-simple" vocabulary for

heavy gamers was 0.843, indicating that the explanatory power was 84.3%. Among the six items, the most relevant feature of "complex-simple" vocabulary observed from the partial correlation coefficient was the lower edge of the game controller body (coefficient: 0.704). This implied that a straight lower edge produced a simple feeling, whereas a convex lower edge elicited a complex feeling. In addition, partial correlation coefficient of casing transparency was 0.581. Transparent casing gave a simpler feeling to heavy gamers, whereas opaque casing generated a complex feeling.

Table 6. The multiple correlation coefficient of "common-unique" and "complex-simple".

items	category	partial correlation coefficient	Common ⇔ Unique	partial correlation coefficient	Complex ⇔ Simple
A grip length	A1 shorter	0.145	-0.066	0.002	-0.001
	A2 longer		0.145		0.002
B chamfering level	B1 lower	0.018	0.018	0.188	-0.164
	B2 higher		-0.005		0.055
C upper edge	C1 straight	0.555	0.555	0.570	-0.821
	C2 arced				0.404
	C3 concave		-0.370		0.411
D lower edge	D1 straight	0.526	-1.031	0.704	1.699
	D2 convex		0.526		-0.334
E panel style	E1 simple	0.576	-0.263	0.280	0.144
	E2 complex		0.576		-0.288
F transparent case	F1 transparent	0.582	-0.677	0.581	0.688
	F2 non-trans		0.582		-0.229
			*multiple correlation coefficient =0.888		*multiple correlation coefficient =0.843

3. Analysis Results of "Obstructed-Smooth" Vocabulary revealed a multiple correlation coefficient of 0.783, indicating that explanatory power reached 78.3%. Among the six items, the partial correlation coefficient did not exceed 0.5, which did not show that any features were more relevant to perceptions of "obstructed-smooth" vocabulary.
4. Analysis Results of "Strenuous-Effortless" Vocabulary for heavy gamers has a multiple correlation coefficient of 0.773 (explanatory power: 77.3%). Among the six items, the most relevant feature of the "strenuous-effortless" vocabulary observed was the body's lower edge (partial correlation coefficient = 0.571), which meant that a straight lower edge elicited a feeling of effortlessness, whereas a convex lower edge produced a feeling of strenuousness.
5. Analysis Results of "Curved-Linear" Vocabulary for heavy gamers revealed a multiple correlation coefficient of 0.882 (explanatory power: 88.2%). The most relevant feature of "curved-linear" vocabulary observed was the upper edge of the game controller body (partial correlation coefficient = 0.758), demonstrating that a straight or arced upper edge made respondents feel that the controller is curved, and an arced shape produced more unique feelings than straight shapes did, whereas a concave appearance elicited a feeling of linearity.
6. Analysis Results of "Stable-Lightweight" Vocabulary for heavy gamers provided a multiple correlation coefficient of 0.894 (explanatory power: 89.4%). The most

relevant feature of this vocabulary combination was grip length, of which the partial correlation coefficient was 0.839. This suggested that a longer grip gave the heavy gamers a feeling of lightness, whereas a shorter grip generated a feeling of stability. Furthermore, the partial correlation coefficient of chamfering level was 0.772, showing that a higher chamfering level gave people a feeling of lightness, whereas a low chamfering level gave people a feeling of stability.

7. Analysis Results of "Unoriginal-Creative" Vocabulary for heavy gamers indicated a multiple correlation coefficient of 0.938 (explanatory power: 93.8%). Among the six items, the most relevant feature of the "unoriginal-creative" vocabulary observed from the partial correlation coefficient (0.691) was the upper body edge. A straight or arced upper edge gave people a more creative feeling; the straight appearance had a greater effect than an arc, whereas a concave appearance elicited a feeling of unoriginality feeling. In terms of lower body edge, the partial correlation coefficient was 0.625, indicating that a straight appearance at the lower edge elicited a feeling of unoriginality, whereas a convex evoked a feeling of creativity. In addition, the partial correlation coefficient of grip length was 0.521, which implies that a longer grip gave heavy gamers a feeling of creativity, whereas a shorter grip elicited a feeling of unoriginality.

8. Analysis Results of "Handmade-Technological" Vocabulary for heavy gamers was 0.924 (explanatory power: 95.4%). Among the six items, the most relevant feature of the "handmade-technology" vocabulary observed from the partial correlation coefficient was grip length, with a coefficient of 0.723. This indicated that a shorter grip gave users elicited a perception of craftmanship, whereas a longer grip elicited a technological feeling. The second most relevant feature was lower body edge, with a partial correlation coefficient of 0.716. A straight lower edge elicited a perception of craftmanship, whereas a convex lower edge evoked a technological feeling.

9. Analysis Results of "Cool-Warm" Vocabulary for heavy gamers revealed a multiple correlation coefficient of 0.914 (explanatory power: 91.4%). The most relevant feature of the "cool-warm" vocabulary observed from the partial correlation coefficient was grip length. The partial correlation coefficient of this item was 0.773, which revealed that a longer grip gave heavy gamers a feeling of coolness, whereas a shorter grip elicited a warm feeling. In addition, the partial correlation coefficient of upper body edge was 0.712, signifying that a straight upper edge produced a cool feeling and an arc or concave upper edge generated a warm feeling.

10. Analysis Results of "Nostalgic-Futuristic" Vocabulary for heavy gamers have a multiple correlation coefficient of 0.943, showing that the explanatory power was 94.3%. The most relevant feature of the "nostalgic-futuristic" vocabulary observed from the partial correlation coefficient is body lower edge, with a coefficient of 0.769. This demonstrate'd that a straight lower edge evoked a feeling of nostalgia, whereas convex lower edge elicited a futuristic feeling. Moreover, the partial correlation coefficient of casing transparency was 0.645, which meant that a transparent casing illustrated a nostalgic feeling, while an opaque casing displayed a future feeling.

5 Design Concept with Kansei Engineering and Affective Computing

After the emotional adjectives and quantitative analysis results of heavy gamers, there are two parts for utilizing emotional attributes to generate design concepts. First part: reconstructing game controller by parametric variables; and second part: generating design concepts based on multiple correlation coefficient.

5.1 Re-constructing Game Controllers by Control Variables

The key features of game controllers are curves of 2D profiles and then constructing 3D models. Based on morphological analysis in Table 5, the fundamental curves to form 2D profile are type A: grip length, type B: chamfering level, type C: upper edge, and type D: lower edge. On contrary, type E and F are panel style and transparent case, which are regardless of 2D profile. Thus, in order to re-construct 2D profile of game controllers by Grasshopper, morphological attributes from type A to D in Table 5 have to be disassembled into parametric variables. There are five parametric variables (i.e. upper edge, lower edge, grip length, chamfering level, and chamfering point in Table 12) were used to re-construct a game controller model. Thus, Grasshopper was used to control parametric variables in Rhino to construct an initial game controller model in Fig. 3.

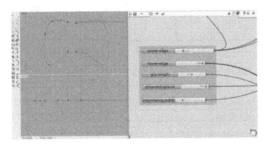

Fig. 3. The initial game controller model re-constructed by Grasshopper.

5.2 Generative Appearance with Emotional Attributes

The Hayashi's quantitative theory type I analysis results of heavy gamers conducted the Stage 2 were applied, and each set of emotional adjectives had a partial correlation coefficient to indicate the effect of appearance components on adjectives. These were used as parameters to construct a game controller model with emotional attributes. Taking the "unoriginal-creative" vocabulary as an example, the emotional value of this vocabulary can be adjusted from 1 (strongly unoriginal) to 5 (strongly creative). Values close to 1 indicated a feeling of unoriginality, whereas values close to 5 indicated a feeling of creativity.

The tendency of vocabulary choice affected the manifestations of grip length, chamfering level, upper edge, and lower edge. The values were determined based on the

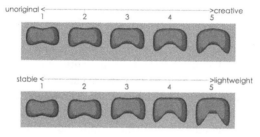

Fig. 4. The changes between "unoriginal-creative" and "stable-lightweight".

quantitative results from experiment in Sect. 3.3 shows that the changes were obvious in "unoriginal-creative" and "stable-lightweight" vocabulary. Moreover, the results will be dynamic based on questionnaires from different groups of users, different countries of people, or numbers of users. Nevertheless, emotional vocabularies form users such as "unoriginal-creative", "stable-lightweight" were used to generate design ideas preferred by heavy gamers (Fig. 4).

6 Conclusions

This study focused on the emotional preferences of heavy gamers in terms of game controller design and employed the generative design system to determine the correlation between the appearance of a game controller and emotional preferences. Heavy user's emotional attributes were analyzed in the initial stage of concept design to produce design concepts preferred by users as a reference for game controller designers.

Ten relevant emotional vocabulary items were obtained from interviews and related studies to conduct experiments related to controller appearance. The emotional vocabulary items were "common-unique," "complex-simple," "obstructed-smooth," "strenuous-effortless," "curved-linear," "stable-lightweight," "unoriginal-creative," "handmade-technological," "cool-warm," and "nostalgic-futuristic." Pictures of game controllers on the market were collected and filtered to 12 samples by experts, and an online questionnaire was constructed using the samples and vocabulary items.

The 12 samples were analyzed through a morphological analysis to obtain five items (i.e., grip length, chamfering level, upper edge of controller body, lower edge of controller body, and panel style). Finally, Hayashi's quantitative analysis was performed, and the results of heavy gamers indicated that grip length could be divided into long and short. The emotional attributes of a long grip was light, creative, technological, and cool; a short grip was considered stable, unoriginal, handmade, and warm. In addition, chamfering level was divided into high and low. The feelings elicited by a higher chamfering level were light and cool; a lower chamfering level was associated with feelings of stability of warmth. Moreover, the upper edge of the controller body was divided into straight, arc, and concave. The straight appearance elicited unique, complex, curved, light, creative, cool, and futuristic feelings; the arc provided unique, simple, curved, stable, creative, warm, and futuristic feelings. The concave appearance elicited feelings of commonalty, simplicity, linearity, stability, unoriginality, warmth, and nostalgia. Furthermore, the lower edge of the controller body was classified into convex and straight appearances.

The convex appearance indicated unique, complex, strenuous, curved, stable, creative, technological, cool, and futuristic feelings, whereas the straight appearance suggested common, simple, effortless, linear, light, unoriginal, handmade, warm, and nostalgic feelings. Finally, the panel style was divided into simple and complex. A simple panel style evoked common, effortless, light, cool, and nostalgic feelings, whereas a complex panel style elicited unique, strenuous, stable, warm, and futuristic feelings.

A game controller model was constructed using the control variables of five appearance features, namely upper edge, lower edge, grip length, chamfering level, and chamfering point. In addition, different emotional attributes (e.g., unoriginal, creative, stable, and light) were included to produce design ideas preferred by heavy gamers. Finally, appearance recommendations of this study are expected to help researchers in this field better understand the relationship between game controller appearances and emotional attributes. This brings designs closer to the needs and preferences of users and changes the process of CAD for concept development.

References

1. Fay, S.: Frog: Form Follows Emotion. Thames and Hudson, London (1999)
2. Krish, S.: A practical generative design method. Comput.-Aided Des. **43**(1), 88–100 (2011). https://doi.org/10.1016/j.cad.2010.09.009
3. Govers, P.C.M., Schoormans, J.P.L.: Product personality and its influence on consumer preference. J. Consum. Market. **22**(4), 189–197 (2005)
4. Lin, S.: The study of touch interface integration on game machine. Master Thesis of Institute of Electronic Engineering National Chin-Yi University of Technology (2012). (in Chinese)
5. Baumann, K., Thomas, B.: User Interface Design for Electronic Appliances. Taylor & Francis, Landon (2001)
6. Evan-Amos: File:N64-Controller-in-Hand.jpg. Retrieved from wikimedia: https://commons.wikimedia.org/wiki/File:N64-Controller-in-Hand.jpg (2011)
7. Nagamachi, M.: Kansei Engineering: a new ergonomic consumer-oriented technology for product development. Int. J. Ind. Ergon. **15**, 3–11 (1995)
8. Shih, Y. L.: A Study on Automatic Generation Design by Knowledge-Based of Parametric Features. Master Thesis of National Yunlin University of Science and Technology (2008). (in Chinese)
9. Chen. C.: Exploring the relationship between the form feature and the style evolution--using loop chair designs as examples. Kaohsiung Normal Univ. J. Sci. Technol. (19), 27–4 (2005). (in Chinese)
10. Neter J., Kuter, M.H., Nachtsheim, C.J., Wasserman, W.: Applied Linear Regression Model. Irwin 3rd edition (1990)
11. Khan, S., Gunpinar, E., Dogan, K.M.: A novel design framework for generation and parametric modification of yacht hull surfaces. Ocean Eng. **136**, 243–259 (2017)
12. Zaichkowsky, L.J.: The personal involvement inventory: reduction revision and application to advertising. J. Advert. **23**, 59–70 (1994)
13. Zwicky, F., Wilson, A.G. (eds.): New Methods of Thought and Procedure. Springer Berlin Heidelberg, Berlin, Heidelberg (1967). https://doi.org/10.1007/978-3-642-87617-2
14. Tung, T.-C., Li, C.-F.: A study of the relation between visual communication design and Kansei engineering. Jo. Des. Res. **3**, 214–221 (2003). (in Chinese)
15. Hayashi, C.: On the quantification of qualitative data from the mathematico-statistical point of view. Ann. Inst. Stat. Math. (2), 35–47 (1950). https://doi.org/10.1007/BF02919500

16. Tanaka, Y.: Review of the methods of quantification. Environ. Health Perspect. **32**, 113–123 (1979)
17. Suzuki, T., Kudo, A.: Recent application of quantification II in Japanese medical research. Environ. Health Perspect. **32**, 131–141 (1979)
18. Hayashi, C.: Quantification method III or correspondence analysis in medical science. Ann. Cancer Res. Therapy **1**(1), 17–21 (1992)

Applying Design Puzzle and Kansei Analysis to Analyze the Emotional Factors in the Bionic Modeling Wooden Toy Design Process by Designers

Chen-Syuan Lin[1] and Teng-Wen Chang[2(✉)]

[1] Graduate School of Design, National Yunlin University of Science and Technology, Yunlin, Taiwan
[2] Department of Digital Media Design, National Yunlin University of Science and Technology, Yunlin, Taiwan
tengwen@yuntech.edu.tw

Abstract. Design is a process where designers apply their experience and knowledge to solve problems. The psychological feelings, emotions, and other Kansei (sensibility) factors that arise during this process indirectly influence the design behavior patterns and the final design outcomes. This research aims to understand designers' Design Puzzles and Kansei sensibilities. Through a qualitative case study using an experimental approach, it utilizes the Design Puzzle framework and Kansei logic within digital media design to analyze whether a designer's design behaviors and Kansei factors have any relevant influences. The results show that designers tend to experience positive emotions during the requirement confirmation, conceptual design, and implementation stages, but face obstacles and negative feelings during task analysis, embodiment, and detailed design stages. **This revealed many latent needs and issues**, which proved highly helpful for improving the design process. It also provided new perspectives and directions for Kansei study.

Keywords: Design Puzzle · Puzzle Rule · Kansei Research · Design Process

1 Introduction

The design process is typically a journey of continuous exploration for designers, involving a tug-of-war between rationality and Kansei. This research goes beyond merely studying the designer's design process; it also investigates the feelings and emotions experienced by designers at various stages of the design process. The study primarily uses Design Puzzle from design computation and the logic of Kansei engineering to analyze designers' design behavior patterns, emotions, and Kansei sensations, aiming to understand the Kansei factors within the design process. It proposes more efficient design thinking and behavioral process recommendations, while also providing new research directions for design processes, design behavior patterns, and Kansei research.

1.1 Background

Design is a problem-solving process in which designers use their own experience and knowledge. From creative conception to the final result, the process is a rather complex cognitive activity. These cognitive activities are the psychological processes of understanding and solving problems through the information perceived by designers, using memory, thinking, reasoning, etc. Howard, Culley, and Dekoninck (2008 [1]) proposed six common stages of the creative process: (A) Establishing the Requirements, (B) Task Analysis, (C) Conceptual Design, (D) Embodiment Design, (E) Detail Design, and (F) Implementation. In each stage, due to the different understandings, definitions, perspectives, and design thinking patterns of designers on issues, their decision-making methods also differ, resulting in different design outcomes.

Chang, T.W. (2004) applied Archea's (1987) Puzzle Rule theory framework from Puzzle-Making to the design field and conducted research on designers' Design Puzzles. He specifically explained the four stages of designers' Design Puzzles: (1) strategy formulation, (2) exploratory observation, (3) trial and error, and (4) integration and verification. Designers integrate and verify their design works to ensure that the design meets the requirements.

From this, we can understand that the actions and feelings in a designer's Design Puzzle behavior also represent an interplay between rationality and Kansei. Just as Kansei Engineering (KE) is used to analyze and optimize design features that satisfy users' perceptions, designers employ a rational attitude to solve problems and achieve objectives on one hand, while utilizing Kansei to create distinctive design outputs on the other. In the field of design, the quality of design outcomes largely depends on the designer's design cognition and behavior. Therefore, in addition to conducting Kansei research on consumers' perceptions of objects, it is equally important to study the Kansei aspects of designers' design processes.

1.2 Research Motivation

Current Kansei Engineering research mainly focuses on users' Kansei towards objects, but designers also interact with various objects or events during the design process, during which they experience various perceptions and Kansei feelings. Since design behavior is quite complex and outcomes differ from person to person, analyzing designers' Kansei using the Design Puzzle and Kansei logic may help unravel the complex design behavior patterns of designers, as well as the complex psychological processes and emotional fluctuations, i.e., Kansei factors, that designers experience during the design process. By understanding the Kansei factors in designers' design processes, it should help designers identify implicit problems or steps that affect design and improve the design process and outcomes.

1.3 Research Purpose

To understand designers' design behavior patterns, this study conducted a design practice observation experiment in the form of a case study. With the theme of "Bionic Modeling Wooden Toy Design," the expected design process of designers during practical design was recorded using qualitative research methods and classified according to the six common stages of the creative process proposed by Howard, Culley, and Dekoninck (2008). The Design Puzzle concept was used to analyze designers' design behaviors. Through interviews, designers' Kansei factors during the product design process were obtained. Finally, the Kansei research logic was used to analyze designers' design behavior patterns. The purpose is to provide designers with more efficient design thinking and behavior flow recommendations during design, as well as to provide a new topic for discussion in Kansei research and design cognition research.

2 Literature Review

2.1 Design Puzzle and Kansei Engineering

The Design Puzzle refers to the interactive puzzle-solving process that designers go through while solving design problems. In his 1987 work "Puzzle-Making," Archea proposed that during the design process, designers would engage in behaviors such as constant exploration, observation, and trial and error to find solutions to design problems. These behaviors are like puzzle pieces in a puzzle game, where the designer has to continuously combine and fit them together to form a complete design work.

Chang, T.W. (2004) applied the Puzzle Rule theory framework to the design field and divided designers' Design Puzzles into the following four stages: (1) Strategy formulation: Designers formulate design strategies based on the goals and constraints of the design problem. (2) Exploratory observation: Designers observe and research to understand the background and requirements of the design problem. (3) Trial and error: Designers find solutions to the design problem through continuous trial and error. (4) Integration and verification: Designers integrate and verify their design works to ensure that the design meets the requirements.

From the above, we can see that during the design process, in addition to rationally solving problems, designers' personal feelings and intuitive judgments, including their understanding of design issues, aesthetic judgments, personal preferences, and even emotional changes during the design process, also affect the final design results like design tools. The entire design process is a process of cognition, Kansei, and emotional changes, which is consistent with the concept of Kansei Engineering research through understanding human Kansei and cognition.

2.2 Puzzle Rule

The Puzzle Rule refers to the rules or conditions that are followed when solving puzzles. These rules describe how to satisfy the puzzle goal based on the information provided by the given hints.

Archea (1987) proposed that designers use four thinking methods to solve design problems: (1) Induction: Deriving general rules or principles from individual observations. (2) Deduction: Inferring specific applications from general rules or principles. (3) Analogy: Finding clues to solve problems from similar situations. (4) Reverse thinking: Thinking about the answer to the problem from the opposite direction.

Chang, T.W. (2004) applied the Puzzle Rule concept to design learning, using specific steps to help students gradually explore and discover new design solutions, thereby enhancing their design abilities and creativity. The five steps of puzzle-solving are: (1) Define the goal: Determine the design purpose to be achieved. (2) Provide hints: Use the design element information generated during the design process to assist in completing the design. (3) Formulate rules: Formulate rules or conditions based on the hints to satisfy the design purpose. (4) Explore solutions: Evaluate different solutions based on the Puzzle Rule to see if they satisfy the purpose. (5) Adjust rules: If the solutions cannot satisfy the design purpose, adjust the rules.

To effectively complete a puzzle, everyone has their own set of Puzzle Rules. By understanding the goal and the hints from the puzzle event, and incorporating personal judgments and intuitive responses to the hints, the goal can be achieved. For designers, the Puzzle Rule is like planning the design process, with each step having a predetermined goal to be completed by the designer. How to determine the end of a step and proceed to the next process step relies on the hints obtained by the designer during the Design Puzzle process. When there are enough hints, the next step can be taken. The relationship between the designer, rule, hint, and goal can be seen in Fig. 1.

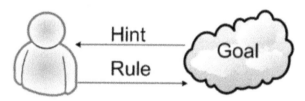

Fig. 1. Relation between hints and rules (Chang, 2004).

2.3 Creative Process

Design behavior is a creative process. This study refers to the six common stages of the creative process proposed by Howard, Culley, and Dekoninck (2008) to help classify the design process of the research subjects. The six stages of the creative process are:

- **(A) Establishing Requirements:** Clarifying the problem and goals, including understanding performance, functionality, user needs, and other relevant factors.
- **(B) Task Analysis:** In-depth analysis of the task, considering details in actual operation.
- **(C) Conceptual Design:** Generating and evaluating various design concepts, including sketches, concept models, or other forms of preliminary solutions.
- **(D) Embodiment Design:** Developing the selected concept into a concrete design, including details of structure, materials, and interactions.
- **(E) Detail Design:** Further focusing on design details, including engineering drawings, technical specifications, and other necessary documents for manufacturing and implementation.

- **(F) Implementation:** Translating the creative design into reality and ensuring that requirements and standards are met. This is an important stage for transforming creativity into actual value.

3 Research Methods

This study refers to Chang, T.W.'s (2004) research concepts, applying the Design Puzzle and Puzzle Rule theory framework from design computing to the design field. The logic concept of Kansei Engineering is used to analyze designers' Kansei thoughts and needs during the product design process, in order to summarize and identify the Kansei factors that influence the designers' design process.

3.1 Research Planning

1. **Selection of research subjects:** This study focuses on individuals who completed a product design graduation project during their university years and have graduated within the last 5 years as the research subjects.
2. **Research tools:** Textual records and post-interviews were adopted. Recording the designers' design process steps facilitated the discovery of the designers' Design Puzzle patterns, while interviews helped understand the emotional changes and Kansei feelings experienced by the designers after actually carrying out the design.
3. **Research analysis:** Qualitative research was employed, and Design Puzzle and Kansei logic were used for analysis to summarize and identify relevant phenomena.

3.2 Research Framework

This study adopted an expert pre-test and case study approach for experiment planning. First, professional designers were invited to participate in a preliminary test to facilitate the planning of the experiment content. Then, three designers were selected as research subjects, with the theme of "Bionic Modeling Wooden Toy Design" for practical implementation. Qualitative research methods were used to record participants' design steps in text form, and interviews were conducted to obtain the designers' Kansei feelings

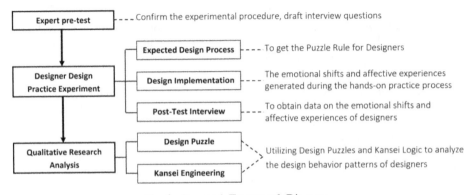

Fig. 2. Research Framework Diagram.

during the creative process. This data was organized and analyzed to facilitate the analysis of the designers' design behavior patterns. Please refer to Fig. 2 for the research framework diagram.

3.3 Process Description

3.3.1 Expert Pre-test

To ensure the smooth conduct of the experiment, this study invited a professional designer with more than 5 years of practical design experience to participate in a pre-test following the process pre-planned for the designers in this research. The main purpose was to confirm the reasonableness of the experiment planning and to refer to the expert's opinions in formulating the interview content. The five steps and content descriptions of the pre-test are as follows:

1. **Design Theme Explanation:** The theme was "Bionic Modeling Wooden Toy Design." An introduction to "what is bionic design" was provided, and photographs of related products on the market were used to explain the preset types and styles of the toys to be completed.
2. **Design Process Planning:** Write down the expected design process steps.
3. **Design Implementation:** Carry out the design implementation according to the planned design steps.
4. **Post-Interview:** Inquire about the emotional feelings during the execution process, step by step, according to the expected design process steps.

3.3.2 Design Implementation Experiment for Designers

Referring to the pre-test results and expert suggestions, the process of the design implementation experiment remained unchanged, but the content was modified. The final experimental steps and content descriptions are as follows:

1. **Design Theme Explanation:** First, explain that the theme is "Bionic Modeling Wooden Toy Design." Then, provide a detailed explanation to the designers about the knowledge required for the theme, including: an introduction to bionic design, the target age group, material limitations, manufacturing method restrictions, and timeline. Finally, use photographs of related products on the market to explain the preset types and styles of the toys to be completed.
2. **Design Process Planning:** Write down the expected design process steps, the objectives of each step, and the key factors for proceeding to the next step.
3. **Design Implementation:** Carry out the design implementation according to the planned design steps.
4. **Post-Interview:** First, introduce the six common stages of the creative process proposed by Howard, Culley, and Dekoninck (2008), and explain the objectives and content to be achieved in each stage. Then, ask the designers about their emotional feelings during the execution of each stage. For example: (A) Establishing Requirements: The main focus of this stage is to clarify the problem and objectives, including an understanding of performance, functionality, user needs, and other relevant factors. How did you feel or what emotions did you experience during these steps? What could be the reasons for this?

4 Experimental Results and Analysis

Before the experiment began, the 8 designers received a detailed explanation of the design theme, including the design scope, bionic modeling concept, and material constraints. First, the designers were asked to list their expected design process, objectives, and the key factors for proceeding to the next step in written form. Then, they carried out the design implementation and completed the product. Finally, interviews were conducted to inquire about their Kansei feelings during the process. Qualitative research analysis was performed using the Design Puzzle concept and Kansei logic from digital media design to analyze the designers' design behavior patterns and Kansei factors.

4.1 Designers' Expected Design Processes

To analyze the designers' expected design processes, this study classified the expected design processes formulated by the designers according to the six common stages of the creative process proposed by Howard, Culley, and Dekoninck (2008): (A) Establishing Requirements, (B) Task Analysis, (C) Conceptual Design, (D) Embodiment Design, (E) Detail Design, and (F) Implementation. Please refer to Tables 1 and 2 for the results of the data classification.

From Tables 1 and 2, it can be observed that among the 8 research subjects' expected design processes, only 3 have all six complete creative processes. Of the remaining 5, 4 lack (D) Embodiment Design and (E) Detail Design. One lacks (A) Establishing the Requirements. For 6 subjects, the steps of (B) Task Analysis plus (C) Conceptual Design account for more than half of their total expected design process steps.

It can be noted that when most research subjects plan their design, they focus on clarifying problems, confirming needs and objectives, proposing design concepts, and evaluating design concepts, hoping that the design will ultimately enter the final production and mass production smoothly. However, they simultaneously lack the structure, materials, and their interactions needed when developing concepts into concrete designs, as well as attention to design details, including engineering drawings, technical specifications, and other process requirements.

The 3 research subjects with all six complete creative processes differ from the other 5 in their background and identity, having richer work experience of about 3 years. Therefore, it can be inferred that designers with more accumulated work experience have more comprehensive planning of the design process.

4.2 Designers' Design Puzzles

This study consolidated the designers' expected design processes, objectives, and key factors for proceeding to the next step. During the experiment, the designers did not fully understand the "key factors for proceeding to the next step" and raised inquiries about it. The study provided an explanation, requesting the designers to identify the key events and content descriptions that would indicate the completion of the current process step and the transition to the next step. Due to space constraints, this study only provides the recording method for Case A (Table 3) as a reference.

Table 1. Expected Design Process and Creative Phase Sorting for Cases A-D.

Case A		Case B		Case C		Case D	
Expected Design Process	Sort	Expected Design Process	Sort	Expected Design Process	Sort	Expected Design Process	Sort
1.Confirm requirements	A	1.Confirm topic content	A	1.Confirm limitations	A	1.User analysis	B
2.Information Gathering	B	2. Data collection	B	2.Data collection	B	2.Questionnaire survey	B
3.Analyze and synthesize data	B	3.Interview survey, questionnaire	B	3.Sketch drawing, proposal classification	C	3.Creative ideation	C
4.Identify the key words	C	4. Sketching	C	4.Size determination	C	4.Market analysis	B
5.Initial sketch ideation	C	5. Sketch modification	C	5.Rough model	C	5.Detail confirmation	B
6. Simple convergence	C	6.Rough model	C	6.Rationality adjustment	C	6.Sketch design and finalization	C
7. Sketching	C	7.Rough model modification	C	7.Confirm styling	C	7.Rough model creation	C
8. Refined draft (1–2 pieces)	C	8. Formal model creation	C	8.Operation confirmation	C	8.Refined model creation	D
9. Rough model	C	9. Model evaluation (expert, user)	C	9.Modify rough model	C	9.Market verification and product optimization	E
10. Finished product creation	F	10. Finished product creation	F	10.Finished product creation	F	10.Finished product creation	F

Through qualitative analysis and comparing the experimental records of 8 cases, the design patterns of the designers are: rationally setting goals and evaluating results through Kansei. The designers' explanations of the goals are relatively rational, while the key evaluations for proceeding to the next step are more Kansei-based. For example, in Case A during the initial ideation phase, the goal was to piece together ideas from keywords and draw concrete images. However, the key to proceeding to the next step was: reaching 10 creative ideas (a self-set number), or until a design the designer liked based on Kansei was drawn. When entering the convergence phase, although the goal

Table 2. Expected Design Process and Creative Phase Sorting for Cases E-H.

Case E		Case F		Case G		Case H	
Expected Design Process	Sort	Expected Design Process	Sort	Expected Design Process	Sort	Expected Design Process	Sort
1. Topic ideation	A	1. Market observation	B	1. Design requirements	A	1. Confirm requirements	A
2. Data collection	B	2. Data collection	B	2. Data collection	B	2. Initial analysis and design strategy	B
3. Organize data	B	3. Topic ideation	A	3. Target group interviews	B	3. Brainstorming and sketch ideation	C
4. Evaluate direction	B	4. Summarization	B	4. Project plan or specifications	B	4. 3D or 2D	C
5. Detailed analysis	B	5. Drafting	C	5. Sketch proposals	C	5. Model creation and testing	C
6. Draft sketching	C	6. Three-dimensional realization (modeling)	C	6. Simple model testing	C	6. Model modification	D
7. Modification and adjustment	C	7. Rough prototype creation	C	7. Target group testing	D	7. CMF and color planning	E
8. Rough model creation	C	8. User testing	D	8. Final decision	E	8. Packaging design	E
9. Rough model confirmation	C	9. Final decision	E	9. Model production outsourcing	E	9. Product finalization and cost estimation	E
10. Physical product creation	F	10. Finished product creation	F	10. Finished product creation	F	10. Enter mass production	F

was to consider cost and structure for modifications, the key to proceeding was through intuition and Kansei, selecting three ideas to develop sketches for (a self-set number).

Analyzing designers' Design Puzzle behavior patterns through the concept of Puzzle Rule, it is found that although goals can be set rationally, designers struggle to determine if these goals have been achieved. They typically rely on their own Kansei feelings (such as intuition, preferences, and sensations) to judge whether the obtained hints are sufficient

Table 3. Case A's Expected Process, **Objectives, and** Key Factors for Next Steps

Expected Design Process Step	Objective	Key Factor for Proceeding to the Next Step
Confirming Requirements	Confirming User Requirements for the Work/Product	Identify key factors you think are helpful, such as: user age, cost, sales locations
Data Collection	Collect relevant existing products on the market, conduct market analysis, categorize styles	Based on existing products, generate ideas and identify key words for product design, such as: cute style, educational…
Data Analysis and Integration	Consolidate the first two stages	Grasp existing market products and your own design direction
Identify key words	Define a more precise product positioning	This will lead to a clear product scope: design style, product attributes, rarity, popularity, price range
Conduct Preliminary Ideation (Sketching)	Combine each key word to create concrete visuals through sketching	Purely develop forms and shapes, approximately draw until about 10 creative ideas emerge or until a design you like appears
Refined Sketches (1–2 pieces)	Consider model production and feasibility for mass production manufacturers, with a clear form and function proposal for clients	Refine aesthetic form, 3D views, orthographic projections, structural explanations
Rough Prototyping (Choose One)	Confirm if the structure is executable. During the process, modify the design based on your own preferences and feedback from others	Create two rough prototypes - one focused primarily on client requirements, and the other based on the designer's/design team's preferences, for the client to select from
Final Product Manufacturing	Proceed with actual production	Actual operation and complete usage flow should have no more than three issues (operation problems, product damage, etc.)

to proceed to the next stage of the process. This research illustrates the designers' Design Puzzle patterns and their interrelationships in Fig. 3.

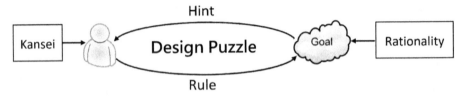

Fig. 3. Designers' Design Puzzle Relationship Diagram.

4.3 Designers' Kansei Factors

After the design practice experiment, this research conducted interviews with three designers, inquiring about their Kansei factors in each design process stage, and organized the interview results into Tables 4 and 5.

Table 4. Designer's Kansei Impressions Record for Cases A-D.

Stages of the Creative process	Case A	Case B	Case C	Case D
(A) Establishing the Requirements	Very happy, feeling a bit troubled and confused	Very happy, feeling troubled, afraid of the past	Eager to try, expectant, wanting to experiment	Feeling trusted and happy, anxious, irritated, unsure
(B) Task Analysis	Very troublesome, feeling pressure	Cautious, impatient, restless	Worried, feeling pressure	Sense of security, feeling elated
(C) Conceptual Design	Relieved, impatient, anxious, sense of urgency, nervous	Very happy, depressed, procrastinating	In a good mood, not troubled, generally positive emotions	A bit annoyed, afraid, worried
(D) Embodiment Design	Troublesome, feeling in need of help	Excited, troubled	Very troublesome, bothered	A little relaxed, feeling of reassurance
(E) Detail Design	Very nervous, pessimistic, curious, expectant	Irritated, relieved and relaxed	Anxious about the process, curious, unsure, unstable	Very troublesome, tedious
(F) Implementation	Fun, confident	Mostly expectant, uneasy	Expectant, excited, confident	Relieved, expectant, worried

Through the analysis of Table 4 and Table 5, which captures the novice designers' descriptions of their emotions and thoughts throughout the creative process, the Kansei results for each creative stage are summarized as follows:

Table 5. Designer's Kansei Impressions Record for Cases E–H.

Stages of the Creative process	Case E	Case F	Case G	Case H
(A) Establishing the Requirements	Excited, nervous and stressed, a bit anxious, struggling between ideal and reality	Grounded, seeking reassurance, feeling powerless	Very excited, puzzled, agitated, anxious, calm	Very happy, wanting to try multiple approaches, excited, dislike, dissatisfied
(B) Task Analysis	Very anxious, urgent, under great pressure, very happy	Motivated, fun, like solving a puzzle	Feeling more relaxed, facing seriously, mentally irritated, anxious	Very focused, facing rationally, not troublesome, irritable, happy steps, clear
(C) Conceptual Design	Very happy, unsure, worried	Nervous, worried, uncertain, feeling out of control, anxious and tense	Very excited and engaged, positive emotions, sense of loss, emotions change with completion level	Complex emotions, fluctuating feelings, very happy, confused, hesitant, mixed positive and negative emotions, uneasy
(D) Embodiment Design	Feeling the need to be cautious, very nervous, treading carefully	Full of curiosity, expectant	Feeling accomplished, satisfied	Feeling calm, confident, slightly satisfied, feeling very stable
(E) Detail Design	Wanting to escape, confused, feeling helpless, wanting to seek help, mixed emotions	Requiring full attention, numb, anxious, nervous	Calm, a bit nervous, emotions leaning towards positive, cautious, afraid of failing at the last moment	Feeling a bit irritated, finding it very troublesome
(F) Implementation	Feeling relieved, like a weight has been lifted, a bit nervous, feeling very happy, finally liberated	Relieved, waiting, problems encountered at this stage don't significantly affect emotions	Expectant, happy, also a bit anxious, worried	Very expectant, unsure, worried, satisfied

1. **Establishing the Requirements:** Happy to receive the design task, feeling a bit expectant and nervous, but finding the subsequent complex processes troublesome.

2. **Task Analysis:** Not good at analyzing large amounts of data, feeling troubled, worried about not being thorough enough which leads to anxiety, stress, and agitation.
3. **Conceptual Design:** When entering this stage, there's a sense of groundedness and reassurance due to having a design direction. Mood is very good when creative inspiration strikes. Facing project timelines creates a sense of urgency and pressure. Additionally, design requirements also generate feelings of restlessness and anxiety.
4. **Embodiment Design:** Feeling confident and relatively relaxed about progressing to this step after the design is finalized. When facing feasibility and practical considerations, it feels very troublesome and confusing, with a sense of needing help.
5. **Detail Design:** Due to lack of engineering expertise and the need to focus on more design details, feeling very nervous, pessimistic, irritated, and in need of assistance. There's a sense of uncertainty, along with anticipation and curiosity about feedback from clients or professionals.
6. **Implementation:** Feeling relieved, relaxed, finding it fun, expectant, and excited. There's a slight uneasiness before the product is completed, but full of anticipation and a sense of achievement regarding the feedback on the final product.

The data shows that designers' positive emotional responses come from: 1. Receiving a design task; 2. Thorough data collection and prior preparation; 3. Having creative inspiration; 4. Proposals being approved; 5. Design finalization and proceeding to detailed design; 6. Completion of the final product. These positive emotions motivate designers to face challenges, solve problems, and achieve goals.

Conversely, designers' negative emotional responses stem from: 1. Complex processes; 2. Lack of proficiency in data analysis; 3. Insufficient thoroughness in processes; 4. Pressure from time schedules; 5. Perfectionist demands on design; 6. Complex practical considerations; 7. Poor communication among various units and personnel; 8. Uncertainty about engineering requirements and detail considerations (engineering drawings, materials, mass production, etc.). When facing negative emotions, designers seek help through: 1. More data collection and analysis; 2. Seeking professional opinions.

Designers experience an interweaving of positive and negative emotions throughout the design process. Understanding these emotional changes can reveal where designers need assistance, thereby allowing for the creation of tailored training courses, tools, and resources to more effectively support their design practice. For example, during task analysis and detail design phases, they may feel confused and helpless about data analysis and engineering requirements, necessitating supportive resources and professional guidance.

This translation maintains the structure and detailed analysis of the original text, accurately conveying the complex emotional landscape of designers throughout their work process, as well as the implications for improving design practice and education.

5 Conclusion

This study employs qualitative research methods, utilizing Design Puzzle and Kansei logic in digital media design to analyze designers' Kansei factors and behavioral patterns when designing bionic wooden toys.

The research reveals that designers, in their expected design process, overemphasize (B) Task Analysis and (C) Conceptual Design while lacking the two critical stages of (D) Embodiment Design and (E) Detail Design, which transform concepts into concrete designs. This finding echoes the reality in most design teaching settings, where there is an overemphasis on creative ideation and a lack of practical engineering-oriented teaching. Moreover, due to limited practical experience, designers often focus excessively on task analysis and concept generation while neglecting the importance of the actual product output process. When assessing whether each process meets the standards, they heavily rely on subjective Kansei factors such as intuition and preferences. This reliance on Kansei evaluation also reflects designers' lack of clarity regarding the criteria for achieving their goals. This aligns with the study's finding that designers' Puzzle Rule features "rationally setting goals and Kansei-based evaluation of results," proving that the primary cause of variations in design outcomes lies in designers' Kansei behaviors and judgments.

By analyzing designers' Kansei feelings during the design process, positive Kansei factors can be identified: 1. Receiving the design task; 2. Sufficient data collection and preparation; 3. Having creative inspiration; 4. Proposal being affirmed; 5. Design finalization and readiness for detailed design; 6. Completion of the final product. Negative Kansei factors include: 1. Complex processes; 2. Lack of proficiency in data analysis; 3. Inadequate processes; 4. Pressure from project timelines; 5. Perfectionist requirements for the design; 6. Complex real-world considerations; 7. Poor communication among personnel from multiple units; 8. Uncertainty in engineering requirements and detailed considerations (engineering drawings, materials, mass production, etc.). Positive Kansei factors provide designers with the motivation to solve problems, while negative Kansei factors suggest that they have encountered issues in the design process and need assistance. The research findings indicate that designers are prone to problems and frustrations in stages B (Task Analysis), D (Embodiment Design), and E (Detail Design), generating more negative Kansei feelings. These emotions and Kansei behaviors imply that designers require support, and if resources and assistance can be provided in a timely manner, design quality can be optimized, and design efficiency can be improved.

The study later discovers that designers' design experience may influence the research conclusions. The research subjects are individuals who have completed a product design graduation project during their university years and have graduated within the past five years. Although they possess product design experience, based on their work experience and data, those with longer work experience tend to have a more complete design process, and their sense of achievement and satisfaction after completing a design comes from positive feedback from clients and the market. On the other hand, those with less experience are less proficient and thorough in handling complex data analysis and engineering considerations. However, once they reach the Implementation stage, they feel reassured and satisfied. Therefore, future research can be conducted to analyze designers with different levels of work experience, providing a more objective analysis to differentiate the Kansei factor differences in the design processes of novice and experienced designers.

This study confirms the feasibility of conducting Kansei research on designers using Design Puzzle and Kansei logic. The research results are helpful in understanding designers' design behavior patterns and uncovering many latent issues from their Kansei factors, which can contribute to improving the design process and developing design education curricula. It also provides new perspectives and directions for Kansei research.

References

Archea, J.: Puzzle-making: What architects do when no one is looking. Principles of computer-aided design: computability of design, pp. 37–52 (1987)

Carliss, Y.B.: Design rules: The Power of Modularity. MIT Press (2000)

Chang, T.-W.: Supporting design learning with design puzzles-some observations of on-line learning with design puzzles. In: Van Leeuwen, J.P., Timmermans, H.J.P. (eds.) Recent Advances in Design & Decision Support Systems in Architecture and Urban Planning, pp. 293–307. Kluwer Academic Publishers, Dordrecht (2004)

Chang, C.-C., Chang, T.-W.: Design Puzzibility: Design idea exploration based on design puzzles with deep learning. Human-Centered Design and User Experience 114 (2023)

Chang, C.-C., Chang, T.-W., Huang, H.-Y., Tsai, S.-T.: Discovering semantic and visual hints with machine learning of real design templates to support insight exploration in informatics. Adv. Eng. Inform. **59**, 102244 (2024)

Howard, T.J., Culley, S.J., Dekoninck, E.: Describing the creative design process by the integration of engineering design and cognitive psychology literature. Des. Stud. **29**, 160–180 (2008)

Huang, K.-L., Chen, K., Chang, J.: Kansei evaluation on the visual and hearing image of interface design. In: Kansei Engineering and Emotion Research International Conference (2010)

Li, M., Dai, Y.: Optimization strategies for the modular resource construction of art gallery's exhibition halls based on kansei engineering. IEEE Access **12**, 27870–27886 (2024). https://doi.org/10.1109/ACCESS.2024.3364751

Nagamachi, M.: Kansei engineering: a new ergonomic consumer-oriented technology for product development. Int. J. Ind. Ergon. **15**, 3–11 (1995)

Yang, L., Chang, T., Lin, C., Chen, H.: Exploring Visual Information with Puzzle Rule–A Design Collage Approach. ISARC2004. Korea (2004)

Cheng, Y.-B., Chang, T.-W., Yang, C.-K.: Generation of design ideas using edos-touch. Des. J. **22**, 169–190 (2019)

Ontology Construction and Sentiment Computation Analysis in Animated Series: A Case Study of Mobile Suit Gundam

Tse-Wei Hsu[✉] [iD]

Southern Taiwan University of Science and Technology, Tainan, Taiwan
dr.fatty.hsu@gmail.com

Abstract. This research involves the construction of structured event templates and the definition of metadata within the animated narrative. By integrating methods such as normalization, authority control, and content analysis, subjects appearing in events are extracted and their relationships defined to form the topological information structure of the story ontology for the animated work *Mobile Suit Gundam 0079*. This data structure, combined with screenshots from the work, is used as the event background and provided to ChatGPT for character sentiment analysis. The study explores whether providing event backgrounds influences ChatGPT's analysis of character emotions in the animation. In the case study, the emotional states of the characters Amuro and Char in episode 41 of *Mobile Suit Gundam 0079* are analyzed under different contextual information.

Keywords: Event Semantics · Sentiment Analysis · Animated Character Emotions

1 Introduction

1.1 Complexity of Character Emotional Layers in Animated Works

In multi-episode animated series, it is often challenging to comprehensively and simultaneously describe the actions and emotions of key characters in relation to their referenced objects within the narrative and emotional space. For instance, in the animated work *Mobile Suit Gundam 0079*, the interactions between main characters Char and Amuro throughout their multiple battles reveal complex and contradictory emotions. They harbor enmity as enemies while also displaying respect and understanding as rivals. Their confrontations on the battlefield involve not only intense combat across the entire battleground but also their individual internal struggles and the emotional conflicts between them. It is often ambiguous whether these battles reflect the broader battlefield scenario, the intertwining of their personal emotions, or merely their inner psychological contest.

    The multilayered emotions of animated characters not only enrich the characters themselves but also enhance the narrative depth of the entire series. As viewers watch these series, they can resonate more deeply with the characters' emotional development. This complexity in emotions transforms characters from mere heroes or villains into beings with genuine human traits and emotional depth. As the storyline progresses, the characters' emotional reactions in different contexts continuously reveal their true thoughts and motivations, allowing viewers to better understand and empathize with them.

1.2 Analysis of Character Emotions in Narratives

ChatGPT is a large language model developed by OpenAI, based on the GPT-4 architecture. Its purpose is to understand and generate natural language. It can simulate human conversations to answer questions, provide suggestions, and perform various natural language processing tasks. ChatGPT is capable of understanding characters' emotional backgrounds and narrative developments from the provided text. It can capture and analyze details of dialogues, events, and character interactions to infer characters' emotional states. Additionally, it can utilize image analysis functionalities to perform content analysis.

To enable large language models like ChatGPT to understand the emotional context of characters in animated series, it is first necessary to analyze narrative events and subsequently interpret the evolving relationship between these events and character emotions. To achieve this, the research must complete several tasks, including the preliminary design of an information model, the creation of a spatio-temporal database with metadata interpretation, and the establishment of authority control for relevant terminology and hierarchical relationships. This preparation is essential for building effective search mechanisms once large datasets are introduced. By employing a spatio-temporal model, ChatGPT can observe the spatio-temporal ontology created by the animated narrative. Conceptual, graphical, and list-based information will be provided to users to illustrate how ChatGPT understands the narrative and the development of character emotions.

Due to the availability of multiple Chinese versions of *Mobile Suit Gundam 0079* from Mainland China, Hong Kong, and Taiwan, there are various translation discrepancies. Beyond authority control, much of the data still requires manual correction to ensure accurate definitions of relationships between entities, proper presentation of events, and to prevent translation-related issues in the ChatGPT testing environment.

1.3 Research Objectives

Based on the above, this research project is divided into three parts: "Establishment of Data Architecture," "Collection of Narrative Data," and "Reconstruction of Animated Narratives," detailed as follows:

1. **Establishment of Data Architecture:** Spatio-Temporal Information Model and Metadata for the Spatio-Temporal Database.

   The spatio-temporal data, i.e., the normalized narrative events, are typically associated with specific character behaviors or particular entities. The creation of metadata aims to structure event information logically, enabling a more comprehensive understanding by the language model. Rigorous metadata definitions are essential for ensuring that all event and character data can be effectively utilized by the computational system.

2. **Collection of Narrative Data**: Analysis and Integration of Animated Events

Throughout the progression of the animated narrative, characters' emotional responses to events and dialogues necessary for narrative explanation—whether close-up expressions or editing logic—affect the language model's comprehension of the story. Collecting these sequential elements is crucial to this research project. Since translation or interpretation discrepancies can arise in naming and definition, authority control is a core task of this project to ensure consistency in the names presented in the database through the interrelation of terms.

3. **Reconstruction of Animated Narratives:** Establishment of Structured Events

Upon completion of the metadata tables for the spatio-temporal database, this research will define the relationships among various tables. By establishing rules and transformations, different elements of spatio-temporal data can be linked into sequences and contexts, forming a complete spatio-temporal information model. A visual interface will be constructed to establish event categories and relationships, presenting the development of animated narratives from different perspectives of specific characters.

2 Case Study

2.1 Overview of the Gundam Series

The "Gundam" animated series has produced over 50 different titles. Additionally, the series has expanded through novels, video games, and even physical model kits, creating a rich and diverse world and timeline. The Universal Century (UC) timeline is a pivotal time axis within the Gundam series and has had a profound impact on the development of contemporary robot animation in a broad sense. Gundam UC timeline works are expressed through various media, including TV anime, animated films, novels, video games, OVA (Original Video Animation), and streaming media. The use of different media platforms has allowed the Gundam UC timeline stories to be presented in a rich and multifaceted manner, achieving cultural influence on a global scale.

Among the UC timeline series, the earliest and the progenitor of all Gundam works is *Mobile Suit Gundam (*機動戦士ガンダム*)*, which originally aired on Japanese television from April 7, 1979, to January 26, 1980. The series consisted of 43 episodes, each approximately 30 min long, including commercials. This foundational series led to the creation of other well-known works such as *Mobile Suit Zeta Gundam (1985,* 機動戦士Zガンダム*), Mobile Suit Gundam: Char's Counterattack (1988,* 機動戦士ガンダム 逆襲のシャア*), and Mobile Suit Gundam Unicorn (2010,* 機動戦士ガンダムUC*).* However, the release times of these subsequent works do not align chronologically with the in-story timeline. Some are set in the distant future or serve as prequels. Many characters span different generational works or appear across multiple series. For this research, the original *Mobile Suit Gundam* will be referred to as *Mobile Suit Gundam 0079*.

The key scenes selected in this study encompass all relevant events involving Amuro and Char, including most of their battlefield interactions and related characters and events. This selection not only integrates the storyline by tracing the narrative progression of both characters but also serves as an essential metric for measuring the protagonists' growth.

2.2 Diversification and Confusion of Events and Characters

Due to *Mobile Suit Gundam 0079* featuring multiple intense battle scenes and recurring character conflicts, these scenarios can often be mistakenly mixed or overlapped in descriptions. For example, the confrontations between Char and Amuro across different battles may involve similar tactics and emotional expressions, leading to potential confusion in specific details when described by ChatGPT.

Furthermore, the main characters Amuro and Char in 0079 experience extensive character development across the entire Gundam series, spanning multiple works such as Mobile Suit Zeta Gundam and Mobile Suit Gundam: Char's Counterattack (1988). In these different generational works, the characters' behaviors and motivations may intersect, causing temporal misunderstandings in their emotional portrayal by both viewers and large language models.

2.3 Limitations in Capturing Emotional Depth

ChatGPT has limitations in capturing the depth of emotions, particularly when considering the complex emotional entanglements between Char and Amuro throughout the Gundam UC timeline. Their relationship is marked by both enmity and mutual recognition, which entails deep psychological motivations and emotional changes that simple narrative descriptions may fail to fully capture. This complexity often leads to an underestimation of the diversity and dynamics of character emotions.

For instance, in a simple test where ChatGPT is asked to analyze a character's emotional state based only on a screenshot (see Fig. 1) without providing background events—such as, "This is a screenshot from episode 41 'A Cosmic Glow' of the anime Gundam 0079, featuring the character Char. Can you analyze the character's emotional state?"—ChatGPT's response tends to focus on the following Fig. 2:

Fig. 1. Char (left) and Amuro's (right) initial reactions to Lalah's death in Gundam 0079

Without event context as a foundation, when ChatGPT analyzes Char's character in the screenshot from episode 41 of *Mobile Suit Gundam 0079*, it primarily relies on

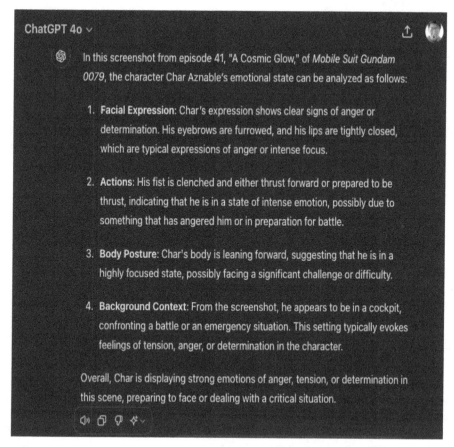

Fig. 2. Analysis of Char's emotions in episode 41 without background events

visual cues and common emotional expression patterns. The analysis methods include facial analysis (eyebrows, gaze, mouth), body posture, environmental clues, and common emotional patterns. This approach focuses on outward emotional indicators to infer the character's emotional state and motivations. While such analysis can provide some initial judgments about emotions, it may lack depth and accuracy, failing to fully capture the complexity of the character's internal emotional changes.

In Fig. 1, Char is experiencing the loss of his lover and comrade, Lalah. The subtle psychological portrayal of Char's feelings for Lalah may not be fully represented in a single textual description, especially since these emotions have multi-layered implications for the character's behavior in the storyline. In another Gundam animated film, *Mobile Suit Gundam: Char's Counterattack* (機動戦士ガンダム逆襲のシャア), Char utters the famous line, "Lalah could have been my mother," indicating that his feelings for Lalah involve both deep romantic affection and a strong maternal attachment. However, since this research focuses solely on *0079*, it does not utilize the additional insights provided by subsequent works to deeply interpret Char's emotions. Therefore, the emotional analysis in this study is primarily based on the data from *0079* and does not cover the developments in later works.

Although providing ChatGPT with relevant background information enables it to recognize Char's complex emotional structure towards Lalah, including his romantic feelings and maternal attachment, and further interpret how this impacts his actions and motivations, ChatGPT still has limited understanding of emotional complexity. For example, ChatGPT can analyze emotional expressions in text but may oversimplify when explaining complex emotions. In cases like Char's intertwining of romantic feelings and maternal attachment, ChatGPT might struggle to fully capture the dynamic changes and deep impacts of such emotional interweaving, potentially simplifying it into emotional opposites and overlooking the importance of this information for narrative progression.

3 Event Semantics

Whitehead (1997) posits that events have a systematizable framework structure, demonstrating that events exist through the integration of individuals and the relationships of time and space involved. The structure of an event is essentially a grid of spatio-temporal relationships. Independent events have their own grids, and there are mechanisms linking events, some of which involve whole-part relationships, while others involve the union of parts within a whole.

Therefore, time records and spatial coordinates serve merely as reference points in a four-dimensional context, but the true subject of observation in a spatio-temporal ontology should be the "event." However, recording spatio-temporal data requires substantial time and resources and risks redundant entries. Excessive spatio-temporal data is not easily retrievable and does not effectively serve as knowledge. Efficiently recording only the segments that change helps reduce memory space consumption and facilitates subsequent tracking of attribute changes (Leslie, Barnes, Binford, and Smith, 2001).

This research normalizes narrative events into data for a spatio-temporal database through content analysis. Each normalized narrative event is subsequently stored in the spatio-temporal database. The elements include Initiator, Action, and Recipient. The Initiator represents the acting character, the cause of initiation, or the origin. The Initiator is well-defined and concrete. Action represents the force of change or the performed behavior, while the Recipient is the object that undergoes the change.

Based on Davidson's (1967) "Davidsonian Event Semantics," this research defines an event entry stored in the spatio-temporal database as composed of "individual," "time," and "action." Each historical event is categorized into Time, Initiator, Action, Recipient, and Complement. The Initiator and Recipient, representing individuals, are akin to subjects, objects, or pronouns in grammatical terms and can refer to a single entity, a composite spatial domain (a collection of individuals), or other types of entities. The relationships between objects include subordinate relationships and inheritance relationships. Action represents a behavior and can be interpreted as a verb used to change the attributes of individuals. An event signifies a turning point where change or transformation into a new state occurs; thus, each event links the involved individuals across different time points.

3.1 Deconstructing Event Narratives Using Five Sentence Patterns

Agentive Sentence (SVO)
The subject is the initiator, and the action is performed by the subject. An agentive sentence includes the initiator (subject, S), the action (verb, V), and the recipient (object, O), forming the pattern [Subject (S) + Verb (V) + Object (O)]. For example: "Amuro (S) + pilots (V) + Gundam (O)."

Experiencer Sentence (SmOV).
The subject is the recipient affected by the action. In an experiencer sentence, the action is not directly performed by the subject on the object but is experienced by the recipient, influenced by the initiator. The pattern is [Sacrificed Lalah (S) + causes (m) + Char (O) + to rage (V)].

Neutral Sentence (SV)
The subject is neither the agent nor the experiencer. In this research, the subject is placed in the initiator column. A neutral sentence includes only the action (verb) and the initiator (subject), where the initiator completes the action alone. It can be a state description or a change in the initiator's attributes, forming the pattern [Subject (S) + Verb (V)]. For example: "Char (S) + retreats (V)."

Attributive Sentence (SVC)
Similar to the neutral sentence, in an attributive sentence, the subject is neither the agent nor the experiencer. The verb is followed by a complement describing an attribute related to the subject rather than another independent entity. The verb used in this type of sentence is a linking verb (such as the verb "to be" or other copulative verbs) connecting the subject to the complement. The pattern is [Subject (S) + Verb (V) + Complement (C)]. For example: "Amuro (S) + becomes (V) + a Gundam pilot (C)."

Double Object Sentence (SVOO)
This sentence pattern involves two recipients. Previously, such sentences were decomposed into two SVO sentences in this research, but the transformed patterns offered weaker interpretive power for the original event. Thus, this pattern was added, forming [Subject (S) + Verb (V) + Object 1 (O1) + Object 2 (O2)]. For example: "Char (S) + provides (V) + false information (O1) + to Garma (O2)."

In the narration of animated stories, a single event may need to be deconstructed into two or more sets of event actions to fully describe the narrative event. Such narrative events, when decomposed into compound event actions, can have the subject of a subsequent event be the entirety of the previous event, forming continuous events. Establishing rules that link all spatio-temporal data is crucial. Beyond the spatio-temporal sequences formed by continuous events, the study attempts to summarize events to create relational links between characters. This research collects events from *Mobile Suit Gundam 0079* and, after defining the rules, links them. It observes the causal relationships derived from each rule and finally integrates the results of these rules to analyze the emotional context of the characters.

3.2 Narrative Analysis and Event Structuring

After consolidating the narrative content of the 43 episodes of the *Mobile Suit Gundam 0079* TV series, a total of 320 structured events were identified. The sentence structure of these events is summarized in Table 1. Through authority control, 293 unique subjects were identified, and 519 relationships were established. To enable ChatGPT to perform a more comprehensive emotional analysis of Amuro and Char in episode 41, 114 events related to Amuro and 39 events related to Char were selected based on the continuity of relationships and entities.

Table 1 Statistics of Event Sentence Patterns

Sentence Patterns	Count
Agentive Sentence (SVO)	193
Experiencer Sentence (SmOV)	8
Neutral Sentence (SV)	78
Attributive Sentence (SVC)	39
Double Object Sentence (SVOO)	2
Total	320

Submitting a similar query to ChatGPT, such as "This is a screenshot from episode 41 'A Cosmic Glow' of the anime Gundam 0079, featuring the character Char. Could you analyze his current emotional state? Additionally, provide structured events related to Amuro," with the structured events summarized in Table 2, resulted in ChatGPT's following analysis of Char's emotional state as in Fig. 3.

Table 2 Structured Events Related to Char

Source	Sentence Pattern	S (Subject)	m (causative verb)	V (Verb)	O (Object)	C (Complement)
#1	SVO	Char		commands	Zaku	to attack White Base
#2	SVO	Char's subordinates		attack	White Base	
#5	SVO	Char piloting Zaku		attacks	White Base	
#5	SVO	Amuro piloting Gundam		confronts	Char	

(*continued*)

Table 2 (continued)

Source	Sentence Pattern	S (Subject)	m (causative verb)	V (Verb)	O (Object)	C (Complement)
#7	SmOV	Char	leads	to attack	the forces	
#7	SVO	Amuro		confronts	Char	engages in fierce air combat
#7	SVO	Amuro		repels	Char	and safely returns to White Base
#10	SVOO	Char		provides	false information	to Garma
#10	SmOV	Char withdrawing support	causes	failure	Garma	
#12	SVO	Char's forces		prepare to attack	White Base	
#12	SVO	Char's forces		attack	White Base	
#12	SVO	Amuro		engages in fierce combat	Char	
#15	SVO	Char leading the forces		prepares to launch an assault	White Base	
#15	SVO	Amuro		engages in fierce combat	Char	
#15	SVO	Amuro with newly discovered abilities		repels	Char and his forces	
#15	SV	Char		reflects	on battle experience and tactics	
#20	SVO	Zeon forces led by Char		attack	White Base	
#20	SVO	Amuro		engages in fierce combat	Char	
#29	SVO	Amuro		confronts	Char	
#34	SVO	Char piloting Red Zaku		challenges	Amuro	

(continued)

Table 2 (*continued*)

Source	Sentence Pattern	S (Subject)	m (causative verb)	V (Verb)	O (Object)	C (Complement)
#34	SVO	Amuro piloting Gundam		engages	Char piloting Red Zaku	
#34	SVO	Amuro		engages	Char	
#34	SV	Char		retreats		
#37	SVO	Amuro		battles	Char	
#37	SVO	Amuro		defeats	Char	
#37	SV	Char		escapes		
#38	SVO	Char		plans	attack strategy	
#38	SVO	Char's forces		attack	White Base	
#38	SVO	Amuro piloting Gundam		confronts	Char's forces	
#38	SVO	Sayla piloting Core Fighter		tries to find	Char	
#38	SV	Char		has a deep conversation		with Sayla
#38	SVO	Char		retreats		
#41	SVO	Amuro		engages in fierce combat	Char	
#41	SVO	Lalah		tries to protect	Char	
#41	SmOV	Sacrificed Lalah	causes	rage	Char	
#41	SVO	Amuro		defeats	Char's mobile suit	
#41	SV	Char		escapes		
#42	SVO	Amuro		clashes	with Char	
#42	SVO	Amuro		defeats	Char	

When conducting a similar operation for Amuro, providing 114 pieces of material related to Amuro, ChatGPT responded as Fig. 4:

Compared to the emotional analysis process without providing structured events, the structured events enable ChatGPT to capture and define Amuro Ray's complex emotions more accurately. In particular, Amuro's remorse for Lalah mainly stems from accidentally causing her death during the battle, resulting in deep self-blame and regret. This incident becomes a central impact on Amuro's emotions, forcing him to confront the consequences of his actions while also intensifying his complex feelings towards Char. Lalah's death becomes an indelible scar between Amuro and Char, deepening their adversarial relationship and filling it with contradictions. The structured events provide critical narrative context for emotional analysis, allowing ChatGPT to more

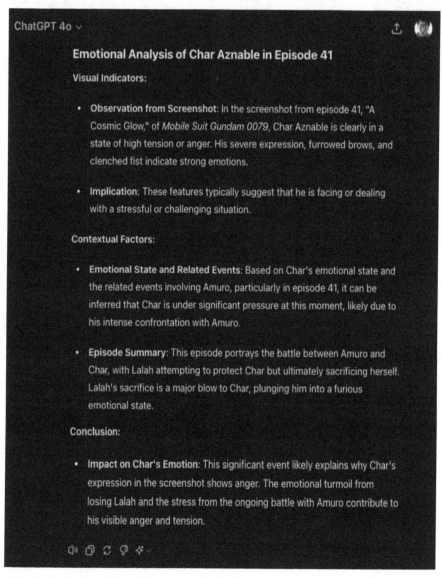

Fig. 3. Analysis of Char's Emotions in Episode 41 with 39 Structured Event Sentences

deeply understand and articulate the character's emotional transitions and underlying psychological motivations.

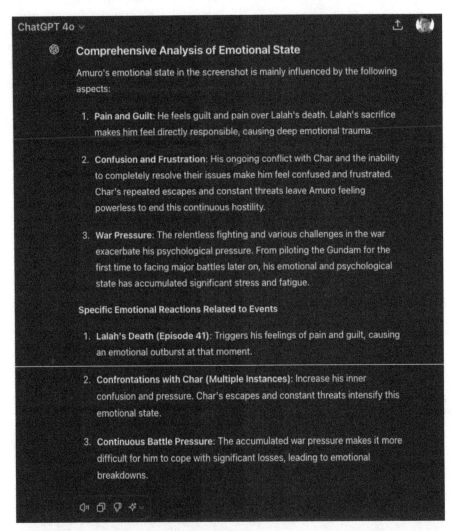

Fig. 4. Analysis of Amuro's emotions in episode 41 with 114 structured event sentences

4 Conclusion and Discussion

In animation, visual and auditory elements such as body language, facial expressions, and voice tones are crucial for understanding emotions. ChatGPT struggles to effectively process these non-textual cues, potentially limiting its ability to fully comprehend and describe the subtle emotional variations of characters. For instance, the tense atmosphere during confrontations between Char and Amuro, or the tone changes in Sayla's conversation with Char, are emotional expressions that may not be fully conveyed through text alone. Therefore, the next phase of research could focus on how to conduct sequential analyses of key scenes in the work, combining continuous character emotions in the story to perform multi-layered emotional analysis of characters in the narrative.

Since ChatGPT can already perform detailed emotional analysis using information collected from the internet, its accuracy in analyzing the stories of Amuro and Char often matches that of real human interpretations of animated characters. Future research should aim to experiment by replacing character names and blocking all information related to Gundam 0079. This approach would prevent ChatGPT from leveraging pre-existing background knowledge or internet data, allowing for more genuine and independent analytical results. This method will better assess ChatGPT's ability to interpret emotions and analyze characters without prior data support, ensuring its analytical capabilities are broadly applicable and reliable across various scenarios.

The existing vocabulary for emotion tagging has not yet been established in this study. Future research will continue to develop and provide a predefined vocabulary for the ChatGPT testing environment, facilitating the judgment of more nuanced emotional changes in animated characters when facing events.

This study employed event semantics to deeply analyze the multi-layered emotional expressions of characters in the anime *Mobile Suit Gundam 0079*, demonstrating how systematic event structures can enhance ChatGPT's emotional understanding capabilities. The findings indicate that by establishing and utilizing structured events, ChatGPT can more accurately capture and interpret the emotional states of characters, especially in complex emotional contexts. Event semantics provided an effective framework for integrating spatio-temporal information and character actions into structured data, thereby improving ChatGPT's accuracy and consistency in the analytical process. However, ChatGPT still faces challenges in processing non-textual information such as body language, facial expressions, and vocal tones, which are crucial for understanding emotions. Future research should explore how to integrate visual and auditory cues from animations for a more comprehensive emotional analysis and test ChatGPT's independent analytical capability without relying on prior background knowledge to enhance its applicability and reliability across various scenarios.

References

1. Davidson, D.: The logical form of action sentences. In: Rescher, N. (ed.) The Logic of Decision and Action. University of Pittsburgh Press, Pittsburgh, PA (1967)
2. Leslie, C., Barnes, G., Binford, M., Smith, S.: A Spatio-Temporal Data Model for Analyzing the Relationship between Property Ownership Changes, Land-Use/Land-Cover and Carbon Dynamics. In: Proceedings of American Congress of Surveying and Mapping Congress. Las Vegas, USA (2001)
3. Whitehead, A.N.: Science and the Modern World. The Free Press, New York (1997). (Originally Published in 1925)
4. Xie, N.: Kongjian Shengchan yu Wenhua Biaozheng: Kongjian Zhuanxiang Shiyu zhong de Wenxue Yanjiu. China Renmin University Press, Beijing (2010)

Image and Media in Kansei Design

A Comparative Study Assessing the Effectiveness of Machine Learning Technology Versus the Questionnaire Method in Product Aesthetics Surveys

Chun-Wei Chen[✉]

Department of Mechanical Engineering, National Chin-Yi University of Technology, Taichung 411030, Taiwan (R.O.C.)
chenschool@yahoo.com.tw

Abstract. In the current market landscape, it is imperative for products not only to meet consumer needs but also to possess an appealing appearance. Consequently, comprehending how consumers perceive the aesthetics of a product becomes crucial. This study aims to fill existing gaps by (1) critically evaluating the effectiveness of current survey tools for assessing consumer product aesthetics, and (2) exploring the application of machine learning (ML) technologies in consumer surveys. Additionally, we seek to (3) compare the questionnaire method and the method employed by ML technologies with conventional survey tools to gauge the practicality of assessing consumer product aesthetics. Our research uncovered two key findings: Firstly, ML technology exhibits remarkable accuracy in predicting consumer perceptions across five representative aesthetic descriptors. Secondly, when comparing the accuracy of the questionnaire method and ML technology in predicting consumer perceptions for these descriptors, ML consistently outperforms the traditional approach. These results indicate that AI, specifically ML technology, not only adeptly forecasts consumer perceptions of product aesthetics but also demonstrates superior performance compared to traditional methods. This study makes theoretical contributions by proposing a survey tool that utilizes artificial intelligence technology to investigate consumer product aesthetics, incorporating practical guidelines for the assessment. Furthermore, it offers valuable insights into consumer product aesthetics and explores the potential impact of technology on this field.

Keywords: AI technology · ML · questionnaire method · consumer · product aesthetic · perception

1 Introduction

Product aesthetics represents a fundamental concept that is intimately intertwined with cognitive acceptance, encompassing two pivotal facets from a design perspective: the visual and tactile impressions a product imparts upon our senses, and the emotional responses it elicits when we perceive its aesthetic allure [1–3]. Product aesthetics is one of the factors for the development of a successful product [4, 5]. Given the robust nexus between emotional gratification and our comprehension of product aesthetics, for product designers and corporations alike, it is paramount to meticulously contemplate these constituents during the product development phase to guarantee its triumphant reception. This elucidates the central role played by consumers' emotional reactions to product aesthetics in the product consumption process [6–9] and highlights the importance of the investigation of survey tools effect for product aesthetics that satisfy consumers.

2 Literature Review of Product Aesthetics Surveys and Their Survey Method

The questionnaire method, often simply referred to as a questionnaire, is a research approach used to gather information and opinions from specific participants [10]. It works by posing written questions on a survey form, which is why it's also known as the "question form method." Here's how it works: Researchers compile a set of questions they want to explore into this survey form. Then, they distribute these forms to respondents who can respond through mail, face-to-face interviews, or follow up interviews. This process helps researchers gain insights into how respondents view a particular phenomenon or issue [11]. A typical questionnaire survey form consists of several sections: a preface, the questions themselves, spaces for answers, and coding sections. These questions come in various types, such as background questions, objective questions, subjective questions, and testing questions [12]. There are two main ways respondents fill out the questionnaire survey form: self-filling and substitute filling. Self-filling surveys are often used for large-scale, one-time data collection efforts. In these cases, respondents usually have a basic understanding of the questions and their content when they fill out the forms. On the other hand, substitute filling involves a survey administrator verbally asking respondents the questions and then recording their answers on the survey form [13]. Despite its widespread use, the questionnaire survey method does have some drawbacks to keep in mind [14]: 1. Lack of Flexibility: If a respondent misunderstands a question, there's no opportunity to clarify or correct it. 2. Limited Insight: Written responses may not fully capture a respondent's true thoughts and feelings. 3. Lack of Control: Researchers cannot control the environment in which respondents complete the survey, potentially impacting their responses. 4. Response Order: Respondents may answer questions out of sequence, affecting the survey's intended structure and possibly introducing response bias.

The recent advancements in Artificial Intelligence (AI) technology have yielded significant progress in the emulation and prediction of human behavior [15]. Machine learning (ML), a distinctive subset of AI [16], is centered on the concept of enabling computers to acquire knowledge from data and enhance their performance without explicit programming [17]. The fundamental mechanism of ML can be elucidated through the following stages: 1). Data Analysis and Model Creation: In this initial phase, algorithms are employed to analyze gathered data and formulate predictive models. These models incorporate elements such as reasoning, knowledge assimilation, and information integration. 2). Model Training: Subsequently, the models undergo a training process to enhance their accuracy and effectiveness. 3). Predictive Capability: With the availability of new data, the established models can proficiently forecast outcomes [18]. In essence, ML empowers computer algorithms to discern intricate patterns and associations within extensive datasets. This newfound capability enables them to make informed decisions and predictions grounded in these identified patterns. Furthermore, ML eliminates the necessity for explicit programming of data access rules, enabling algorithms to autonomously learn and validate data [19]. ML technology encompasses a plethora of learning models, including supervised learning, unsupervised learning, semi-supervised learning, and reinforcement learning, depending on the specific nature of the data and the desired outcomes [20–21]. Its primary applications encompass object classification, item identification, outcome prediction, and comprehensive decision-making [23]. When confronted with complex and unpredictable datasets, researchers have the option to select specific ML algorithms or combine multiple algorithms to achieve the most precise predictions. This approach enhances survey methodologies, reduces errors, and minimizes the requirement for extensive human resources and time investment.

The difference between ML and questionnaire methods offers a nuanced perspective on understanding and forecasting differences in human behavior. The technical principles underlying both approaches reveal intriguing similarities and distinctions. The questionnaire survey method, a traditional yet effective tool, involves a sequential process: designing questions, collecting answers, compiling data, employing statistical analysis, and deriving conclusions. Conversely, ML's predictive prowess follows dis-tinct steps: preparing questions for model training, having subjects input data for training, training the neural network model, and using the trained model to predict outcomes based on new data. Notably, both methods share the commonality of utilizing questions and answer options to collect behavioral data. However, their differences are profound. ML, driven by the objective of constructing a predictive model, collects behavior data for training purposes, while the questionnaire method seeks to understand behavioral characteristics. Technologically, ML endows the ability to learn and predict human behavior dynamically, a capability absent in the static rules of the questionnaire survey method. This adaptability renders ML more versatile in predicting a spectrum of human behaviors. In essence, while both methods leverage questions, the divergent purposes and technological capacities underscore the superiority of ML in the realm of predicting human behavior.

3 A Framework of Research

This study endeavors to present a novel approach for the analysis of consumer product aesthetic perception utilizing AI, specifically ML, and evaluate its viability (Fig. 1). In the course of our research, we conducted an exhaustive inquiry into this domain by employing two distinctive methodologies. The initial stage of our inquiry entailed the deployment of a meticulously designed structured questionnaire to systematically examine and predict consumer perceptions regarding the aesthetic attributes of products. In a subsequent phase, the latter part of our investigation harnessed ML technology to pursue identical objectives.

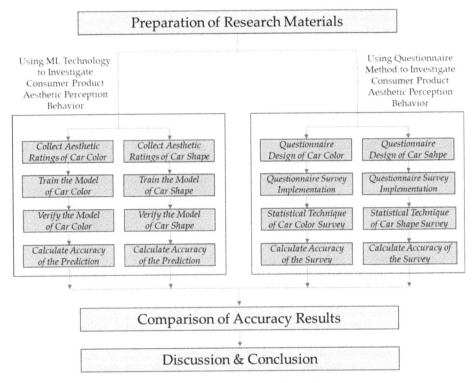

Fig. 1. The framework of research

4 Research Methodology

4.1 Using ML Technology to Investigate Consumer Product Aesthetic Perception Behavior

4.1.1 Preparation of Research Materials

In this research endeavor, we employed ML techniques to prognosticate the perceptual evaluation of the aesthetic attributes associated with distinct automobile models. To facilitate this undertaking, we meticulously curated a dataset encompassing a total of 160 high-fidelity images of automobiles. This dataset was categorized into two distinct subsets, each serving a unique purpose in our modeling efforts. The first subset comprises 60 high-resolution color photographs, meticulously selected to train models responsible for capturing and interpreting the nuances of product color aesthetics (refer to Fig. 2). These images span a diverse range of car brands, including but not limited to Toyota, Honda, Kia, Hyundai, Ford, Mazda, BMW, Mercedes-Benz, Skoda, Volvo, Infiniti, Nissan, Volkswagen, Porsche, Chevrolet, and Audi. Conversely, the second subset encompasses 60 black-and-white photographs of automobiles, thoughtfully curated to train models focusing on the discernment of product shape aesthetics (Fig. 3). These monochromatic images also represent a wide array of automotive brands, thereby ensuring the comprehensiveness and robustness of our model training process. Subsequently, we employed an additional set of 20 color car photographs to validate the efficacy of the trained model pertaining to product color aesthetics (refer to Fig. 4). Furthermore, we utilized a separate set of 20 black-and-white automobile images to validate the trained model responsible for product shape aesthetics (refer to Fig. 5). The validation images employed in our study constitute a pivotal element, facilitating the evaluation of the generalization and precision of our ML models. Collectively, this comprehensive dataset and methodological approach encompassing color and monochromatic images from a multitude of automobile manufacturers facilitated the development of models capable of discerning and predicting consumers' perceptions of car aesthetics with scientific rigor and precision.

This investigation meticulously manages extraneous variables, including "brand awareness," "brand recognition," and "brand familiarity," which have the potential to exert influence on the perceived aesthetics of the product. The approach to controlling these variables is outlined as follows: 1). Brand Neutrality in Sample Selection: Subsequently, in the process of sample selection, product samples characterized by strong brand styling or those emblematic of the brand were deliberately excluded. This meticulous curation of samples was undertaken with the intent of preventing subjects from gaining access to any brand-specific information through the research samples, thereby upholding the integrity of the study's controls. 2). Brand Information Deprivation: In preparation for the assessment, advanced image processing technology was applied to obfuscate the brand logo on the product samples used in the study. This deliberate obscuration was executed to ensure that the participating subjects were not privy to any direct brand-related information from the research samples prior to evaluation.

In order to assess the aesthetic attributes of the depicted automobiles in Fig. 2 and Fig. 3, we meticulously curated a collection of descriptive aesthetic adjectives, as delineated as modern, traditional, smooth, technological, stable. These adjectives were employed as evaluative tools by participants in scrutinizing the visual aesthetics of the vehicles.

4.1.2 Training and Verifying of ML in Predicting Consumer Product Aesthetic Perception Behavior

The paper describes a process for training and verifying a ML model to predict consumer evaluations of product aesthetics. Here is a step-by-step breakdown of the process:

1). Step 1: Define Adjective Vocabularies and Input Photos
   Start by defining five descriptive aesthetic adjectives. Then, input 120 car photos (Fig. 2, Fig. 3) into the ML system.
2). Step 2: Collect Aesthetic Ratings
   Engage individuals with prior experience in automobile procurement to provide aesthetic evaluations for a set of 120 automotive images, employing a lexicon of five adjectival vocabularies. These evaluations are to be expressed on a numerical continuum spanning from 1 to 5, wherein a rating of 1 point indicates "extremely unaesthetic," 2 point conveys "unaesthetic," 3 point designates "neutral," 4 point signifies "aesthetic," and 5 point represents "highly aesthetic." The aesthetic rating tasks encompass the following:

Fig. 2. 60 color cars photos for training models of product color aesthetics

A. Evaluating Car Color Aesthetics: In this task, participants are tasked with the responsibility of appraising and assigning ratings to five discrete aesthetic attributes associated with the coloration of automobiles. These attributes are prominently featured in each photograph within Fig. 2 of our ML system. The resulting evaluations will be recorded and stored within the ML system's database.

B. Assessing Car Shape Aesthetics: In this task, participants will undertake the assessment and rating of five distinct aesthetic dimensions pertaining to the shapes of automobiles, as depicted in each photograph within Fig. 3 within our ML system. The assessments rendered by participants will be meticulously recorded within the ML system, contributing to a comprehensive dataset for subsequent analysis.

Fig. 3. 60 black-and-white cars photos for training models of product shape aesthetics

3). Step 3: Train the Model

After the acquisition of aesthetic ratings for a dataset comprising 60 automobile images, it is imperative to commence the configuration of the ML model's parameters for subsequent training. This configuration entails the utilization of a deep neural network architecture, which is instrumental in handling complex image data. We employed the VisLab®'s AI model training system, developed by 3DFAMILY Company, in our research to facilitate image recognition. Key considerations during this configuration process encompass the specification of image attributes, encompassing dimensions such as image width (size 512 pixel), image height (size 512

Fig. 4. 20 color cars photos for verifying the trained model of product color aesthetics

Fig. 5. 20 black-and-white cars photos for verifying the trained model of product shape aesthetics

pixel), and image depth (size 3 pixel). Additionally, the training regimen necessitates the determination of critical hyperparameters, including the number of iterations (set at 10), the batch size (set to 6), and the learning rate (1X10–4). Subsequent to the meticulous parameter setting, the training phase of the ML model is initiated. This phase is characterized by iterative optimization, with each iteration involving the exposure of the model to a batch of training data and the subsequent adjustment of its internal parameters to minimize prediction errors. The desired convergence criterion for this training endeavor is the attainment of a prediction accuracy level exceeding 93.5%. Upon achieving the specified accuracy threshold, the trained model is ready to be preserved within the ML system. This preservation process encompasses the archival of the model's architecture, weights, and learned features. These artifacts collectively encapsulate the model's acquired knowledge and predictive capabilities, rendering it a valuable asset for subsequent applications and inferences within the domain of aesthetic image rating and analysis.

4). Step 4: Verify the Model

Process 40 automobile images (referenced as Fig. 4 and Fig. 5) as input data for the ML system's validation procedure. Extend an invitation to the same set of participants to perform aesthetic assessments on these 40 car images utilizing an identical lexicon of five adjectival descriptors and the established 1–5 point rating scale. These rating tasks replicate the aesthetic evaluation processes described within the "Collect Aesthetic Ratings" stage. Subsequently, employ the pre-trained model, which has been previously saved, to make predictions pertaining to the aesthetic scores for the aforementioned 40 photographs. Following this phase, a comparative analysis is executed between the actual aesthetic ratings assigned to the images and the predictions generated by the saved, trained model, particularly in regard to the 20 images. The assessment includes the automated computation of prediction accuracy for the trained model. This critical step seeks to evaluate the efficacy of ML in predicting consumer evaluations of car images.

4.2 Using Questionnaire Method to Investigate Consumer Product Aesthetic Perception Behavior

4.2.1 Research Material Preparation

In the context of this investigation, we have employed a questionnaire-based approach to elucidate the manner in which consumers discern the aesthetic attributes of products. To accomplish this, we have selected five salient descriptive aesthetic adjectives as modern, traditional, smooth, technological, stable, and subsequently integrated images of automobiles depicted in Figs. 3 and 4 to serve as our experimental stimuli.

4.2.2 Questionnaire Design

The assessment instrument devised for the investigation into consumer perceptions regarding the aesthetic attributes of a product comprises a comprehensive document spanning 21 pages. The structure of this instrument is logically organized, encompassing distinct sections, namely the preamble, the inquiry segment, the responses section, and the coding division. The preamble, residing on the initial page, serves the pivotal function of acquainting survey participants with the questionnaire's over-arching objectives and furnishing them with explicit instructions. Subsequently, the substantive inquiries are prominently presented across pages 2 through 21. Each of the inquiry pages conforms to a uniform template (see Fig. 6 as reference). They each prominently feature one of the 20 color car illustrations as delineated in Fig. 3. Additionally, these pages incorporate a series of five queries aimed at evaluating the aesthetic qualities of the depicted automobile. Furthermore, accompanying each color car image is one of the 20 black-and-white car depictions outlined in Fig. 4, together with an analogous supplementary set of five aesthetic evaluation questions. Respondents are hereby mandated to judiciously appraise the aesthetic allure of each depicted vehicle on a calibrated ordinal scale, with the following categories: 1 point: "very unattractive," 2 point: "unattractive," 3 point: "neutral," 4 point: "attractive," and 5 point: "very attractive."

4.2.3 Questionnaire Survey Implementation

To investigate consumer perceptions of product aesthetics, we implemented a systematic methodology (Fig. 8), which was structured into four distinct phases:

1). Step 1: Selection of Respondents

In this initial stage, we outline how we chose the participants for our survey. We've opted for a total of 80 respondents, which is a reasonable sample size. These 80 respondents are car buyers from four Taiwanese groups for car appre-ciation and car buying. Additionally, we ask demographic information about the respondents, including their gender distribution, average age, and the common characteristic of having experience with car purchases. This demographic information was used for analyzing the survey results.

2). Step 2: Questionnaire Distribution

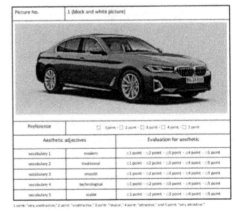

A. Question Example for surveying consumer perceptions regarding the product color aesthetic

B. Question Example for surveying consumer perceptions regarding the product shape aesthetic

Fig. 6. Examples of Questions in Questionnaire for investigating into consumer perceptions regarding the aesthetic attributes of a product

In this step, we distributed 80 questionnaires to the selected respondents, and we received 78 completed questionnaires in return. The response rate for this survey did not reach 100%, but 78 out of 80 responses was considered a high response rate. When analyzing the data, we need to account for any missing responses.

3). Step 3: Statistical Technique

In this phase, data underwent a preliminary processing step encompassing categorization and application of descriptive statistical techniques, inclusive of measures like the computation of central tendencies. This pivotal stage is indispensable in facilitating the extraction of insights from the dataset and inferences of substantive import from the survey's findings.

4). Step 4: Validation of Data Analysis Results

This ultimate stage involves a comparative analysis of the outcomes derived from questionnaire data against data garnered via an alternative methodology, specifically, the aesthetic evaluations provided by the study participants for each automobile image depicted in Fig. 4 and Fig. 5. This comparative evaluation serves as a means to gauge the precision and dependability of our questionnaire in probing into consumer product aesthetic discernment. Its pivotal function lies in corroborating the research findings and ascertaining the alignment of our questionnaire with real-world perceptual phenomena. In this context, we juxta-pose the results derived from the questionnaire-based investigation with the actual aesthetic appraisals of the automobile images presented in Fig. 4 and Fig. 5, as reported by the study subjects. This comparative scrutiny serves as a critical mechanism for evaluating the accuracy and consistency of our questionnaire in exploring the aesthetic perceptions associated with consumer products. Its fundamental role is to validate the research findings and ensure the congruence of our questionnaire with real-world perceptual experiences.

5 Results

Figure 7 displays a comparative chart that highlights the differences between survey results obtained through ML technology and those gathered through the questionnaire method. The chart reveals two key findings: 1). In the realm of predicting consumer perceptions of product shape aesthetics, ML technology demonstrates greater predictive accuracy across vocabulary1 to vocabulary2 when compared to the questionnaire method. 2). Similarly, when it comes to predicting consumer perceptions of product color aesthetics, ML technology also exhibits higher prediction accuracy across vocabulary1 to vocabulary2 in contrast to the questionnaire method. In essence, these results indicate that ML technology excels in achieving higher accuracy when investigating two critical aspects of consumer product aesthetics: shape perception and color perception. This implies that AI technology, specifically ML, surpasses the traditional questionnaire approach in accurately assessing consumer perceptions of product aesthetics and does so with greater efficiency. Consequently, this study's hypothesis is substantiated and confirmed.

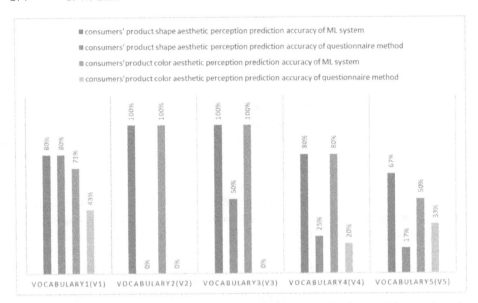

Fig. 7. A comparison diagram of survey effects between the ML technology and the questionnaire method

6 Conclusions

The successful development of a product is contingent upon a multifaceted array of critical determinants, which encompass functionality, usability, aesthetics, brand coherence, and the user experience. For both product designers and corporations, it is imperative to meticulously consider these elements during the product development phase to ensure its triumphant reception. Given the strong association between emotional satisfaction and our understanding of product aesthetics, practitioners in the field of product design and researchers specializing in this domain require a reliable tool for the effective quantification of consumer emotional contentment with regard to product aesthetics, spanning a diverse range of products. This paper endeavors to introduce a method for studying consumer product aesthetic perception behavior utilizing artificial intelligence (AI) technology, specifically ML, and assess its feasibility. Our empirical observations have yielded discernible insights, which are encapsulated by the following key findings: 1). ML Precision: ML technology has exhibited an exceptional level of precision in predicting consumer perceptions of product aesthetics, particularly with respect to five representative aesthetic descriptors. 2). Comparative Analysis: A comparative analysis of predictive accuracy between the traditional questionnaire-based methodology and the cutting-edge ML technology consistently favored the latter. This result underscores the superiority of ML technology as a predictive tool in the context of consumer perceptions related to the aforementioned aesthetic descriptors. 3). Efficacy of AI: These compelling outcomes underscore the idea that AI, particularly ML technology, serves as an effective instrument not only for predicting consumer perceptions of product aesthetics but also for probing this aspect with a high degree of proficiency.

References

1. Li, H., Xu, J., Fang, M., Tang, L., Pan, Y.: A study and analysis of the relationship between visual—auditory logos and consumer behavior. Behav. Sci. **13**, 613 (2023)
2. Naeini, S., Mostowfi, S.: The role of pleasure criteria in product design: an integrated approach in ergonomics and hedonomics (a review). Res. Psy. Beha Sci. **3**, 39 (2015)
3. Zhang, J., Park, S., Cho, A., Whang, M.: Recognition of empathy from synchronization between brain activity and eye movement. Sensors **23**, 5162 (2023)
4. Li, X., Niu, Y., Xu, J.: Factors affecting purchase intention of Hanfu: considering product identification, cultural motivation, and perceived authenticity. Behav. Sci. **13**, 689 (2023)
5. Pieter Desmet,: Measuring emotion: development and application of an instrument to measure emotional responses to products. In: Mark Blythe, Andrew Monk, (ed.) Funology 2. HIS, pp. 391–404. Springer, Cham (2018). https://doi.org/10.1007/978-3-319-68213-6_25
6. Hsiao, K.A., Chuang, W.J.: A study on pleasure, aesthetic, attraction and visual fixation time with related to product shapes. J. Des. **22**(4), 1–20 (2017)
7. Ye, S.: Do I destroy product beauty? effects of product aesthetics and usage process on consumption affect. Ming Chuan University, Graduate Institute of Business management (2018)
8. Shi, J., Honjo, T., Zhang, K., Furuya, K.: Using virtual reality to assess landscape: a comparative study between on-site survey and virtual reality of aesthetic preference and landscape cognition. Sustain. **12**, 2875 (2020)
9. Preston, V.: Questionnaire survey. In: International Encyclopedia of Human Geography, pp. 46–52(2009)
10. Gill, M.: The purpose, design and administration of a questionnaire for data collection. Radio. **11**, 131–136 (2005)
11. Williams, A.: How to … Write and analysis a questionnaire. J. Orth. **30**, 245–252 (2014)
12. Petra, M. B.: Trisha, G.: Hands-on guide to questionnaire research: selecting, designing, and developing your questionnaire. BMJ **328**, 1312–1315 (2004)
13. Kmetty, Z., Stefkovicsb, Á.: Assessing the effect of questionnaire design on unit and item-nonresponse: evidence from an online experiment. Int. J. Soci. Res. Meth. **25**, 659 (2021)
14. Asada, M.: Towards artificial empathy. Int. J. Soc. Robot. **7**, 19 (2015)
15. Huddleston, S.H., Brown, G.G.: Informs Analytics Body of Knowledge (2022)
16. Alizamir, M., Kim, S., Kisi, O., Zounemat-Kermani, M.: A comparative study of several ML based non-linear regression methods in estimating solar radiation: case studies of the USA and Turkey regions. Energy **197**, 117239 (2020)
17. Osisanwo, F., Akinsola, J., Awodele, O., Hinmikaiye, J., Olakanmi, O., Akinjobi, J.: Supervised ML algorithms: classification and comparison. Int. J. Comput. Trends Technol. **48**, 128 (2017)
18. Zöller, M.A., Huber, M.F.: Benchmark and survey of automated ML frameworks. ArXiv: 1904.12054 (2021)

Cognitive Research on AI-Assisted Generation of Animations with Diverse Artistic Styles

Chia-Ling Chang(✉)

National Taitung University, 369, Sec. 2, University Rd., Taitung, Taiwan
idit007@gmail.com

Abstract. In recent years, AI generation technology has experienced explosive growth, encompassing both text and images. As a result, more and more people are learning and utilizing these tools. Animation, a consistently popular medium for conveying information, transcends gender, age, ethnicity, and cultural boundaries. Compared to a single art style, short videos that combine audio and visual elements are particularly well-suited for integrating diverse artistic styles into a cohesive whole. This study begins by comparing three well-known AI video generation tools: "Runway Gen-2," "Genmo," and "Moonshot." The comparison focuses on their strengths and weaknesses across several dimensions, including "image quality," "database completeness," "video duration," "usability," and "auxiliary functions." The research also explores the public's growing preference for the integration of diverse artistic styles within a single piece, as opposed to single-style works. To investigate this, the study applied AI-assisted technology to generate 20 short animations in diverse artistic styles, with cups as the central theme. The objective was to examine the young audience's perceptions and acceptance of AI-generated, multi-style animations. Data was collected through online questionnaires, analyzing participants' perceptions and levels of acceptance. This research aims to apply diverse artistic styles and AI animation techniques to the practical creation of short videos. The findings provide insights for developing short animation advertisements that could improve the effectiveness of commercial promotions for the new generation.

Keywords: Diverse Artistic Styles · AI Generation · Cognitive · Animation · Short Videos

1 Introduction

With the rapid advancement of technology, the application of artificial intelligence (AI) in the field of art has garnered significant attention, particularly in exploring how diverse artistic styles can be integrated into AI animation. The exploration in this field aims to expand the creative scope of audiovisual arts, offering audiences richer artistic experiences. The introduction of diverse artistic styles is expected to enrich the visual expression of videos, break through the constraints of traditional animation styles, and provide a more creative space for expression. The application of AI technology has already achieved a series of successes in the field of art. For instance, Gatys, Ecker, Matthias

(2016) proposed a new algorithm that can separate and recombine the content and style of images. Through an optimization process, a content image is combined with a style image to generate an image that retains both content and style. For images of specified styles, the study shows the application effect of the algorithm on images of various styles (such as Van Gogh's "Starry Night" and Picasso's paintings), proving the effectiveness and diversity of the algorithm. And the generated image is visually highly similar to the original style image, while retaining the basic structure of the original content image [1]. The application of diverse artistic styles in AI animation not only helps enrich the visual effects of artistic works but also promotes communication and integration among different cultures and artistic fields. This cross-cultural impact could become a new opportunity for the development of art, driving global artistic growth [2]. It is necessary for us to reflect on the impact of the application of diverse artistic styles in AI animation on creators and audiences. Is it merely a showcase of technology, or does it genuinely enrich the experience of artistic creation and appreciation?

This study aims to explore the application of diverse artistic styles and forms in AI animation short videos, propose technical realizations of integrating diverse artistic styles into AI animation short video creation, and deeply analyze their impact on visual presentation and aesthetic experience. It analyzes the influence of different artistic styles on the visual effects of AI animation short videos, explores their differences in emotional expression, information transmission, and audience engagement to provide creators with direction and strategic recommendations for creation.

2 Literature Review

2.1 AI-Generated Images and Animations

In the late 1960s, Harold Cohen began developing AARON, becoming one of the early significant artificial intelligence art systems. AARON is a typical representative of symbolic artificial intelligence art, and its method of generating images is based on symbolic rules. Cohen's intention was to encode drawing behavior using artificial intelligence, and the initial version of AARON created simple black and white drawings. Subsequently, Cohen continuously developed AARON's drawing skills through his own painting activities, enabling it to paint with special brushes and dyes without the need for human intervention [3].

Since 2014, Generative Adversarial Networks (GANs) have become a commonly used tool for AI artists. This system adds new images through a "generator" and evaluates the success of the generated images through a "discriminator." Recent models adopt vector quantization generation and are trained through GANs and contrastive text-to-image generation [4].

DeepDream, introduced by Google in 2015, utilizes convolutional neural networks to search for and enhance patterns in images using visual illusion algorithms, creating intentionally over processed images. With the advent of DeepDream, some companies have also begun to introduce applications that can transform photos into the style of famous art collections.

In recent years, some AI drawing programs have been able to generate various images through text prompts. For example, Open AI's DALL-E, Google Brain's Imagen and Parti (released in May 2022), and Microsoft's NUWA-Infinity. The input forms of these systems can include text, images, keywords, and/or configuration parameters. By specifying prompts for specific artistic styles, such as "in the style of {name of an artist}," or broad choices of aesthetic/artistic style elements [5].

This study compiles and analyzes five AI rendering programs/software that are currently both popular and convenient. The advantages and disadvantages of each are summarized in Table 1. After analysis, this study concludes that Moonshot has significant advantages and convenience in AI rendering.

Table 1. Analysis of pros and cons of AI rendering software (Compiled by the study).

Name	Advantages	Disadvantages
Moonshot	• Basic functions are free, many additional features, easy to use • Supports Chinese	• Lower image quality, requires additional resolution enhancement • Relatively incomplete database
Midjourney	• Leading market share with a robust database • Can adjust scale and resolution of works • Can generate derivative works based on specific AI artwork	• More expensive • No Chinese interface
Stable Diffusion	• Completely free with unlimited usage • Can be operated online via web pages	• Requires web setup, more complex to us • No Chinese interface
Novel AI	• Specializes in anime-style generation • Can generate images from images	• More prone to distortions when generating characters • More likely to encounter copyright issues • No Chinese interface
Stablcboost	• Provides preset AI drawing styles, lens and painting forms, lighting variations	• Many setting options, higher learning cost • Slower rendering speed • No Chinese interface

This study further organizes and compares three AI tools capable of generating dynamic images from pictures, detailing their advantages, disadvantages, and prices, as shown in Table 2.

Table 2. Analysis of AI Tools for Generating Dynamic Images (Compiled by the study).

Name	Advantages	Disadvantages
Runway Gen-2	• Comprehensive database • Customizable camera settings • Converts existing images into animations • Web-based operation • Magic brush for partial edits • Preserves original image appearance when generating animations • Has a material library and examples • More parameters for camera movement customization	• Limited free usage per day • Simpler operation • Supports Chinese input commands • Has a command library for reference • Easy operation via Line
Genmo	• Completely free with unlimited usage • Web-based operation • Usable by directly inputting text • Adds effects to existing videos via the web if they are insufficiently rich • Supports camera movement • Automatically generates text descriptions matching provided images • Supports 6-s videos • Highest image quality among the three (1280*2304) • Comprehensive database	• No support for Chinese • Disrupts original composition • Results may differ from the provided image • Higher likelihood of generating failed or mismatched videos
Moonshot	• Limited free usage per day • Simpler operation • Supports Chinese input commands • Provide a command library for reference • Easy operation via Line	• Lower image quality, requires additional resolution enhancement • Relatively incomplete database • Only generates GIF files • Inconsistent rendering, often producing distorted or mismatched animations

2.2 The Rise of Short Videos

With the analysis of big data and changes in audience preferences, shorter and more attention-grabbing short videos have emerged. These videos swiftly delve into key topics or intriguing subjects, featuring vivid colors to captivate viewers. Consequently, within just one year in 2021, posts about short videos have grown by over 470%. Short videos, defined as videos under 15 min in length, have seen a dramatic rise in popularity. Initially, social media platforms predominantly focused on sharing text or long-form content, such as Facebook. However, from 2020 to 2023, the usage rates of short video apps like TikTok and Snapchat, or apps incorporating short video features like YouTube and Instagram, have surpassed other software in Europe, America, and the

Australia-New Zealand region. Currently, short video content covers a wide range of topics, including lifestyle entertainment, fashion trends, movie clips, as well as commercial advertisements and e-commerce promotions [6].

Nowadays, more and more creators are choosing to upload their daily short clips or edited videos to short video platforms. Their concise and rich themes allow viewers to quickly grasp the content amidst their busy lives, while also providing entertainment and visual impact. By 2022, the average viewing rate for short videos on YouTube and Instagram has exceeded the typical 1–3% for regular videos. Consequently, many companies have identified new business opportunities with the rise of short videos. They choose to place their advertisements in short videos to increase revenue and brand awareness. This expansion of the short video market encourages more individuals to engage in this market for creative purposes [7]. E-commerce short videos use short videos as carriers, users experience sequential transitions from the "viewing" scene to other scenes, with the visual styles of short video content playing an important role in influencing user purchasing decisions [6]. Summarize the format and duration requirements of videos on mainstream short video platforms available on the market, as shown in Table 3. This will serve as a reference for determining the ideal length of our short videos. Finally, this study adopts 60 s as the length for creating animations.

Table 3. Video Durations on Short Video Platforms.

Platform	Duration
IG reel	60 s
YouTube short	60 s
Douyin	15 s - 15 min
Tiktok	15 s - 15 min

2.3 Diverse Artistic Style Animation and Applications

In recent years, the materials used in animation creation have become increasingly diverse, including hand-drawn, puppetry, image synthesis, 2D, 3D computer animation, and various other techniques. This diversity has changed the way creators operate and how viewers appreciate animations. The influence of animation is not only reflected in its commercial success but also through media technology, which strengthens the communication and immersion of animation culture, exerting a profound impact on the cultural mediation of animation [8]. The length of animated videos also affects the artistic expression. While single artistic style works are suitable for feature-length animations to maintain narrative emotions, the fusion of diverse artistic styles in short animated videos is an innovative approach. For example, "Gorillaz: Journey to Plastic Beach" is a 15-min animation produced by the British band Gorillaz. Through short episodes featuring virtual characters representing the band, it tells a story that runs through their album content. Its animation style is diverse, combining rough hand-drawn, 2D drawing,

3D modeling, and live-action elements, showcasing richness in artistic expression and creativity [9].

Another animation work presented in diverse artistic styles is "MATE?" (https://www.youtube.com/watch?v=ri4MqY0zvFM) produced by the Argentine animation studio Buda. This 2-min short video animation, created by 23 animators from South America, showcases various patterns and vibrant colors on cups, forming a local mate tea culture. Collaboration with multiple animators with different styles results in a fusion of distinctive brushstrokes and styles, enriching the visual presentation of the animation. According to the findings of Lyu et al.'s (2021) study, participants can still identify different artistic styles, especially textures, after watching AIGC films composed of four style groups (including Fauvism, Expressionism, Cubism, and Renaissance) The recognition of strokes and strokes is greater than that of colors, and viewers may prefer samples with higher recognition at the semantic and effectiveness levels [10].

In summary, this research aims to explore the visual impact of different artistic styles on AI animation short videos, understand the differences in emotional expression, information transmission, and audience engagement brought about by diverse artistic styles. Through a questionnaire survey targeting audiences, the study aims to analyze their preferences and perceptions of diverse artistic styles and their impact on the acceptance of AI-generated animation short videos, providing creators with direction and strategic recommendations for creation.

3 Process of Research

This study primarily uses AI-generated short animation clips in diverse artistic styles as research samples. Participants are invited to watch these samples and complete a questionnaire, based on which the acceptance and preference perceptions of viewers regarding the characteristics of different AI-generated styles are analyzed. The theme of the AI-animated short clips is set as a cup for two reasons: firstly, it references the 2022 mate tea project with diverse artistic styles, which successfully demonstrated how diverse artistic styles can be integrated into a marketing video. Secondly, cups are objects that have existed throughout history and across cultures. Regardless of the style, AI should be able to generate reasonable images, reducing the discordance between objects and styles.

3.1 Steps for AI-Generated Animations

1. Determining Types and Number of Artistic styles:
   Referring to significant styles and schools in art history, incorporating both Eastern and Western artistic styles as well as new modern digital trends, styles that were overly similar were eliminated, ultimately establishing 20 artistic styles. These include: Gothic, Impressionism, Rococo, Fauvism, Cubism, Abstract, Pop Art, Pixar, Sketch, Van Gogh, Street Graffiti, Picasso, Pixel Art, American Comics, Ukiyo-e, Bronze, Blue and White Porcelain, Ghibli Studio, Cyberpunk, and ink painting.
2. Generating Style Images Using AI:
   Using cups as the theme, an AI rendering tool was used to generate images in the aforementioned 20 styles. The primary reasons for choosing Moonshot include: (1) a large

number of example keywords making it easy to use, (2) simple operation requiring only adding the Moonshot LINE official account, eliminating complicated registration processes, (3) official discussion and tutorial areas with the website providing a gallery for referencing others' keyword use, (4) mostly free features with affordable advanced plans, (5) unlimited monthly rendering, and (6) a Chinese interface supporting Chinese keywords, facilitating future feedback.

3. Selecting Images Matching the Preset Styles:
   Three art teachers with relevant backgrounds were invited to discuss multiple generated images for each of the 20 styles, deleting those that were overly similar or had obvious errors (e.g., a cup with two handles on one side). Ultimately, one representative image for each style was retained, resulting in 20 images as shown in Table 4.
4. Creating GIFs in AI-Generated Styles:
   Using AI tools to generate dynamic images for each of the aforementioned 20 style representative images. This study chose Genmo due to its comprehensive database, free access, and unlimited generation capability. The generated GIFs have better quality and appropriateness.
5. Selecting Suitable and Reasonable GIFs:
   Three art background teachers were invited to review and select the 20 generated GIFs based on their compatibility with the artistic styles and the dynamic coherence of the cup subject and the background.
6. Considering the Order of GIFs and Compiling into a Short Animation:
   The 20 GIFs obtained above were compiled into a short animation using video editing software. The order of the GIFs was random, but visually similar GIFs were spaced apart to enhance viewer recognition. To meet short film specifications and viewer preferences, the total length of the video was edited to be within one minute.
7. Adding Background Music:
   Background music is an important factor influencing the animation experience. This study created two versions of the background music: a soft version and a lively version, ultimately completing the fully AI-generated animation.

4 Research Results and Analysis

4.1 Questionnaire Analysis

This study collected a total of 110 valid questionnaires. The first part of the questionnaire included four questions about the participants' basic information, which are summarized as follows:

Gender: 44 males (40%) and 66 females (60%), indicating a higher willingness to participate among females. Age distribution: 20 participants aged 16–20 (18.2%), 44 participants aged 21–25 (40%), 13 participants aged 26–30 (11.8%), 8 participants aged 31–35 (7.3%), and 25 participants aged 36 and above (22.7%). Background in art and design: 40 participants (36.4%) had an art and design background, while 70 participants (63.6%) did not. Experience with AI drawing software: 59 participants (53.6%) had used AI drawing software, and 51 participants (46.4%) had not. The most commonly used AI drawing software among the participants was Midjourney, followed by NovelAi, Stable Diffusion, Moonshot, and StableBoost.

After watching the diverse style experimental animation generated in this study, the participants completed a questionnaire to understand their preferences, appropriateness,

Table 4. AI-generated images in diverse styles (Organized by this study).

recognition, and conveyance of applying diverse artistic styles in short animations. The questionnaire also encouraged participants to spontaneously create and share their AI-generated artworks. The questionnaire included four questions on preference, and the statistics are shown in Table 5 and Table 6.

Table 5. Statistics on the Preference of Animations with diverse artistic styles.

Question	Degree				
	Strongly agree	Agree	Neutral	Disagree	Strongly disagree
Would this diverse styles animation attract you?	30 (27.3%)	38 (34.5%)	31 (28.2%)	9 (8.2%)	2 (1.8%)
Do you like this animation?	22 (20%)	48 (43.6%)	32 (29.1%)	7 (6.4%)	1 (0.9%)
Does this animation have a unique style?	32 (29.1%)	55 (50%)	14 (12.7%)	8 (7.3%)	1 (0.9%)

Unit: person

To understand which animation elements attract the participants, a multiple-choice format was used to let respondents select their choices. Among the 110 respondents, the elements of diverse style animations that were considered attractive in order were: color (57.3%), lighting (52.7%), lines (42.7%), and detail (32.7%). The researchers believe that color and lighting are the most direct and obvious elements that can be visually distinguished, while the elements of lines and detail may require a background in art and design to better appreciate the variations.

Table 6. Statistical Summary of Elements Attracting You in this Animation.

Elements in Animation	Frequency (Multiple Selections)	Percentage
Color	63	57.3%
Light and Shadow	58	52.7%
Lines	47	42.7%
Detail	36	32.7%
Others	4	3.6%

Total respondents: 110 people

In terms of appropriateness, there are three questions, and the statistics are shown in Table 7. According to the data, 79% of people agree or strongly agree that they want to continue watching this type of devise style animation. Furthermore, 83.6% of respondents believe that devise style animations are more suitable for release as short videos rather than feature films. Additionally, 83% of respondents think that devise style animations in the form of short videos are more appealing than feature films.

Table 7. Statistics on the Appropriateness of Devise-Style Animation.

Question	Degree				
	Strongly agree	Agree	Neutral	Disagree	Strongly disagree
Do you want to continue watching this devise style animation?	27 (24.5%)	49 (44.5%)	2 5(22.7%)	8 (7.3%)	1 (0.9%)
Do you think devise style animations are suitable for release as short videos?	45 (40.9%)	47 (42.7%)	16 (14.5%)	2 (1.8%)	0 (0%)
Do you think devise style animations released as short videos are more appealing than as feature films?	41 (37.3%)	50 (45.5%)	18 (16.4%)	1 (0.9%)	0 (0%)

Unit: person

In order to assess whether participants could distinguish diverse artistic styles, three questions were asked regarding recognition, with statistics shown in Table 8. 69% of participants believed they could recognize certain specific artistic styles. Within the experimental animations, researchers designed 20 different artistic styles. The most easily recognizable style was oil painting (70.9%), followed by blue and white porcelain (58.2%), ink painting (45.5%), sketching (44.5%), and pixel art (41.8%). These top five styles are more strongly felt, and most of them were previously studied in high school art classes, making them more familiar to participants. In contrast, the least easily recognizable styles were Rococo, Gothic, Fauvism, and Street, with recognition rates all below thirty percent.

Table 8. Recognition Statistics of diverse artistic styles animations.

Question	Degree				
	Strongly agree	Agree	Neutral	Disagree	Strongly disagree
Can you recognize certain specific styles among these devise styles?	27 (24.5%)	49 (44.5%)	26 (23.6%)	8 (7.3%)	0 (0%)

Unit: person

Regarding the conveyance of information in the videos, statistics are presented in Table 9. 54.5% of participants agreed that diverse styles animations are suitable for conveying plots. However, upon further examination of what kind of plots the animations conveyed, among the 94 respondents who selected "Strongly Agree," "Agree," or "Neutral," 53 people (56.3%) believed that the short videos conveyed the concept of "visual impact," while 10 people (10.7%) thought the videos conveyed advancements in AI image generation technology. Three people (3%) thought the video was a cup advertisement, but 28 respondents provided other different keywords. For example, many mentioned seeing teacups or cups, but could not provide more descriptive phrases. In summary, while most participants agreed that diverse styles animations are suitable for conveying plots, a closer examination revealed that their responses still largely focused on the visual and concrete aspects, without truly mentioning plot details. This highlights a current limitation of AI animations.

Table 9. Conveyance of information of diverse artistic styles Animations.

Question	Degree				
	Strongly agree	Agree	Neutral	Disagree	Strongly disagree
Do you think devise styles animations are suitable with light-hearted music?	25 (22.7%)	35 (31.8%)	34 (30.9%)	16 (14.5%)	0 (0%)

Unit: person

Finally, to encourage participants to engage in this AI-generated image experience, the questionnaire provided the keywords used by the researchers at the end, allowing participants to use AI image generation software to create and upload their own AI-generated images with a cup as the theme. Among the 110 participants, only 25 uploaded images, resulting in a total of 54 images, of which 50 successfully depicted a cup, while 4 had no relation to a cup, as shown in Fig. 1. Among the 50 successfully generated images, participants with an art background produced 28 images, while those without an art background produced 22 images. This shows that the ratio of the number of people to the ratio of images generated is not significantly different. In terms of style, it can be seen that those with an art background generated a wider variety of styles compared to those without. Both groups produced popular styles such as sketches, blue and white porcelain, and pixel art, as listed in the previous tables. However, those with an art background also produced additional styles such as Impressionism, color block style, and comic style, which appeared less frequently or were not included in our research. Conversely, images from participants without an art background tended to be more realistic or lacked a clear style.

Fig. 1. Participants' Experiences with Image Generation.

5 Conclusion and Recommendations

The primary aim of this study was to explore the impact of applying diverse artistic styles and forms to AI-generated short animations and videos. The main findings of this study are summarized as follows:

1. More participants found diverse style animations to be unique rather than simply likable or attractive.

    Analysis of the questionnaires revealed that among those attracted to or fond of the animations, thirty percent of participants were unable to make a selection. However, among those valuing uniqueness, a significant proportion made positive selections, indicating that diverse style animations possess distinctiveness.

2. Diverse styles animations paired with lively musical styles are preferred over softer musical styles.

    Questionnaire analysis showed that among those who preferred lively musical styles, the proportion of participants favoring softer styles decreased by twenty percent, while those deeming them unsuitable increased by ten percent, indicating that lively musical styles are more suitable for diverse styles animations.

3. AI-generated diverse styles animations are still immature in conveying plotlines effectively.

    According to questionnaire analysis, participants' responses regarding plot comprehension remained limited to the visual aspects, indicating that AI-generated diverse styles animations have yet to effectively convey narrative elements to participants.

4. Incorporating elements such as color and lighting into animations can enhance viewer engagement.

    Survey results showed that over fifty percent of participants believed that color and lighting would attract them, suggesting that these elements should be incorporated into animation production to capture viewers' attention.

Participants with an art background produced images with more varied and distinct styles compared to those without. Analysis of generated images showed that participants with an art background produced a greater variety of less familiar styles and fewer monochromatic or repetitive teacup designs, indicating that individuals with an art background are better able to accurately express their desired types and exhibit a greater diversity of styles.

Acknowledgements. We would like to extend our sincere gratitude to all the participants for their enthusiastic participation and valuable feedback. Additionally, we would like to thank three students, Xiao-Zhang Chen, Qing-Hong Jiang, and Jia-Yu Yao, for their assistance with AI image generation and questionnaire data organization, which greatly contributed to the smooth execution of this research. This study received partly financial support from the National Science and Technology Council (NSTC 113-2635-H-143-001-).

Disclosure of Interests. The authors have no competing interests to declare that are relevant to the content of this article.

References

1. Gatys, L., Ecker, A., Bethge, M.: A neural algorithm of artistic style. J. Vision **16**(12), 326 (2016). https://doi.org/10.1167/16.12.326
2. Choi, Y., Uh, Y., Yoo, J., Ha, J.-W.: StarGAN v2: diverse image synthesis for multiple domains. In Proceedings of the IEEE/CVF Conference on Computer Vision and Pattern Recognition, pp. 8188–8197 (2019)
3. McCorduck, P.: AARON's Code: Meta-Art, Artificial Intelligence, and the Work of Harold Cohen. W. H. Freeman and Company, New York (1991)
4. Goodfellow, I., et al.: Generative adversarial nets. In: Proceedings of the International Conference on Neural Information Processing Systems (NIPS 2014), pp. 2672–2680 (2014). https://papers.nips.cc/paper/5423-generative-adversarial-nets.pdf
5. The New York Times. Are A.I.-Generated Pictures Art? https://reurl.cc/ey3M8L. Accessed 14 Aug 2024
6. Feng, J.: Review of research on E-commerce short videos. Front. Human. Social Sci. **4**, 235–242 (2024)
7. Brennan, M.: Attention Factory: The Story of TikTok and China's ByteDance Paperback. Independently published, Traverse City (2020)
8. Li, S.-H.: A study on media technology development and cultural mediation: taking Japanese anime culture as an example. In: Proceedings of the Conference on Digital Networks, Media Technology, and Cultural Studies, pp. 77–88. Wenhe Publishing, Taipei City (2009)
9. Young, A.: Watch: Gorillaz-"Journey to Plastic Beach". Consequence sound. https://consequence.net/2010/07/watch-gorillaz-journey-to-plastic-beach/, last accessed 2024/8/15
10. Lyu, Y., Lin, C., Lin, P., Lin, R.: The cognition of audience to artistic style transfer. Appl. Sci. **11**(7) (2021). https://doi.org/10.3390/app11073290

A Study About Visual Movement of Applying Proximity of Gestalt Psychology on Straight Line

Tsu-Min Hsiang[✉] and Wei-Ming Liao

Department of Design, National Taiwan Normal University, Taipei, Taiwan
c4203490@gmail.com

Abstract. Lines generated by the movement between two Points guide the overall visual direction, thereby producing Visual Movement. Through Aesthetic Principles and Gestalt Psychology, it has been discovered that "Straight Line" can perfectly express rhythmic composition. Furthermore, the relationship between rhythmic composition and Visual Movement is closely intertwined, as different formations can generate Visual Movement. The rhythmic compositions related to Straight Lines include Length, Thickness, Density, and Radiation. Among these, the condition for generating rhythm is "Distance", and a certain quantity of repetition is required. To examine the importance of Length, Thickness, and Distance on the impact of Visual Movement, this study conducted Quantitative Analysis through experimental design. Experiment tasks were designed using Paired Comparison Methods to identify how images formed by Straight Lines might generate Visual Movement. The results reveal that 1. Density and Length both influence Visual Movement, with Density showing a higher significance level, thus confirming Distance as the primary factor. 2. Thickness is not the primary factor affecting Visual Movement of Straight Lines, but in specific contexts where it interacts with Density and Length, it still has some degree of influence. 3. There are interactions and multiplicative effects between Density and Length; as Straight Lines become longer and denser, the sense of Visual Movement becomes stronger. The concept of Visual Movement is bound to appear in any pictorial object. The study aims to understand which elements in the composition of straight lines enhance their sense of movement. As a foundational theory for future Visual Movement.

Keywords: Straight Line · Visual Movement · Visual Composition · Gestalt Psychology · The law of Proximity

1 Introduction

In Principle of Form, Line is one of fundamental concepts. All shapes originate from a point, and when two points are connected to form a line, it can then expand into a plane. According to Lee (2016), Line will guide the viewer's visual direction and the structure of the visual image. Factors such as Length, Thickness, Density, Verticality or Horizontality, Angles and Curves of lines all influence visual composition and the direction of the image. During the process from point to line, motion trajectories appear. The position and formation of lines created through movement within an image influence the viewer.

Viewers might imagine the direction in which the line could move, leading to a concept of Visual Movement both visually and psychologically. Wang (2021) mentioned that our understanding of movement is not a projection of past experiences onto current visual experiences, but rather an inherent visual phenomenon of the object itself. Therefore, the Visual Movement inherent in the elements that compose the image is related to human psychological perception.

Wang (2021) discussed the concepts of Balance and Grouping Principles in visual composition. Balance involves concepts of distance, gravity, and direction, while Grouping Principles focuses on size, shape, color, direction, and texture. Yeh (2002) pointed out that the concept of Grouping Principles can be seen in Gestalt Psychology. Grouping Principles include similarity, proximity, length, space, closure, and continuity. Thus, these principles are related.

In summary, lines, situated between points and planes, possess directionality and are more likely to guide the viewer's visual direction and generate movement. Objects inherently have Visual Movement due to the psychological effects induced by our vision. In visual composition, the concepts of balance and grouping, both part of Gestalt psychology, can reveal the viewer's psychology through visual image phenomena. Compared to a complex painting, graphic designers often encounter simpler images like logos with geometric shapes. However, there has been less in-depth discussions of Visual Movement about them. There are many interrelated concepts between Aesthetic Principles and Gestalt Psychology. The concept of rhythm in Aesthetic Principles suggests that geometric shapes, like lines, can create Visual Movement effects, but it does not further explore the necessary distance and conditions to achieve this effect. Gestalt Psychology discusses Grouping Principles related to objects, such as similarity and proximity, which are also important factors in Aesthetic Principles, yet their relationship to Visual Movement has not been explained.

The three theories mentioned above are fundamental for designers learning about design, primarily focusing on the Visual Movement of static overall images. They do not delve into the discussion of geometric shapes such as lines. However, the concept of Visual Movement is bound to appear in any object with a visual element.

As a consequence, the study will explore Visual Movement of Straight Lines through the proximity theory of Gestalt Psychology. By using the theoretical background of Gestalt Psychology and Aesthetic Principles, the study will conduct experimental designs to investigate how the Visual Movement of Straight Lines are generated. This includes variations in the number, distance, and thickness of lines. The study aims to explore the causes and conditions of Visual Movement, providing an important reference for visual design. The purpose of this study are as follows:

1. To summarize the relationship between Gestalt Psychology's proximity and the Visual Movement of straight lines through literature analysis.
2. To explore the relationship between the length, density, and thickness of straight lines and Visual Movement using Paired Comparison Methods.
3. To analyze the influencing factors of straight lines Visual Movement and their interrelationships.

2 Straight Line's Theories Related to Composition and Visual Movement's

2.1 The Composition of Lines and Straight Lines

Line is the trajectory formed by the movement from a Point in an image to another Point within the same image, having a start and end. Its role in the image is the extension of the point. Chen (2018) mentioned that in geometry, a line only has length and direction, without the concept of width. However, visually, viewers will automatically perceive elongated objects as lines, giving these lines a sense of area. Lin (2007a: 2007b) noted that strictly speaking, if the width-to-length ratio of a line exceeds 1/5, it no longer qualifies as a line. Therefore, to form a line, the length-to-width and area ratios must be considered, and it should not be too thick. Straight Line has a fixed direction and lacks curve or bend characteristics. Yoshioka (2021) stated that an imaginary line is formed between two points, providing a visual guiding effect, and the shortest distance between them is a straight line.

From the above, it is clear that lines possess the characteristic of movement, with the start and end two points connected by a straight line, giving them directionality. The two endpoints guide Visual Movement, directing the visual flow within the image. Due to their directional and length properties, lines can take different forms such as Straight Line, Curved Line, or lines of varying thickness. Among these, straight lines are relatively fixed, categorized into three types: Vertical, Horizontal, and Diagonal. Vertical Line represents the pull of upward and downward forces, Horizontal Line represent the pull of left and right forces, and diagonal lines depend on their directional pull. Additionally, by varying length or spacing, a richer visual presentation can be achieved.

2.2 The Visual Movement of Straight Line in Aesthetic Principles

Lin (2007a: 2007b) mentioned that lines are the most practical elements for expressing the beauty of rhythm, and Aesthetic Principles are systematic theoretical rules developed from the accumulated experiences of human aesthetic perception, analyzed by numerous art scholars. As mentioned in Sect. 2.1, lines can achieve diverse compositions through various forms of variation. To attain overall balance in a composition, Aesthetic Principles can be utilized to assist in achieving this goal.

Regarding Aesthetic Principles, since they are based on the accumulation of aesthetic experiences, scholars have varying interpretations. The study has compiled these interpretations into Table 1, merging those with similar meanings, and categorizing them into ten principles: Simplicity, Repetition, Gradation, Balance, Proportion, Symmetry, Rhythm, Harmony, Contrast, and Unity.

Among the aforementioned Aesthetic Principles, the one most directly related to movement is Principle of Rhythm. Asakura (1991) believes that rhythm and movement are closely related. Rhythm is a form with a sense of movement. Yang et al. (2022) mentioned that rhythm in design refers to the arrangement of elements such as shape or color in repetitive, overlapping, or gradational ways to convey a sense of periodic and orderly movement. Lin (2007a: 2007b) noted that rhythmic form is a type of rhythm and movement, with the expression of lines being the most versatile. This demonstrates

Table 1. Aesthetic Principles proposed by various scholars.

Scholar	Era	Aesthetic Principles
Ohchi (Wang, trans.)	1975	Repetition, Harmony, Gradation, Symmetry, Contrast, Unity, Rhythm, Balance, Proportion
Lin	2007	Order, Repetition, Gradation, Balance, Proportion, Rhythm, Contrast, Harmony, Emphasis
Chen, Lee, Lin	2019	Repetition, Gradation, Symmetry, Balance, Harmony, Contrast, Proportion, Rhythm, Unity
Yoshioka Toru	2021	Repetition, Alternation, Rhythm, Gradation, Symmetry, Balance, Contrast, Harmony, Dominance and Subordination, Unity, Proportion
Yang, Cheng, Huang, Chen, Chuang	2022	Simplicity, Contrast, Gradation, Proportion, Repetition, Symmetry, Balance, Harmony, Rhythm, Unity

that rhythm and movement are closely connected, achieved through the repetitive and continuous arrangement of the same elements, which can evoke a sense of varying speeds.

However, although rhythm involves the repetition and continuity of the same elements, it still requires specific rules to form a shape with a sense of movement. Yoshiko (2021) explained that rhythm takes on different forms based on the arrangement of its elements, as illustrated in Fig. 1. The Figure shows that: a. has no rhythm, b. has rhythm, c. has no rhythm, d. has rhythm.

Fig. 1. Rhythmic Arrangement Patterns (Adapted from Yoshiko 2021).

Wang (2019) classified and analyzed the composition of rhythm, primarily through the principles of repetition, gradation, and proportion. These principles were varied and combined, such as gradation being divided into shape, size, position, direction, and color. She further categorized them into nine types: shape, vertical, horizontal, size, distance, spatial sense, density, speed, and radiation.

The study re-analyzes the rhythmic types related to straight lines mentioned above, which include Shape Length, Shape Thickness, Distance Density, and Linear Radiation. It was found that the condition for producing rhythm is distance, and a certain amount of repetition is required. For example, the types of distance density and linear radiation, as shown in Table 2, generate rhythm based on the distance between objects and their positional relationships.

Table 2. Types of Rhythmic Composition Related to Straight Lines.

Type	Figure	Description
Shape Length		Through the repeated arrangement and gradation of long and short straight lines, the original shape can be altered. However, a certain distance with a gravitational effect must be maintained between the lines to produce a rhythmic effect.
Shape Thickness		Through the repeated arrangement and gradation of thick and thin straight lines, the original shape can be altered. However, a certain distance with a gravitational effect must be maintained between the lines to produce a rhythmic effect.
Distance Density		By varying the spacing between lines to be sparse or dense, a rhythmic effect can be achieved. Typically, proportional relationships such as arithmetic progressions, geometric progressions, or Fibonacci sequences are used to determine the distances between lines.
Linear Radiation		Creating a rhythmic effect through radial patterns can be achieved by increasing the density of lines as the quantity of straight lines increases. This arrangement visually creates a radiating or diverging effect from the center outward.

Through the aforementioned types of rhythmic compositions involving straight lines, it is evident that rhythm itself possesses dynamism. This is because under certain conditions such as shape, space, and time constraints, "distance" becomes the primary factor influencing the direction of movement.

2.3 The Visual Movement of Straight Line in Gestalt Psychology

Liu (2011) pointed out that a complete work visually presents dynamic balance. Our vision tends to group together objects with similar perceptual properties, following the Grouping Principles of Gestalt Psychology.

Yeh (2002) mentioned Wu (1986) explanation of Grouping Principles as a method of using perceptual grouping. We assign relationships such as connections and sequences to the stimuli we receive. Through the "similarity principle," we determine the degree to which stimuli resemble each other and thus their degree of mutual belonging. Grouping Principles has been developed into many laws (principles) through scholarly research, and this study organizes them into Table 3.

Table 3. Principles of Gestalt Psychology proposed by various scholars.

Scholar	Era	Principle of Aesthetic Construction
Bruce, V., Green, P. R. & Georgeson, M. A	1996	Proximity, Similarity, Common Fate, Good Continuation, Closure, Relative Size (Surroundedness, Orientation, Symmetry), The Law of Prägnanz
Maurice Hershenson	1999	Proximity, Similarity, Closure, Good Continuation, Common Fate
Yeh, C. H	2002	Prägnanz, Closure, Proximity, Similarity, Continuity, Familiarity, Closed Forms, Objective Set, Common Fate, Common Movement, Symmetry, Simplicity
Liao M. H	2022	Similarity, Proximity, Continuity, Closure, Symmetry, Figure-ground

From the above, it is known that similar objects, such as those visually similar, similar in color, similar in spatial position, and similar in movement direction, are grouped together according to Grouping Principles. Therefore, straight lines are also visually grouped due to our visual tendency towards balance.

The principle related to distance in the aforementioned Grouping Principles is "proximity," According to Yeh (2002), the proximity principle signifies that under equal conditions in a stimulating context, elements that are closer to each other in both time and space will form a group and be perceived together. In the Sect. 2.2 on Aesthetic Principles, it was mentioned that rhythm is generated by distance and requires a certain amount of repetition. This psychological aspect also influences visual perception, causing similar elements to group together. From this, we can understand that Aesthetic Principles and Gestalt Psychology influence each other, impacting visual perception through psychological processes. This also explains within the types of rhythmic construction that if distances change, the rhythm disappears; therefore, maintaining a certain distance is necessary to sustain its rhythmic effect.

3 Experimental

3.1 Experiment Background, and Procedures

In Sect. 2, the relationship between straight lines, Aesthetic Principles, and Gestalt Psychology was explained. It was found that "distance" is the most important factor influencing Visual Movement of straight lines. However, the specific distance and conditions required to generate Visual Movement are not detailed in the literature on Aesthetic Principles and Gestalt Psychology. The study references the concepts of Visual Movement of Straight lines in Aesthetic Principles, as mentioned in Table 2, which include Shape Length, Shape Thickness, and Distance Density. Therefore, this study will design experiments to explore how proximity in Gestalt Psychology affects and determines the Visual Movement of straight lines.

This experiment will use Paired Comparison Methods to design questions and conduct pairwise comparison tests. The experiment consists of four stages: the first stage involves identifying the experimental scope and limitations through literature review; the second stage involves experimental design, preliminary testing, and revisions; the third stage involves conducting the experiment and inviting scholars or designers with a design background to participate in the tests; and the fourth stage involves collating the experimental records and performing data calculation and analysis.

3.2 First Stage: Experiment Scope and Limitations

Through literature review, we identified the fixed and variable ranges. The fixed range includes the carrier and drawing size of the patterns, with changes only applied to the patterns themselves. The variable elements can explore Distance, Quantity, Thickness, and Length to determine the proportions most likely to generate Visual Movement. This study needs to control the quantity of each variable to better observe the impact of each variable. Therefore, Distance, Thickness, and Length are each set to three levels, with values increasing or decreasing in arithmetic progression, and placed at the center of the frame. The variable aspects of this study are listed below:

Quantity. Based on Yoshiko's (2021) basic design textbook, "quantity" is set to 15 lines. Creating Visual Movement requires a certain number of lines. Even numbers tend to be too stable due to symmetry, while odd numbers lack symmetry, making imbalance more apparent and showcasing Visual Movement. Therefore, the experiment is designed to vary using odd numbers and is conducted on a 17 × 17 grid (each grid being 5mm), ensuring patterns do not touch or exceed the outer frame.

Distance. Distance must exhibit a certain proportional relationship to perceive Visual Movement. Therefore, this study utilizes an arithmetic progression to create variations in density using grids as the basis for line spacing. The first variation involves 15 straight lines spaced at 1.5 grids → 1 grid → 0.5 grid → 0.5 grid → 1 grid → 1.5 grids → 1 grid → 0.5 grid → 0.5 grid → 1 grid → 1.5 grids → 1 grid → 0.5 grid → 0.5 grid, cycling through this interval pattern. The second variation decreases by 0.25 grids, while the third variation decreases by 0.5 grids.

Thickness. According to Lin (2007), the description of line thickness relative to its length indicates that once the width-to-length ratio exceeds 1/5, it begins to deviate from the conditions of a line. Therefore, the proportion of line thickness should not exceed this range. The experiment uses an Arithmetic sequence of 1pt, 3pt, and 5pt for varying line thicknesses.

Length. Length varies based on the drawn pattern, ensuring that the width-to-length ratio does not exceed 1/5. Therefore, the shortest length is set to 5 grids, and an Arithmetic sequence is used for the line lengths, ranging from 5 grids to 15 grids, with increments of 5 grids (i.e., 5 grids, 10 grids, 15 grids).

3.3 Second Stage: Experimental Design, Testing, and Revision

First, pattern drawing was initiated based on the conclusions drawn from First stage. The experimental patterns were designed using a 17 × 17 grid. Each pattern consists

Table 4. Experimental Pattern.

Thickness	Pattern	Pattern	Pattern
T1(1pt)	D1L1T1	D1L2T1	D1L3T1
T1(1pt)	D2L1T1	D2L2T1	D2L3T1
T1(1pt)	D3L1T1	D3L2T1	D3L3T1
T2(3pt)	D1L1T2	D1L2T2	D1L3T2
T2(3pt)	D2L1T2	D2L2T2	D2L3T2
T2(3pt)	D3L1T2	D3L2T2	D3L3T2
T3(5pt)	D1L1T3	D1L2T3	D1L3T3
T3(5pt)	D2L1T3	D2L2T3	D2L3T3
T3(5pt)	D3L1T3	D3L2T3	D3L3T3

of 15 lines, and they were cross-designed according to Density (D), Length (L), and Thickness (T). The patterns were labeled using codes D, L, and T to represent the three variables. Each pattern was numbered sequentially starting from D1L1T1, resulting in a total of 27 different patterns as listed in Table 4.

After completing the experimental patterns, the interface design for the patterns was conducted. This experiment used the BENQ PD2700Q 27-inch monitor for testing, designing the interface based on its monitor size and 2560 × 1440 px resolution. For the size of the patterns, according to Liao (1994), who quantitatively studied color simultaneous contrast, the size of the color areas for pairwise comparison experiments was set to 1.5 × 1.5 cm. Therefore, the patterns were scaled to this specified size, adjusting the width variable accordingly. Regarding experiment background colors, following Kaufmann and O'Neill (1993), experiments showed that a black background led to confusion with unsaturated colors, while a gray background improved the screen. Thus, the experiment used color code #8a8b8c as the background color.

Finally, the patterns from Table 4 were used to create questions for pairwise comparison. The patterns were categorized into three groups based on their thickness (T1, T2, T3), with each group consisting of 9 patterns. Subsequently, each set of 9 patterns was further divided into pairs in a crosswise manner, resulting in 36 questions per group (excluding left-right considerations). Therefore, a total of 108 questions were created across T1, T2, and T3 groups. Next, the items were encoded and randomized using a random number generator (RNG) utilizing built-in cryptographic functions (crypto.getRandomValues). This process generated item numbers randomly, ensuring that the 36 questions for T1, T2, and T3 were shuffled in a non-repeating manner. This setup enabled the pairwise comparison experiment between patterns to proceed in a randomized fashion.

The experimental questions from T1, T2, and T3 will undergo a crossover sorting classification into 6 groups: A (T1 → T3 → T2), B (T2 → T1 → T3), C (T3 → T2 → T1), D (T2 → T3 → T1), E (T3 → T1 → T2), and F (T1 → T2 → T3). This categorization is implemented to prevent fixed sequences during the experiment, thereby ensuring that the experimental screen order does not bias the results.

3.4 Third Stage: Conducting the Experiment

Once the above experiment design is completed, preliminary testing will be conducted. After testing, adjustments, and revisions are finalized, the experiment will officially commence. The experiment started in January 2024 and concluded in June.

Experimental Location. To ensure the accuracy of the experiment, a fixed experimental location will be used where all participants will be tested in the same experimental space. The seating arrangements for both observers and participants will be planned and organized.

Participants. The experiment was publicly announced online to recruit participants. From January to May 2024, forms were widely distributed to design groups and school communities. There're 36 participants, evenly split between 18 males and 18 females, ranging in age from 18 to 70 years old. Participants are allocated based on gender within each group. This experiment aims to serve as a reference for designers in future creations. Therefore, all participants have received at least one semester of training in basic design courses at school. Currently, participants include working students, designers, and educators.

Experimental Steps. The experiment consists of three parts: T1, T2, and T3, each comprising 36 questions, totaling 108 questions. There are 6 sets of crossover classifications, arranged according to the order in which participants signed up for the test. Each experiment duration is approximately 40 to 60 min, varying depending on each participant's testing conditions, with a 10-min break between each part.

3.5 Fourth Stage: Data Analysis for the Experiment

Experimental data is stored in the content management system, where participants' selected answers and response times are recorded during the experiment. This study employs ANOVA using SPSS statistical software to examine whether these factors can generate Visual Movement. Theoretically, variations in these factors should create corresponding perceptions of movement under certain conditions.

Initially, the experiment uses density (D), length (L), and thickness (T) as three factors. A three-way ANOVA is conducted to test these factors individually—D, L, and T. However, due to the experimental design, D and L were not cross-compared with T in previous item setups. Therefore, in the analysis of the T factor, changes in the mean are not apparent. Consequently, T is singled out for a one-way ANOVA to determine its specific effects. Next, the study employs a two-way ANOVA to examine the combined effects of D and L two factors, exploring whether their interaction has a cumulative impact on overall Visual Movement.

4 Experimental Results and Analysis

4.1 Three-Way ANOVA for Straight Line's Density, Length, and Thickness

In the experiment, there are 108 questions, derived from different levels of density (D), length (L), and thickness (T). Hence, a three-way ANOVA is appropriate for this experiment.

For D, L, and T, the experiment uses a 95% confidence interval. However, because T was not cross-compared with D and L in the three-way ANOVA, conclusive results regarding its interaction effects cannot be drawn from this analysis. Therefore, in addition to the three-way ANOVA, T was separately analyzed using a one-way ANOVA to examine its effects within a 95% confidence interval.

First, the scores given by 36 participants for the 27 patterns in Table 4 are collected. In the experiment, patterns corresponding to groups T1, T2, and T3 are labeled as S01 to S27. Next, calculate the scores for patterns S01 to S27 within each respective group (T1, T2, T3) using SPSS statistical software to determine their average scores.

Following that, a three-way ANOVA was conducted, and the results are presented in Table 5. From the analysis in Table 5, it was found that for D, $P = 0.050$, just reached the conventional significance threshold of 0.05. For L, $P = 0.099$, although very close to 0.05, suggesting a marginal significance level. To examine whether this marginal significance is due to errors or other factors, Fig. 2 was used to visualize the trends.

In Fig. 2, D, L, and T were examined separately. For instance, for D, the average scores of D1, D2, and D3 across the 108 questions were plotted, showing an upward trend, indicating "higher line density leads to higher perceived movement." Conversely, for L, the plots of L1, L2, and L3 showed a downward trend, suggesting "longer length leads to higher perceived movement." Since T was not cross-compared with D and L, its values could not be calculated here. However, based on the clear trends observed in Fig. 2, it suggests that D and L's effects might be significant due to consistent patterns, potentially beyond mere error margins. Therefore, based on the trends observed in Fig. 2, D and L could be considered significant factors influencing perceived movement, despite the marginal P-values initially observed in the ANOVA analysis.

Table 5. Three-way ANOVA for D, L, and T (Rounded to the third decimal place).

Source		Type III Sum of Squares	df	Mean Square	F	Sig
D	Sphericity Assumed	88.154	2	44.077	3.125	**0.050**
	Greenhouse-Geisser	88.154	1.067	82.636	3.125	0.083
	Huynh-Feldt	88.154	1.073	82.172	3.125	0.083
	Lower-bound	88.154	1.000	88.154	3.125	0.086
L	Sphericity Assumed	121.463	2	60.731	2.383	**0.099**
	Greenhouse-Geisser	121.463	1.124	108.039	2.383	0.128
	Huynh-Feldt	121.463	1.136	106.949	2.383	0.127
	Lower-bound	121.463	1.000	121.463	2.383	0.132
T	Sphericity Assumed	0.000	2	0.000		
	Greenhouse-Geisser	0.000				
	Huynh-Feldt	0.000				
	Lower-bound	0.000	1.000	0.000		
D * L	Sphericity Assumed	24.920	4	6.230	3.833	**0.005**
	Greenhouse-Geisser	24.920	2.261	11.020	3.833	0.021
	Huynh-Feldt	24.920	2.425	10.274	3.833	0.019

4.2 One-Way ANOVA for Straight Line's Thickness

In Three-way ANOVA, no conclusions were drawn for T. This is because, in the prior experimental design, D and L were not cross-compared with T. Therefore, T will now be analyzed separately using a one-way ANOVA to determine its specific effects. First, all scenarios will be listed, and a one-way ANOVA will be conducted. For example, under the D1L1 scenario, the average score for T1 will be calculated to examine the impact of all T factors on the perceived movement.

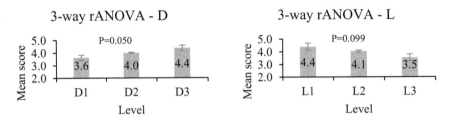

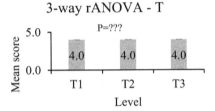

Fig. 2. Three-way ANOVA for D, L, and T individual analysis chart.

Overall, the scores for the T factor across various scenarios do not show a consistent trend. However, in certain scenarios, such as D1L2 and D3L3, there appears to be an upward trend, suggesting that there may be more nuanced factors at play. Excluding these two scenarios, the trends in other scenarios are not clear. Additionally, the one-way ANOVA results for T do not yield a P-value less than 0.05, indicating that T is not a significant factor.

4.3 Two-Way ANOVA for Straight Line's Density, Length

In Table 5, the three-way ANOVA includes interaction analysis, meaning it examines whether the effects of combining any two factors are additive. However, since T could not be calculated in the three-factor ANOVA, interactions involving T are not considered. Only the interaction between D and L is examined, and the result shows a P-value of 0.005, indicating significant interaction. The relevant data is as described in Table 6 below. Therefore, we will further explore the interaction between D and L.

Table 6. Three-way ANOVA for D*L (Rounded to the third decimal place).

Source		Type III Sum of Squares	df	Mean Square	F	Sig.
D * L	Sphericity Assumed	24.920	4	6.230	3.833	**0.005**
	Greenhouse-Geisser	24.920	2.261	11.020	3.833	0.021
	Huynh-Feldt	24.920	2.425	10.274	3.833	0.019

First, the average scores for the interactions between T1, T2, and T3 and D1, D2, and D3 are listed in Table 7. This data is then used to create Fig. 3 to examine the results under interaction effects. The results indicate a multiplicative effect between D and L. Therefore, in future experiments, when designing patterns that incorporate D and L, participants should be shown patterns with significantly different combinations, such as "long and dense" and "short and sparse." Theoretically, the participants' perceptions should differ greatly between these two scenarios.

Table 7. Three-way ANOVA for D, L, and T (Rounded to the third decimal place).

Mean	L1	L2	L3
D1	11.4	11.0	10.4
D2	13.4	12.1	10.4
D3	14.9	13.4	11.1

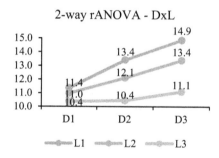

Fig. 3. Two-way ANOVA for D, and L analysis chart.

5 Conclusion

1. To summarize the relationship between Gestalt Psychology's proximity and the Visual Movement of straight lines through literature analysis: research shows that "straight lines" best express rhythm. The rhythm formation in Aesthetic Principles and the proximity principle in Gestalt psychology can create Visual Movement. The rhythm-related characteristics of straight lines include Shape Length, Shape Thickness, Distance Density, and Linear Radiation. The conditions for generating rhythm include distance and a certain degree of repetition.
2. To explore the relationship between the length, density, and thickness of straight lines and Visual Movement using Paired Comparison Methods: Based on Sect. 4 of the experimental analysis regarding density (D), length (L), and thickness (T).
    a. Through three-way ANOVA, it was found that both density (D) and length (L) significantly impact Visual Movement, with density (D) showing a higher level of significance. This confirms that distance is the primary factor influencing Visual Movement in straight lines.

b. One-way ANOVA indicates that thickness (T) is not a primary factor influencing Visual Movement of straight lines. However, in specific conditions, it still exerts some degree of influence.
3. To analyze the influencing factors of straight lines Visual Movement and their interrelationships: Through two-way ANOVA, it was determined that there is an interaction effect between density (D) and length (L). Specifically, when a straight line tends towards being "long and dense," the perception of movement becomes stronger and exhibits an amplification effect. Thus, there is a multiplicative effect between D and L. From the above, it is evident that both density (D) and length (L) influence Visual movement. Density (D) shows a higher significance level, confirming that "distance" is the primary factor affecting Visual Movement in straight lines. Moreover, the interaction and multiplicative effects between D and L indicate that as the straight line becomes longer and denser, the perception of Visual Movement increases.

References

Bruce, V., Green, P.R., Georgeson, M.A.: Visual Perception, 3rd edn. The Psychology Press, East Sussex (1996)
Hershenson, M.: Visual Space Perception. The MIT Press, Boston (1999)
Ohchi, H.: Basis of Design. (Wang, H.H., trans.) Ta Lu Book, Taipei (1975)
Wang, H.H.: The Psychology of Art. Artist Magazine, Taipei (2021)
Wu, S.M.: Psychology. Sanmin, Taipei (1986)
Lin, C.H.: Basic Design-2 Dimensional Composition Theory. New Wun Ching, Taipei (2007a)
Lin, C.H.: Basic Design Theory-The Creativity Thinking of Form and Composition. Chuan Hwa Book, New Taipei (2007b)
Chen, M.Y., Lee, M.L., Lin, H.Y.: Fundamental Design Practice. Tiked Books, New Taipei City (2019)
Asakura, N.: Fundamental Problems of Creating in the Two-Dimensional Space. (Lu, C.F., trans.) North Star Books, New Taipei (1991)
Yang, C.T., Cheng, C.G., Huang, R.Y., Chen, C.L., Chuang, T.C.: Design Introduction. Chuan Hwa Book, New Taipei (2022)
Liu, S.L.: Yi Shu Xin Li Xue-Yi Shu Yu Chuang Zao, 5th edn. Artist Magazine, Taipei (2011)
Liao, M.H.: Yin Yue Zhong Di Zhi Jue Xian Xiang. Silk Road 82, 34–41. Taipei Chinese Orchestra, Taipei (2022)
Kaufmann, M.C.: O'Neill: colour names and focal colours on electronic displays. Ergonomics **36**, 881–890 (1993)
Barmley: Paired Comparison Methods. In: Qualifications and Curriculum Authority, pp. 246–300 (2008)
Chen, H.K., Lin, P.C.: A study of applying visual principles of visual movement in visual communication design. J. Natl. Taiwan College Arts **74**, 81–98 (2004)
Chen, H.K., Chang, C.Y.: The influence of visual balance on visual movement and its applying research for visual principle. J. Natl. Taiwan College Arts **74**, 75–88 (2003)
Wang, Y.L.: Research and creation of the application of rhythm to chinese character moving posters design. Master's thesis, National Taiwan Normal University (2019)
Lee, Y.L.: A discussion on the meaning of line drawing and its applications of line element. Master's thesis, Hsuan Chuang University (2016)
Liao, W.M.: Quantitative research of simultaneous contrast of colorzeng. Master's thesis, National Chiao Tung University (1994)

Lin, C.C.: Dian Xian Mian De Hui Tu Gui Ji Yan Jiu. Master's thesis, National Taiwan Normal University (2004)

Chen, J.R.: The creation and study on visual forming in tabletop game design. Master's thesis, National Taiwan University of Arts (2018)

Yeh, C.H.: A study on the effect of attention in product interface structure - applying gestalt psychology. Master's thesis, National Yunlin University of Science & Technology (2002)

Yoshioka Toru Homepage. https://www.yoshiokatoru.com/. Accessed 10 Dec 2023

Application of the Immersive Virtual Training

The First Immersive Simulation Cave System to the First Responder Training in Taiwan

Hao-Yang Chen[1], Pey-Yune Hu[1(✉)], and Lien-Shang Wu[2]

[1] Kun Shan University, No. 195, Kunda Road, Tainan City 710303, Taiwan
t093000397@g.ksu.edu.tw
[2] Kaohsiung Normal University, No. 62, Shenjhong Road, Kaohsiung City 824004, Taiwan

Abstract. The Virtual Reality Technology has become one of the important digital trends in decades. In nowadays the VR technology engages with people's common lives gradually and the variety of the digital experience is though out many fields. Multimedia technologies become an inevitable trend that benefits our life, and furthermore, saves lives. The study is based on a practical VR application which had been co-established by National Chung-Shan Institute of Science and Technology and Kun Shan University. And the purpose is seeking enhancement for the first responders' training programs of the national army. We built a cave automatic virtual environment (CAVE) simulation system to simulate three disaster scenarios that projects surrounding motion picture with 12 4k high-definition projectors. To strengthen the immersive experience, the system also works with 4 high fidelity human patient simulators (HPS), 4 earth quake simulation platforms, a heating generator and 5.1 surround sound. It is the first CAVE application built for the first responders training programs in Taiwan. With the advantages of the VR technologies and physical effects, the system would trigger personal's stress and pressure similar to real situations. Furthermore, it enhances first responder's assessment and preparation that the regular drills are hard to achieve.

Keywords: Virtual training · CAVE system · First responder · Disaster simulation · Visual effects

1 Instructions

1.1 Application Technology Profile

Digital technologies have extended our imagination in many fields. It delivers innovation and diversity of problem-solving. Virtual Reality (VR) technology is one of the good examples that people have keep expand its potential to apply the advantages to benefit our live. In recent years, VR technology has emerged as a transformative force across multiple sectors, including entertainment, education, healthcare, and more. Characterized by its ability to create immersive, interactive, and highly realistic digital environments, VR is redefining how individuals engage with digital content and each other. This academic paper delves into the multifaceted impact of VR technology, analyzing its evolution, applications, and implications for the future.

The concept of VR can be traced back to the mid-20th century with the development of early flight simulators and computer graphics. However, it is the advancements in computational power, graphics processing, and sensory technologies in the past decade that have propelled VR into mainstream consciousness. VR CAVE (Cave Automatic Virtual Environment) is a type of immersive virtual reality system that allows users to experience and interact with digital environments in a room-sized cube. The term "CAVE" was originally coined by researchers at the University of Illinois at Chicago in the early 1990s. The Cave Automatic Virtual Environment (CAVE) represents a pioneering development in the field of immersive virtual reality (VR), offering a unique platform for experiencing and interacting with digital environments. Developed in the early 1990s by researchers at the University of Illinois at Chicago, the CAVE system is designed to project stereoscopic 3D images onto the walls and floor of a room-sized cube, thereby creating an encompassing virtual environment around the user (Cruz-Neira et al. 1992).

A VR CAVE typically consists of a cube-shaped room where three to six internal walls (in some cases including the floor and the ceiling) function as projection surfaces. High-resolution projectors display images or motion pictures on these surfaces, some models work with stereoscopic glasses which provide users with a compelling sense of depth and immersion (Cruz-Neira et al. 1993). The integration of motion sensor systems enables real-time interaction and navigation within the virtual space, as the system adjusts the perspective based on the user's movements, thus enhancing the feeling of presence within the digital environment. Base on the limitation of real time motion capture technology, the CAVE installation comes with motion sensor allow only one person to experience at a time.

1.2 Application Technology Profile

The applications of VR CAVE technology span a diverse range of fields. In scientific research, it facilitates the visualization of complex data sets, such as molecular structures, astronomical data, and geological formations, enabling researchers to gain new insights through immersive exploration (Leigh et al. 1997). In engineering and industrial design, CAVEs are employed for virtual prototyping and simulation, allowing designers to evaluate the ergonomics, aesthetics, and functionality of products before physical prototypes are constructed (Mapes and Volino 2001). The medical field benefits from VR CAVEs through applications in surgical training, where practitioners can practice procedures in a risk-free environment, and in therapeutic settings, where immersive environments are used for exposure therapy and pain management (Riva et al. 2019). Firefighters training drills are essential to elevate firefighters' perception and help to do preparation toward danger scenarios. With the advantages of VR CAVE system, some specific dangerous drills such as training in enclosure fire, would be considered as a good approach to training firefighters. (Grabowski 2021) Immersive simulation has been used for training first responders since 2008. A CAVE system designed for observing first responders' behavior in disaster scenes was built at the University of Michigan. Group of qualified specialists included a senior emergency medicine (EM) attending physician and an EM chief resident that both have substantial additional experience in disaster response, tactical EM, fire/rescue/emergency medical services (EMS) operations, and out-of-hospital education. The team developed the disaster scenarios with representatives from law enforcement, fire, and EMS (Wilkerson et al. 2008). The scenario is about a bombing disaster scene in a football stadium, 12 victims need to be proper approached according to the criteria. The research team evaluated 14 participates performance in the installation and a structured interview was conducted to help the researches to understand how well the simulation triggered participants anxious as engaging with actual scenario. And the study results prove that the VR CAVE installation is effective to benefit first responders training.

1.3 The Applications of VR CAVE Technology in Field Training and Education

Virtual Reality (VR) technology has significantly impacted various sectors by providing immersive and interactive environments. The Cave Automatic Virtual Environment (CAVE), a sophisticated VR system, has shown particular promise in training and educational applications. Such teaching methods and learning experience will be better than traditional apprenticeships, which are taught directly by teachers (Vanderbilt 1990). The realistic and immersive technology of virtual reality is a digital interactive media solution that allows learners to be more immersed in the environment (Visch et al. 2010). This paper examines the utilization of VR CAVE technology in these domains, highlighting its methodologies, benefits, and implications for the future of training and education.

- **Applications in Medical Training**

    VR CAVE technology has proven particularly beneficial in medical training. It provides a realistic and interactive platform for practicing complex surgical procedures, emergency response techniques, and routine medical tasks in a risk-free environment. Riva et al. (2019) emphasize the use of VR CAVE in surgical training. Detailed anatomical models and interactive simulations enable surgeons to practice and perfect procedures before performing them on actual patients. This enhances the surgeons' skills and improves patient safety by reducing the likelihood of errors during real surgeries.

- **Applications in Industrial and Engineering**

    The industrial and engineering sectors utilize VR CAVEs to train personnel on complex machinery and processes. These virtual environments allow engineers and technicians to gain hands-on experience without the associated risks of real-world training. (Mapes, and Volino 2001) discuss the application of VR CAVE systems in industrial training. They highlight that these systems offer a cost-effective solution for simulating hazardous environments and training personnel in safety procedures. The immersive nature of CAVEs ensures that trainees can experience realistic scenarios, leading to better retention of knowledge and skills.

- **Applications in Military and Defense Training**

    The military and defense sectors have adopted VR CAVE technology for training purposes. Soldiers and officers use CAVE systems to simulate combat scenarios, strategic planning, and mission rehearsals in controlled parameters. Leigh (1997) underscore the importance of VR CAVEs in military training. The ability to simulate diverse terrains especially the models function as flight simulators and combat situations enhances the preparedness and effectiveness of military personnel. The detailed and interactive simulations help trainees develop critical skills and improve their performance in real-world operations.

- **Applications in Educational Settings**

    VR CAVE technology has also found applications in educational settings, providing students with immersive learning experiences that enhance understanding and engagement. In subjects like history, biology, physics, and geography, CAVEs offer students the opportunity to explore virtual reconstructions, conduct virtual experiments, and interact with educational content in a meaningful way. Johnson et al. (2016) highlight the use of VR CAVEs in education, noting that immersive environments can significantly improve student engagement and comprehension. By providing a hands-on learning experience, CAVEs help students grasp complex concepts more effectively than traditional teaching methods.

- **Applications in Firefighters drill Settings**

    Enclose fire scenario is an important subjects to fire fighter trainings. As one of the most dangerous tasks for firefighters, it is crucial that effective training applications help to develop strategies toward actual situation. VR CAVE and HMD (Head Mounted Display)-based virtual training simulators have already conducted to elevate training quality. The research conducted by Andrzej Grabowski (2021) reveals both of CAVE and HMD-based VR are effective tools for training firefighters in enclosure fire scenarios. The findings support the integration of VR training simulators into firefighter training programs complementing traditional training methods to prepare trainees for real-life challenges.

- **Applications in First Responders Training**

    It is essential to deal with challenges of live rescued in any disaster scenario to first responders. Training of high-acuity, low-frequency events like mass casualty incidents is the most important of all training programs. Traditional training programs are hard to simulate disaster scenarios which is convincible enough to trigger anxiety and tension. VR CAVE technologies can be convincing options to help first responder personnel doing preparation. The realistic and stressful scenarios provided by the CAVE environment help bridge the gap between theoretical knowledge and practical application. While challenges such as cost and technical complexity remain, the benefits of enhanced realism, safety, and repetitive practice make immersive VR simulations a valuable addition to traditional training methods. The CAVE application in University of Michigan has been proved as a good model of integrate VR technologies into first responder training (Wilkerson et al. 2008).

1.4 The Advantages and Challenges of VR CAVE Applications

The VR CAVE applications in training and education have several advantages as following points:

1. **Immersion Experience**: The capabilities of CAVE environments provide an immersive training and learning experience, triggering senses that traditional methods can't achieve for enhancing skill acquisition and knowledge retention.
2. **Safe Environment**: Participants can go through tasks and experiments in a risk-free environment, reducing the likelihood of accidents.
3. **Collaboration**: VR CAVE would be considered as a better model compared to other VR technologies in multiple user participation. Participants can interact within the same virtual environment, facilitating teamwork and collaborative learning exercises.

There are challenges also that the limitations VR CAVE has been down played, and some are quite obvious. Despite the advantages above, VR CAVE technology also presents challenges, including high costs of installation and maintenance, technical complexity, and the need for significant physical space (Cruz-Neira et al. 1993). The complexity of system integration is another barrier to application development. Compare to other VR technologies such as HMD VR (head mounted display), the CAVE system without VR function impose a less cognitive and physical loads (Grabowski 2021).

1.5 Conclusion

In general, Immersive VR simulation, particularly using CAVE technology, can be an effective tool for training first responders. The realistic scenarios provided immersive experience help to bridge the gap between theoretical knowledge and practical application. Despite the challenges like cost and technical complexity, the benefits of enhanced realism, safety, and repetitive practice make VR CAVE simulations a valuable addition to traditional first responders training methods.

2 Pre-production Overview

2.1 Installation

The Application is the first immersive simulation system with VR Cave Automatic Virtual Environment (CAVE) for training the first responders of National Defense R.O.C in mass casualty events. The simulation utilized a high-resolution CAVE integrated with installations which enhance the immersive experience, 4 high-fidelity human patient simulators (HPS), 4 earth quake simulation platforms, a heating generator, a 10 blades heavy duty turbo fan and 5.1 surround sound system. The Installation is in the Medical Training Center of Medical Affairs Bureau Ministry of National Defense. First of all, three disaster scenarios are considered to be the most critical challenges to first responders in Taiwan, and which are earthquakes in urban areas, landslides in mountain villages and severe traffic accidents in highway tunnels to be setup as simulation training subjects. The VR experience environments are based on real sites. All scenarios refer to true incidents which are generally regarded as severe tragedies in Taiwan as Table 1 (Fig. 1).

2.2 Participants

Two multimedia servers show control software and hardware tools for real time projection mapping. Twelve 4K projectors are driven by the two servers that project surround

Application of the Immersive Virtual Training 309

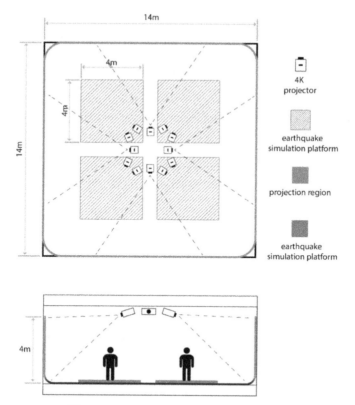

Fig. 1. The configuration of the Installation.

Table 1. The profile of scenarios and reference.

Disaster scenario	Environment	Reference		
		incident	Date/ location	casualty
earthquake	Taipei City	921 earthquake	Sep. 21, 1999 /Taiwan	2417 deaths 29 missing 11,305 wounded
landslides	mount–ain area in Nantou county	Bazhang River tragedy	July 22, 2000 /Chiayi county	4 deaths
traffic accident	Hsuehshan Tunnel	Vehicle fire accident in Hsuehshan Tunnel	May 7, 2012/ Hsuehsa-n Tunnel	2 deaths, 31 wounded

motion pictures on a room-size wall surface as 4 m high and the dimensions is 14 m by 14 m. The servers process not only rendering real time sequence, but also to function as hardware integrate system capable of controlling projection images and physical effects simultaneously. The project was supervised by Aeronautical Systems Research Division of National Chung-Shan Institute of Science & Technology (NCSIST), and also in charge of application development with Visual Communication Design Department in Kun Shan University (KSUVCD). A project manager and a technical supervisor from NCSIST cooperated with an associate professor who has expertise in new media from KSUVCD and a production coordinator as the core team of production management. The development initiated in mid-2020. For the matter of the 3 years long-term project which many experts with different professional back ground engaged the production pipe-line, multi preproduction tasks were conducted for a better understanding of the project development in full scope, and also the risks management in every perspective. The production team participated the development in the latter part of the preproduction. The production group consists of three expert groups, a technical director, a visual effects and animation team, and a green screen production crew which was directed by an associate professor from Motion Picture Department in Kun Shan University (KSUMPD), and whose expertise is visual effects. There were also senior first responders of Tainan City Government Fire Bureau and three military instructors from the Medical Training Center of Medical Affairs Bureau Ministry of National Defense participated the project as exercise advisors. In the final stage of development, there were 3 performance tests and 2 pressure tests had been conducted in the proper order.

2.3 Pipeline and Pre-production

The term pipeline can be realized as a well-defined set of processes for achieving a certain result. It is also a linear workflow with irreversible processes [14]. A large scale project development like this one, a good pipeline design with precise consideration would not only make the workflow organize, but also well risk management and efficient workflow. Several preparation procedures and working protocols needed to be done before the production. Creating the Pipeline was one of the essential tasks in the pre-production period and also considered to be the most important job. The subject studies and field research are the key points in the very first stage. Figure 2 is the flow framework of the pre-production.

2.4 Development Workflow

Development workflow is sequence of processes that teams engaged in the develop follow to create, manage, test and deploy software or projects. It ensures that tasks are completed in an organized and efficient manner, often incorporating best practices to maximize productivity and minimize risks. The tasks in the phase were not only about the preparations to finish the project in time, there were far more meaning beyond than that. The expectations of any possible challenge, outlined a draft of the application, the production network establishment, cloud-base project management, resource planning, corresponding solutions to unexpected situation, development criteria, coordination system to multi-disciplinary working, and etc. The project involves teams consisting of experts

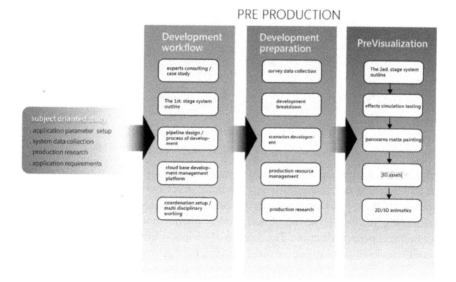

Fig. 2. Pre production workflow.

with various expertise, working separately under different infrastructure and multiple servers. It's essential that teams have enough support to unify working structures by a cloud base platform. Production criteria was set up simultaneously on the early-stage. The implementation of the naming convention was beneficial to organize output data and make it traceable as Fig. 3. Tree structure was considered as the fundamental data organization on the cloud file server. The experts from various fields related to the project benefited a lot by sharing profession knowledge and experience. The consulting experts experienced in rescued operation are from the Tainan City Government Fire Bureau and military instructors from the Medical Training Center of Medical Affairs Bureau Ministry of National Defense. Dacoms Technology Co. a media company has rich experience in creating VR CAVE entertainment contents supported technical consulting. In the support of simulation scenarios creation, an experienced film visual effects and animation company, WhiteDeer Animation Studios, that serves international film industry in computer graphics visual effects and animation supported the project as consulting and also participated the production. In the matter of efficiency and collaboration production process, production conventions were setup as the fundamental parameters of production management. Cloud-based tools helped to built-up organized workflow.

2.5 Development Workflow

The phase of process focused on the data collection and detailed the production technically. The survey data collection in Fig. 2 is about the physical data of the sites, integrating different types of spatial data, geometry measurement, and the disaster records relate to the simulate scenarios. Documentary of the locations with photography and video recording was used frequently. The high resolution panorama as Fig. 4 was created by 20 photos with 35mm aperture lens, it clarified uncertain spatial issues to later processes

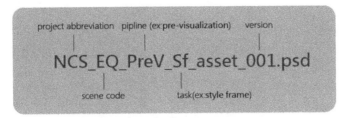

Fig. 3. Production file naming convention.

recreating the environment in digital 3D space. Integrating visual resource and survey data is an important task to digitalize environment in further procedure. Scenarios development in this stage is scriptwriting. In this project, it came with a certain format to easy understanding, the events, acting, description of phenomenon, and notes for the further production and post production. The feature of scripts is timeline, every events and acting should be described in time records. Scenario scripts would break down in time list in some particular cases, especially the complicated scenes. Creating convincing computer graphic simulation that enough to trigger participants anxious are sophisticated processes. Creating high quality application with finite resource in this case is challenge. Well resource management did help that efficient and effective deployment and allocation of the project's resources when and where they were needed. The resources included financial resources, human resources, technological resources, and time. It's quite an experience to learn that effective resource management is crucial for maximizing productivity, reducing waste, and ensuring that projects are completed on time and within budget.

Fig. 4. The panorama photo of the highway entrance of the Hsuehshan Tunnel as visual reference.

2.6 Pre Visualization

Pre Visualization often represents not only the best way to develop a sequence but the best way to collaboratively link the variety of departments, technologies, and the points of view that have to come together in a modern production to bring the sequence to life [14]. It was considered as the most important communication process in the early phase. The project director visualized the concepts and planned elements in the scenarios. Furthermore, analyzed every scene that would be taken over by the post production team. The development of the system is sophisticated and elaborate, all the tasks we had done in the phase to make sure every process would go smoothly for further development. A requirement for clarity in the visual draft came from the project director and production teams. The director visualized the concepts and planned elements in the scenarios. Furthermore, analyzed every scene that would be taken over by the post production team. Visualization also benefited the production team in understanding every shots for the visual effects compositing in the post production. There were three main processes for visualization, which are scenario matte painting, storyboarding, and animatic. The communication frameworks of creating visual contents were deeply relied on the three tasks. The panorama photos which were created in the preparation period had been post edited by photo compositing method for picturized ideas (Fig. 5). Matte painting is the popular technique has been conducted in the film industry. In this project, pre-production team matte pained three concept arts to represent the disaster scenarios and be references to 3D digitalization. The Fig. 5 shows the matte painting of traffic accident scenario of the simulation and the Fig. 6 high lights the key elements in the scene should be recreated to 3D animation and visual effects.

Fig. 5. The digital matte painting of the scenario.

Fig. 6. The key animation assets are highlighted for the further production.

Storyboarding specifies how the scenarios go in every possible detail. In this case, shot lists were created by the director before storyboarding. Storyboards were conducted to plan visual elements in scenes to be shot or produced in the further processes (Fig. 7). Due to the amount of live-action footage filming as requirements, the director worked on the specific details to shots and kept a high degree of flexibility. The pre-production team visualized the concept to screen precisely by going through the storyboards over and over, narrowing down the problems which might have occurred in the following filming. The CG elements created illusions to produce immersive senses convincing enough to the first responders have been considered as important methods to the system. Most of the shots composited CG elements with live-action footage in post-production. Good storyboarding was also a good way to understand the full scope of the development and accounting. On the other hand, the storyboards were very important for visual communication. For the scenes which needed to be sophisticated approached, the storyboards might be not enough to deliver the concept completely, or for some challenging shots in the production lately, the pre-visualization team animated the shots as advanced versions which called animatic to post production teams to render the assignments as well as they can by 2D or 3D animation.

Fig. 7. The storyboard represents some detail of shots in the Hsuehshan Tunnel car accident scenario.

3 Development Overview

3.1 Production

The key idea of virtual production is the composition of virtual and real scene elements or different virtual elements (Grau et al. 2000). Virtual studio is usually taken as a technique of combining live action in a studio with a keyed background image. Creating a green screen-based virtual studio that combines CGI (Computer-Generated Imagery) and live-action involves several steps and technologies, such as matching the camera movements with the virtual environment for creating seamless and believable composites. 2. Use soft, even lighting for both the green screen and the subjects. This reduces shadows and ensures the green screen is uniformly lit. Typically, the camera films one or more actors in a controlled environment, these actors usually perform in a green area without seeing any real solid set around them. This setup requires actors to imagine their surroundings and interact with virtual elements that will be added in post-production. To help actors perform and react to the position and plot in a green screen-based virtual studio, several techniques can be employed:

1. **Real-Time Compositing Screen**: Using a big screen or multiple screens positioned around the green screen area to display the virtual environment in real-time as Fig. 8 shows can significantly help actors visualize their surroundings and interact more naturally with virtual elements. The green screen must be evenly lit to ensure a consistent color that can be easily keyed out in post-production. Use multiple light

sources to eliminate shadows and hotspots. Light the actors and props separately from the green screen to avoid color spill. Use backlighting to create separation and prevent green reflections on the subjects.

2. **Interactive and Physical Reference Points**: Place markers on the green screen for motion tracking. These markers help in aligning CGI elements with the live-action footage in post-production. Place physical markers or lightweight props on the green screen floor and set. These serve as reference points for actors to interact with, representing virtual objects or positions in the CGI environment. Ensure that markers are placed consistently and are easily distinguishable from the green screen. Inform actors of any movement restrictions to avoid breaking the illusion of the final composite. Ensure they stay within the green screen area. Actually placed the makers in the footage filming was not for camera motion tracking, they were used as guide marks for the performers. Even though the screen monitor provided the real time compositing video delivered the basic spatial idea reflected to the digital space, but visual guide still necessary to stage acting.

3. **Stage Direction**: Provide clear, detailed instructions about movements, interactions, and timing. Direct actors on where to look and how to interact with objects or characters that will be added later. Use placeholders or reference objects to help actors visualize the scene. Movement Restrictions is also about inform actors of any movement restrictions to avoid breaking the illusion of the final composite. Ensure they stay within the green screen area.

4. **Rehearsals and Adjustments**: Use playback features to review and analyze rehearsals and takes. This allows for immediate adjustments and fine-tuning of performances. In the other point of view, the phase of works are something about testing. Stage acting needs well preparation before camera rolling (Fig. 9).

5. **Lighting and Camera Setup**:

   - **Camera Setup**: In this case, production crew deployed two 4K high resolution cameras in the scene. So each shot had two different angles so the camera director had options to the good. Specific survey data recording is necessary to the post production. Survey data is a film term indicates the parameters refer to camera setup such as the high of the camera tripod, the distance from the camera to the shooting subject, film aspect ratio, aperture, and etc. Certain visual effect producing which relate to camera movement such as camera tracking and match moving might not work effectively without camera survey data. Even through, most of shots were filmed in still camera which means shots without camera movement, but the production crew still made the survey data as well.
   - **Lighting**: The green screen must be evenly lit to ensure a consistent color that can be easily keyed out in post-production. Use multiple light sources to eliminate shadows and hotspots. Sufficient lighting to the green screen backdrop is good to visual effects artists go the compositing jobs more easily. The green screen color spill would be a problem if the filming subjects don't have enough distance to the green backdrop. Light the actors and props separately from the green screen is also good to avoid color spill. Use backlighting to create separation and prevent green reflections on the subjects (Box 2020).

6. **Special Effects Make-Up**: Producing realistic compositing sequence is essential to great immersive experience. All visual elements in quality and convincing enough to trigger participants' sensation. Both computer graphics and live-action footages need to be as good as possible. The application simulates three disaster scenarios and which one has casualties with different types of wounds. In the realm of performances, special makeup played a significant role in transforming performers into injured characters (Fig. 10). Here were two key methods of special make-up that been applied in the production.

   - **Prosthetics**: Realistic prosthetics were used to create wound appearance, such as scar, blood, scratch, fracture scald, and amputating. Prosthetics were made from various materials such as silicone, latex, and foam latex, ensuring a convincing on-screen transformation.
   - **Feature Transformation**: Portraying different age groups on stage or screen demanded an alteration in the physical appearance of the performers. Wrinkle application, paler makeup tones, and specific hairstyles are some of the techniques employed to create a convincing age-appropriate character. For the character diversity in the scene, different dressing delivered the variety of occupations and personalities.

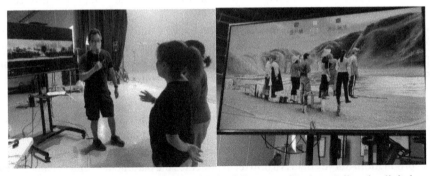

Fig. 8. Real-time compositing monitor provided a full scope of understanding the digital spatial issue. The film crew and performers might have clear ideas about shooting green screen footages for the compositing visual effects.

3.2 Production

In general, post-production is the process about integrating visual and audio elements into systematic assets. It is the final stage of the development, and amount of sophisticate tasks execute by several groups that their expertise differs from each other. The production management becomes extreme important to coordination, supervising the pipe line, quality control, progress review, and much more. The challenge is in proportion, the key factor to achieve the goal are every professional group full fill responsibilities well with organize workflow which established in pre-production. Unlike ordinary motion picture, the projection animation and visual elements render in real time and have been controlled

Application of the Immersive Virtual Training   317

Fig. 9. Rehearsals were not only for the performers, the production team also had meeting and testing frequently to be advanced camera outputs.

Fig. 10. Special make-up elevated the quality of live-action footages. Produced convincing atmosphere in the later visual compositing tasks.

by the Pandoras Box media system which consists of three hardware workstations and a software application. In the final stage, the technical director integrated visual elements and edited compositions of scenarios in the Pardoras Box application. The length of three simulation scenarios is around 10 min to 13 min and the resolution of the wall projection is 36864 × 2160 pixels. The post-production development processes of the application had went through as following:

1. **Ingest and Organization:**

    - **Footage Review**: All green screen footage was reviewed and logged. Footage review is a quite complicated and tedious task. In the studio filming, production crews filmed every shot in many takes in case the post production team might need. But it would be unnecessary the post production team put every footage in compositing. The director went through all the footage to make the selection.
    - **Organization**: Resource files were organized into a coherent structure for easy access. The production management system used by the post production team was the main platform to organize the resource Files which included the footages. In post-production, the visual effects director plays a critical role in bringing the project to life. By combining creative vision, technical expertise, and strong leadership, the VFX director ensures that the final product is visually stunning

and seamlessly integrates all visual effects elements. This process transforms raw footage into the polished, immersive experiences that captivate audiences.

2. **Visual Effects Compositing:**
   - **Shot Breakdown**: Each visual effects shot was broken down into its components, and organized by shot lists which related to production management deeply as huge amount of compositing works to be done. Visual effects artists worked with the coordinator to fit in the production team.
   - **Compositing**: This involves combining live-action footage with CGI elments. Compositing always be considered to be the most important process in the visual effect. The visual effects artists used Adobe After Effects to completed the jobs. Most of the video clips shot from the green screen studio had been passed on to the composting process. The first step of the compositing was removed the green background, leaving only the actors and props. This allows the background to be rep-laced with CGI or other footage and the process called keying. Refine the keying process to ensure clean edges and avoid any green spill. This might involve additional rotoscoping and edge cleanup (Brinkmann, R., 2008) .In this case, the final integration was done in the Parndors Box application, so the all of the compositing element were outed put with alpha matte. All of the time consuming jobs turn out to be convincing illusions which is good enough to give the first responders an immersive sense while they experience the system.
   - **3D environment Creation**: 3D models, environments, and animations were created using Autodesk Maya. Nearly 1000 3D assets to build up the disaster scenarios. The 3D artists modeled assets with reference collected in pre-production period. The process of realistic 3D assets creation is a quite complicated task, it need time to craft and digitalize the scenarios even before the production started. For better rendering performance in massive 3D environment, the post production team used Unreal Engine 5 to environment creation (Fig. 11). Benefited by the powerful 3D render algorithm, the 3D engine has capability to render high realistic quality in real time just like this case. The crews were able to review the scenes without going through the time consuming 3D render tasks.
   - **Simulation and Dynamics**: Realistic simulations of natural phenomena (e.g., smoke, fire, water) are created. The project needs lots of particle effects to produce realistic simulation such as burning vehicles, mass explosion, bridge collapse, city building topple down, flooding river, mudslide ,and etc. Visual effects artists rendered the phenomena by Houdini which is a popular particle effects simulation engine in the visual effects industry (Fig. 12). To increased productivity, many effects shots were done by compositing with effects stock footage for time saving matters.

3. **Audio Engineering:**
   - **Sound Design**: Crafting unique sounds that define the film's sonic identity. This can include everything from the subtle to the surreal, giving the film its unique auditory signature.
   - **Capture the Raw Sonic Material**: During production, the focus is on capturing the best possible on-set audio. This includes dialogue, ambient sounds, and live

sound effects. The quality of these recordings is paramount, as they form the backbone of the post-production process. The audio engineer collected the sounds in the studio during the live action filming.
- **Foley and Sound Effects**: Custom sound effects are recorded to match specific actions in the film. Creating and recording custom sound effects to enhance the realism and impact of the film. Sound effects manipulate the participants' senses as an essential method to raise their anxiety levels. The sound of victims' cry for help, bystanders engaging rescues in the disaster scenes, crowd's shouting, police sirens and etc., numerous sound clips were processed as much important as visual elements. The audio engineer analyzed footage and produced simulated sound of circumstances simultaneously in a foley stage.
- **Mixing**: Crafting unique sounds that define the film's sonic identity. This can include everything from the subtle to the surreal, giving the film its unique auditory signature.

4. **Editing in the Media System:**
   - **Integration**: It is the final stage of simulation development. The media server is capable of controlling motion picture projection and physical effects simultaneously. The methodology of non-linear editing the video clips is similar than common editing software. The media server supports GPU-accelerated and has high hardware expandability that has full capacity to drive the application. To achieve 360° immersive video projection on 14 m x 14 m space, there are 12 4k resolution projectors controlled by 3 media servers. The technical director worked closely with the project director and the supervisor for the final integration. The operation interface as Fig. 13.
   - **Rough Cut**: Rough cuts of the simulation scenarios, arranging scenes in the correct order and cutting down the footage to the necessary length. Doing rough cut was about the preview delivered specific ideas for testing.
   - **Composition Layout**: The panorama matte paintings as Fig. 5 became the main reference to layout visual elements in the system. There are around 70 to 100 visual elements in every scenario.
   - **Timeline Editing**: Timeline editing indicates the events happen in sequence to simulation scenarios. There are around 10 to 20 event sequences to simulate the disaster circumstance expect to trigger participants' tensions to simulate real situation.

5. **Final Review and Testing:**
   - **System Tests**: The director, supervisor, technical director, and other key stakeholders review the near-final product, making last-minute tweaks and adjustments. Before inauguration of the application, several tests had conducted to discover any defect that may cause the performance dragging. Also found out any possibility to take the application to its maximum potential. In general, it's a process of trial and error. Timeline Editing: Timeline editing indicates the events happen in sequence to simulation scenarios. There are around 10 to 20 event sequences to simulate the disaster circumstance expect to trigger participants' tensions to simulate real situation.

- **Drills**: As the Fig. 14 shows, before the official inauguration, the first responders of military hold drills in the application. In the perspective of system development, consider running the application with actual drills considers as system stress testing. Gathering feedbacks from the first responders as importance reference improved the system. Ensuring the quality meet the project's standards. Elevating the training quality by the immersive simulation application is the purpose of the development, so developing the application fulfils the improvement of first responder training programs is the priority.

Fig. 11. The post production team used Unreal Engine 5 to integrate 3D environment. A large scale integration of 3D assets as well as this one, crews were able to have a comprehensive review on the simulated disaster scenes.

Fig. 12. The visual effects team used particle simulation engine Houdini to simulate the collapsing bridge caused by flood.

Fig. 13. The media server not only bridges the hardware installations and digital contents also has complete control over the entire workflow.

Fig. 14. The first responders of military hold an exercise in the CAVE system.

4 Conclusion

The implementation of the Cave Automatic Virtual Environment (CAVE) system for first responder training in Taiwan represents a significant advancement in utilizing immersive simulation technology. This project, a collaborative effort between National Chung-Shan Institute of Science and Technology and Kun Shan University, demonstrates the potential of VR CAVE to enhance the realism and effectiveness of training programs for critical scenarios. By simulating disaster environments such as urban earthquakes, mountain landslides, and tunnel traffic accidents with high fidelity through 12 4K projectors, high-fidelity human patient simulators, and physical effects like earthquake platforms and surround sound, the CAVE system offers an unparalleled training experience. The application of VR CAVE technology in first responder training has shown to improve the assessment and preparation of individuals in ways that traditional drills cannot match. The immersive experience not only triggers stress responses similar to real-life situations but also allows for repetitive practice in a safe, controlled environment. This enhances the skills and readiness of first responders, ultimately contributing to better outcomes in actual emergency scenarios. Despite challenges such as high costs and technical complexity, the benefits of VR CAVE systems, including enhanced realism, safety, and the ability to provide detailed, interactive training scenarios, make them a valuable addition to traditional training methods. The successful implementation and positive feedback from performance and pressure tests suggest that this technology holds significant promise for future applications in various fields beyond first responder training, including medical, industrial, and military training. In conclusion, the CAVE system developed for Taiwan's first responder training programs exemplifies the transformative power of virtual reality technology. It sets a benchmark for immersive training solutions, paving the way for broader adoption and further innovations in VR-based training and education.

References

Box, H.C.: Set Lighting Technician's Handbook: Film Lighting Equipment, Practice, and Electrical Distribution, 2nd edn, pp. 131–159. Routledge, London (2020)

Brinkmann, R.: The art and Science of Digital Compositing: Techniques for Visual Effects, Animation and Motion Graphics, 2nd edn, pp. 206–228, Morgan Kaufmann, Burlington (2008)

Cruz-Neira, C., Sandin, D.J., DeFanti, T.A.: Surround-screen projection-based virtual reality: the design and implementation of the CAVE. In: Whitton, M.C. (ed.) Proceedings of the 20th Annual Conference on Computer Graphics and Interactive Techniques, vol. 2, pp. 135–142. ACM Press, New York (1992)

Cruz-Neira, C., Sandin, D.J., DeFanti, T.A., Kenyon, R.V., Hart, J.C.: The CAVE: audio visual experience automatic virtual environment. Commun. ACM **35**, 64–72 (1993)

Grabowski, A.: Practical skills training in enclosure fires: an experimental study with cadets and firefighters using CAVE and HMD-based virtual training simulators. Fire Saf. J. **125**, 103440 (2021)

Grau, O., Price, M.C., Thomas S.G.A.: Use of 3D techniques for virtual production. In: Videometrics and Optical Methods for 3D Shape Measurement. SPIE, vol. 4309, pp. 40–50 (2000)

Jeffrey Okun, V.E.S., Susan Zwerman, V.E.S. (eds.): The VES Handbook of Visual Effects: Industry Standard VFX Practices and Procedures, 1st edn. Routledge, London (2020)

Onyesolu, M.O., Eze, F.U.: Understanding virtual reality technology: advances and applications. In: Schmidt, M. (ed.) Advances in Computer Science and Engineering, vol. 1, pp. 53–70. IntechOpen, London (2011)

Leigh, J., Johnson, A.E., DeFanti, T.A.: Issues in the design of a flexible distributed architecture for supporting persistence and interoperability in collaborative virtual environments. In: Crawford, D. (ed.) Proceedings of the 1997 ACM/IEEE conference on Supercomputing, pp. 1–14. Association for Computing Machinery, New York (1997)

Mapes, D.P., Volino, P.: An inexpensive CAVE virtual environment. IEEE Comput. Graph. Appl. **2**, 58–61 (2001)

Riva, G., Wiederhold, B.K., Mantovani, F.: Neuroscience of virtual reality: from virtual exposure to embodied medicine. In: Wiederhold, B.K. (eds.) Cyberpsychology, Behavior, and Social Networking, vol. 22, pp. 82–96. Mary Ann Liebert, New York (2019)

Vanderbilt, T.C.A.T.G.A.: Anchored instruction and its relationship to situated cognition Anchored instruction and its relationship to situated cognition. Educ. Res. **19**, 2–10 (1990)

Visch, V.T., Tan, E.S., Molenaar, D.: The emotional and cognitive effect of immersion in film viewing. Cogn. Emot. **24**(8), 1439–1445 (2010)

Wilkerson, W., Avstreih, D., Gruppen, L., Beier, K.P., Woolliscroft, J.: Using immersive simulation for training first responders for mass casualty incidents. Acad. Emerg. Med. **15**(11), 1152–1159 (2008)

Applying the Story Context Analysis Method in Teaching Immersive Media Development

Tzu-Wei Tsai(✉), Su-Ting Tasi, and Shao-Han Liao

National Taichung University of Science and Technology, Taichung, Taiwan
wei@nutc.edu.tw

Abstract. Teaching immersive media development involves integrating technical skills with creative storytelling, offering unique opportunities for engaging user experiences. This study explores the integration of the story context analysis method into the teaching of immersive media development. The technique involves dissecting narratives to understand their components, themes, and emotional arcs and to create immersive media experiences. The study reviews the literature on augmented immersive media, the story methods, and the case analysis of augmented immersive media. The technique involves deconstructing narratives, identifying themes, and applying insights to immersive media projects. Practical workshops are conducted to integrate the method into teaching, focusing on concept design, display design, and content production. Through case studies, practical exercises, and collaborative projects, students gain a structured approach to understanding and applying storytelling principles in immersive media development. The study highlights the significance of interdisciplinary collaboration, stressing its role in empowering students to create compelling, immersive experiences. This approach enhances the learning experience and prepares students for the real-world challenges of the evolving landscape of immersive media.

Keywords: Story Context Analysis · Immersive Media · Teaching Method · Emotional Design

1 Introduction

Immersive media, such as virtual reality (VR) and augmented reality (AR), offer unique storytelling and user engagement opportunities. Teaching immersive media development requires systematically integrating technical skills and creative storytelling. In this context, the Story Context Analysis Method plays a pivotal role. This method involves dissecting narratives to understand their components, themes, and emotional arcs and using this understanding to create immersive media experiences, underscoring the significant impact of this method in the field. This study is dedicated to exploring the integration of the Story Context Analysis Method into teaching immersive media development.

2 Literature Review

2.1 Augmented Immersive Media

Through augmented immersive media, one must gain visual and kinesthetic stimuli or surprising interaction, understand the meaning of the narrative, and arouse them to participate in visual, behavioral, and narrative fun to have a pleasant experience. (Zhao and Tsai 2018) Tamaki (2021) proposed using intuitive, surprise, and story design to analyze creativity and people's immersion in media. The immersive experience also allows people to obtain three levels of emotional design: sensory, behavioral, and reflective experiences entirely in a specific situation (Norman 2004; Chen 2020). Therefore, metaverse technologies or environments, such as AR, VR, and MR, create near-realistic situations so that users can fully immerse in the experience, generate resonance connections, and achieve immersive emotions. Immersion sense includes three states.:

1. Perceived immersion: From visual, auditory, or narrative storylines, sensory perception gains enjoyment and aesthetic pleasure, such as watching performances and movies.
2. Behavioral immersion: Focusing on interactive operations smoothly and competently, one achieves the "flow" phenomenon proposed by psychologist Csikszentmihalyi.
3. Physical immersion in space: One is immersed in it when entering a surrounded space. Immersion participation allows participants to gain spiritual and ideological reflection, such as in the exhibition of Team Lab or a circular exhibition space over 180°.

In addition, a significant correlation exists between participation behavior and empathy (Zhao and Tsai 2018). The immersion projection scale has a variety of participants' immersion experiences in positive, challenging, ability, immersion, and flow aspects. For example, the projection scales include window type (90°), surround type (120–180°), and field type (above 360°). Furthermore, various levels of interaction involvement bring various psychological feelings to people (Tsai and Tsai 2009). The intelligent interactions of augmented immersive media include environmental information (people flow or position) and somatosensory information (body movements, image manipulation). The following is a case study of some immersive media in conditions such as projection scale, intelligent interaction, Opto-kinesthetic, and experience, as shown in Table 1.

2.2 The Creative Thinking of the Scenario-Based Analysis Method

The story context analysis method systematically dissects narratives to understand their underlying components, themes, and emotional arcs. This method examines the context in which a story unfolds and the elements contributing to its structure and meaning. By analyzing various aspects of a narrative, such as characters, plot, setting, and themes, the story context analysis method seeks to uncover the deeper layers of meaning and significance within a story. Key components of the story context analysis method include:

1. **Deconstruction**: Deconstruction involves breaking the narrative into its fundamental elements, including characters, plot points, conflicts, and resolutions. This process

Table 1. A case study of augmented immersive media.

Immersive media	Analysis items
 2019 Taichung Flower Expo Discovery Hall	projection scale: field type (above 360 degrees) Intelligent interaction: location of people flows Opto-Kinaesthetic: abstract moving images Experience: visual pleasure and interaction
 Team Lab	Scale: field type (above 360 degrees) Interaction: image interaction Opto-Kinaesthetic: abstract, concretive motion graphic Experience: visual pleasure
 Exploring a wondrous landscape	Scale: field type (120-180 degrees) Interaction: body movements Opto-Kinaesthetic: concrete motion graphic Experience: behavior pleasure
 Precious Moments for Mikimoto Jewelry	Scale: window (90 degrees) Interaction: b crown flow Opto-Kinaesthetic: concrete motion graphic Experience: narrative pleasure

helps identify the building blocks of the story and how they contribute to its overall structure.
2. **Theme Identification**: Theme identification involves identifying the narrative's central themes or messages. Themes often reflect universal human experiences or societal issues and provide insight into the story's deeper meaning.
3. **Emotional Arcs**: Emotional arcs involve mapping the emotional journey of characters and audiences throughout the story. They consist of identifying critical emotional beats, such as moments of tension, triumph, tragedy, and resolution, and understanding how they contribute to the overall emotional impact of the narrative.

4. **Worldbuilding**: Worldbuilding examines the world or environment in which the story occurs. It includes considerations of the setting, culture, history, and rules governing the story's universe and how these elements shape its narrative and characters.
5. **Application**: The application is to apply insights gained from the analysis to various purposes, such as creating immersive media experiences, developing characters or plotlines, or understanding the cultural significance of the narrative. This step involves translating theoretical understanding into practical applications.

The story context analysis method provides a structured framework for understanding and interpreting narratives across different mediums, from literature and film to interactive media. By examining the context, themes, and emotional dynamics of a story, this method enables more profound insights into the meaning and impact of narratives on individuals and society.

2.3 Freytag's Pyramid

Freytag's Pyramid is a structural framework for dramatic storytelling, commonly used to describe the progression of a narrative. It was developed by the 19th-century German novelist Gustav Freytag, who analyzed the structure of ancient Greek and Shakespearean dramas. The Pyramid consists of five key stages:

1. **Exposition:** This is the story's introduction, where the setting, characters, and central conflict are established.
2. **Rising action:** After the initial setup, the story's tension builds as complications arise, leading to the development of the central conflict.
3. **Climax:** This is the turning point or the most intense moment of the story, where the protagonist faces the peak of the conflict.
4. **Falling Action:** Following the climax, the story begins to resolve the conflicts, leading towards the conclusion.
5. **Denouement (Resolution):** This is the story's final stage, where loose ends are tied up, and the conflict is resolved, providing closure.

Freytag's Pyramid is a classic narrative structure emphasizing the progression from calm beginnings through intense drama to a satisfying conclusion (Fig. 1).

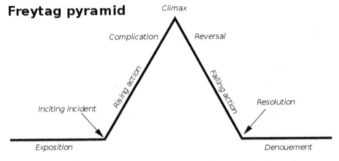

Fig. 1. Freytag's Pyramid (Hanson 2014).

3 Method

Integrating the story context analysis method into teaching immersive media development involves the following approaches in the study:

1. **Case Studies**: Students analyze and study immersive media project cases, examining how to apply storytelling techniques to create engaging experiences. They deconstruct narratives within these projects and identify the strategies employed in the cases.
2. **Practical Exercises**: Assign practical exercises where students deconstruct existing stories and propose adaptations for immersive media platforms. Guide students in creating storyboards, interactive scripts, and prototypes that integrate narrative elements with immersive technology.
3. **Collaborative Projects**: Facilitate collaborative projects where students work in teams to develop immersive media experiences. Emphasize the importance of interdisciplinary collaboration, with students from diverse backgrounds contributing their expertise in storytelling, design, programming, and audiovisual production.
4. **Critique and Feedback**: Provide constructive feedback on students' projects, focusing on how effectively they utilize storytelling principles within immersive media. Encourage peer critique sessions where students analyze and provide feedback on each other's work, fostering a culture of continuous improvement.

3.1 The Three Stages of Workshops

Execute three workshop sections to proceed with the research methods.

Session 1: Concept Design Workshop for Augmented Immersive Media

Objectives: The session aims to help participants understand the basic concepts of innovative augmented immersive media, decide themes and situations for businesses, local art performances, or space experiences, and propose innovative ideas through group discussions.

Steps:

1. Teaching Unit I: Curator Narrative Survey. Demonstrate the example of the theme, situation, and issue of augmented immersive media. Introduce creative thinking to survey the curatorial narration. Group discussion for the theme, concern, and problem, and complete the Curator Narrative Survey.
2. Teaching Unit II: Participants present the curatorial narration plan and design using the "Story Arc" and "Hero's Journey" methods.

Session 2: Intelligent Augmented Immersive Media Display Design Workshop

Objectives: In-depth exploration of the three intelligent perception modes and augmented immersive projection scale. Teamwork uses divergent and convergent thinking to propose innovative interactive display ideas for specific businesses, local art performances, or space experiences.

Steps:

1. Introduce referred video examples to explain the concept of intelligent perception modes and augmented immersive projection. Group discussions on innovative interactive display ideas
2. Teaching Unit III: Extend the results from Unit I and Unit II from the previous workshop and use the script method/situational story method to design and complete the storyboard drawing.

Session 3: Intelligent Augmented Immersive Media Production Workshop

Objects and Steps: Learn tools such as Resolume Arena and Touch Designer to transform the creative ideas from the previous workshop into augmented immersive media practical content.

4 Result

One representative outcome is described as follows.

4.1 Case Study

First, one group of students chose the exhibition "On the Oceanic Edge" (Fig. 2) as the case study target and analyzed its story context.

Curatorial Discourse
This exhibition explores the origins of nature and the evolution of life from the Cambrian era, spanning hundreds of millions of years. It features marine life like trilobites, fish, and cephalopods, highlighting the gradual emergence of aquatic biodiversity. With over 300 meticulously collected paleontological fossils from 35 years of Science Museum research, the exhibit offers a close-up and striking visual experience of life's evolution. It showcases the transition from sea to land, illustrating organisms' rich and complex history through detailed displays.

Curatorial Narrative Structure
The exhibition starts with a detailed look at flattened insect fossils and traces their evolution. As visitors move through the exhibit, they can observe the changes in fossil layers, extending downward from stacked slopes to show the evolutionary path. The display transitions from insect and plant fossils to the metamorphosis of ancient organisms, covering the evolution from early fish to modern life. This rich presentation helps the audience understand the layered nature of biological evolution and highlights the fundamental changes in organism evolution across different periods.

Venue Creativity
The exhibition features marine life suspended with steel wire or string and blue-tinted light tubes at the bottom of display cases to simulate seawater, creating an immersive marine atmosphere. This design enhances the experience with lively on-site illustrations, adding interest and excitement. Combining lighting, decoration, and illustrations offers a unique visual effect, boosting audience engagement and participation.

Exhibition Media Interaction

There are interactive screens that can change pages through gestures, projection screens displayed on glass, and physical exhibits, which account for many. In addition to the interactive media screen, a giant shark mouth is created with a big model to allow the audience to experience the feeling of changing perspective. This display is not a multimedia application, but it also achieves the effect of experiencing interaction.

Fig. 2. The case study of "On the Oceanic Edge."

4.2 Design and Planning of Works

As shown in Table 2, the "Freytag's Pyramid" of the story structure can be used to consider the narrative construction creatively.

Table 2. Using the "Freytag's Pyramid" to create the narrative construction.

	Story Narrative	Exhibition Design
Exposition	The exhibition spans hundreds of millions of years of evolution, unveiling the origins and extinctions of nature and the generational shifts in living organisms	Through animated presentations and interactive elements, the history of evolution is revealed through story-driven dialogues
Rising action	Starting with biological fossils and gradually uncovering the evolution and changes in the marine environment	Starting with biological fossils and gradually uncovering the evolution and changes in the marine environment
Climax	Discover how fossil displays gradually evolve from insect and plant fossils to the metamorphosed fossils of ancient organisms	Using interactive games, observe the evolution of organisms through movement or gestures
Falling action	Tracing the evolution from large fish species to modern organisms	Using touch interaction to select and move different fish species onto fossil images
Denouement	Immersed in the awe and tranquility of marine visuals	Create ocean-themed set designs, short videos, and animations to showcase the movement of water waves and the migration of fish schools

Description of the Work

With the increasing pollution of the oceans in modern society, this immersive exhibition aims to express the current beauty of the sea through an interactive installation. By simulating the movement of fish shoals and the ocean environment, the exhibition provides an opportunity to experience and appreciate the purity and beauty of the ocean, highlighting the importance of maintaining the cleanliness and ecology beneath the water's surface.

Many individuals have gradually forgotten the ocean waves' tranquility, purity, and vastness over time, which is a significant loss. The ocean, present since the Earth's formation, is indispensable. Its existence has fostered numerous precious forms of life and countless possibilities, similar to the air surrounding the environment. Although its impact may not always be visible, the ocean's influence is deeply integrated into daily life. This exhibition aims to evoke beautiful memories buried deep within the consciousness, offering an immersive experience in the deep blue sea and sharing the joy of such immersion.

Presentation of Creative Proposal (Implementation)

The project aims to showcase ocean scenes through immersive projection. Participants can interact with the projection by touching it, allowing them to dance with shoals of fish or admire shimmering waves. Standby states will be implemented for varied visual displays when the system is inactive. The projection will feature additional effects, such as swaying and twinkling, to enhance the visual experience. Serene background music will further deepen the immersion. By moving their right hand and stepping left and right in front of the sensor, participants will make the shoal of fish follow their movements, offering a shared experience of the ocean's cleanliness and beauty.

Design Specifications

The augmented projection space is located in the immersive projection studio at NCTU, with a surrounding scale of 120–180 degrees. Visitors interact with the projection by standing two steps in front of the sensor and waving their right hand, directing the shoals of fish to swim in response. The visuals include underwater scenes, shimmering ocean waves, and subtle twinkling lights. This setup allows visitors to collectively experience the ocean's tremors and beauty, creating an immersive virtual sea experience where their gestures control the movement of the fish.

Production

The artwork will be produced using software such as TouchDesigner and Resolume Arena and hardware equipment such as a Kinect sensor and a projector.

Evaluation:

Apply the story context analysis method to evaluate the project. Shoot the exhibition scene and analyze the narrative image for the visitors' behavior and interaction. The shooting time is about an hour, and the shooting intervals are about 10 to 20 min. Illustrate the records and analysis in Table 3.

4.3 Evaluation Results

Observe audience behavior and responses to evaluate the outcome using qualitative and quantitative methods. The findings suggest the following adjustments:

Table 3. Shows the evaluation of the project using the scenario-based method.

Picture	Shooting Time	Illustrate Visitor Behavior
	Time: 11:40	The visitor tried to approach the machine and waved his right hand to interact successfully.
	Time: 11:50	Visitors get too close to the equipment, causing the fish to gather in one place due to poor sensing.
	Time: Day 2, 11:45	Visitors stand in place, immersed in the exhibition space, listening to surround music.
	Time: 12:00	Visitors began to move around, waving their left and right hands to interact with the fish.
	Time: 12:10	Visitors gather in groups, causing the fish to converge again due to induction errors and poor contact.
	Time: 12:25	Visitors lined up and took turns interacting with each other.

Lighting: Ensure all light sources in the projection environment are turned off, leaving only the light on the head screen. It will enhance the immersive experience by reducing distractions.

Fish Model Size: Slightly elongate the fish model to create a greater sense of depth and distance, improving the visual experience from far to near compared to the original size.

Background Simplicity: Reduce the number of layers in the background, retaining only water waves and fine glitter. It will help maintain focus on the interactive elements and avoid visual clutter.

These modifications aim to enhance the overall impact of the interactive projection and improve visitor engagement.

Audience Participation Feedback

Thirty audiences completed an aesthetic experience measurement consisting of 12 questions rated on a scale of 1 to 5 points. The results, presented in Table 4, indicate positive responses overall. The mean scores reflect a favorable reception of the interactive projection's aesthetic qualities.

Table 4. Audiences' Aesthetic Experience Response Result.

Items	Mean	SD
I feel charming/like it:	4.519	0.643
Makes me joyful	4.519	0.802
It makes me calm and relaxed	4.519	0.753
It feels novel and interesting	4.630	0.629
Feels challenging	3.704	1.171
Captivates and engages me	4.370	0.792
Inspires me	3.963	0.940
It makes me feel relaxed	4.370	0.839
Impressively attractive	4.370	0.967
Evokes deeper meaning	3.926	0.958
Moves me emotionally	4	1.074
Energizes me	4.333	0.679
Mean	4.27	

5 Conclusion

Integrating story context analysis methods into teaching immersive media development provides a structured approach to understanding and applying storytelling principles within immersive experiences. This method involves deconstructing narratives to identify key themes, mapping emotional arcs, and utilizing these insights to enhance immersive media projects. By applying these techniques, students can create compelling and engaging experiences that effectively resonate with audiences.

Educators can facilitate this process through various pedagogical strategies. Case studies offer students real-world examples of successful storytelling in immersive media, illustrating how narrative techniques are employed in practice. Practical exercises enable students to experiment with storytelling elements in their projects, fostering hands-on learning and creativity. Collaborative projects encourage teamwork and diverse perspectives, further enriching the storytelling process. Additionally, constructive critique provides valuable feedback, helping students refine their narrative approaches and improve their immersive media creations.

By leveraging these methods, educators can empower students to harness the power of storytelling in the dynamic and evolving field of immersive media. This approach enhances students' technical skills and deepens their understanding of how narrative can shape and elevate immersive experiences.

Acknowledgments. This study was funded by the National Science and Technology Council in Taiwan (Grant NSTC 112–2410-H-025–034-). We also thank all the participants for their help in this study.

References

Bakdash, J.Z., Augustyn, J.S., Proffitt, D.R.. Large displays enhance spatial knowledge of a virtual environment. In Proceedings of the 3rd Symposium on Applied Perception in Graphics and Visualization, pp. 59–62. ACM (2006)

Bailenson, J.N., Blascovich, J., Guadagno, R.E.: Self-representations in immersive virtual environments. J. Appl. Soc. Psychol. **38**(11), 2673–2690 (2008)

Brockmyer, J.H., Fox, C.M., Curtiss, K.A., McBroom, E., Burkhart, K.M., Pidruzny, J.N.: The development of the game engagement questionnaire: a measure of engagement in video game-playing. J. Exp. Soc. Psychol. **45**(4), 624–634 (2009)

Brooks, K. There is nothing virtual about immersion: Narrative immersion for VR and other interfaces. Alums. Media. MIT (2003). Edu/brooks/storybiz/immersiveNotVirtual.Pdf. Accessed May 2007

Brown, E., Cairns, P.: A grounded investigation of game immersion. In: CHI'04 Extended Abstracts on Human Factors in computing SYSTEMS, pp. 1297–1300. ACM (2004)

Hallnäs, L., Redström, J.: Slow technology - designing for reflection. Pers. Ubiquit. Comput. **5**, 201–212 (2001). https://doi.org/10.1007/PL00000019

Chen W.F. Playful design: changing the wonderful experience of millions of people; exciting work and life! Shen-gun Publisher (2020)

Heavy, M.: Immersive technologies: the best way to deliver a memorable experience (2020). https://www.heavym.net/immersive-technologies/

Hanson, A.A.: 7-Step Freytag's Pyramid (2014). https://writingitch.com/2014/08/21/7-step-freytags-pyramid

Norman, D.: Emotional Design. Basic Books (2004)

Tamaki. Experience Design: Creative Thinking. Peace Publisher (2021)

Tokey, K., Loup-Escande, E., Christmann L., Richi, S.A.: questionnaire to measure the user experience in immersive virtual environments. In: VRIC 2016: Proceedings of the 2016 Virtual Reality International Conference (2016)

Triberti, S., Chirico, A., La Rocca, G., Riva, G.: Developing emotional design: emotions as cognitive processes and their role in the design of interactive technologies. Front. Psychol. **8**, 1773 (2017)

Tsai, T.W., Tsai, I.C.: Aesthetic experience of proactive interaction with cultural art. Int. J. Art Des. **2**, 94–110 (2009). (NSC 94–2622-E-025–001-CC3)

Wang, C.H.: University students' perception of shapes and colors in dynamic and static images (2018)
Witmer & Singer. Version 3.01. September 1996. Immersive Tendencies Questionnaire, Revised by the UQO Cyberpsychology Lab (2004)
Woolman, M.: Motion design: moving graphics for television, music video, cinema, and digital interfaces (2004)
Zagola, N.: Engagement Design: Designing for Interaction Motivation. Springer, Heidelberg (2019)
Zhao, I.L., Tsai, T.W.: The effect of projection scales on participants' immersive experience. In: Taiwan Digital Media Design Society International Symposium, Taipei, Taiwan (2018)
https://taicca.tw/article/6b239982
https://arplaza.co/article/what-immersive-experience/
https://www.kingone-design.com/blog/ImmersiveExperienceDesign

Innovative Design for Cultural Sustainability

A Study on Qualia's Experiential Model from the Concept of Service Innovation Design: A Case of 7-Eleven Service Experience in Taiwan

Yi-Hang Lin[1] and Po-Hsien Lin[2(✉)]

[1] Graduate School of Creative Industry Design, National Taiwan University of Arts, Taipei, Taiwan
[2] Graduate School of Creative Industry Design, National Taiwan University of Arts Professor, Taipei, Taiwan
t0131@ntua.edu.tw

Abstract. The shift in the consumer landscape in Taiwan's retail industry is due to technological advances and service differentiation. The added value of culture and creativity not only enriches lifestyles, but also changes consumers' experiences and perceptions of products. This study uses the case study method to propose a research framework, combining the theories related to the three levels of qualia and service innovation, to explore in depth the correlation between the development of convenience superstores and qualia design, using 7-Eleven as a case study. The results are summarized as follows: (1) 7-Eleven has successfully shaped its differentiation and uniqueness after 2019 through "brand co-branded theme stores", and through the qualia analysis, it has found that the service contains moving stories and creative designs, and has finally realized the service innovation process of "shaping differences and establishing identity"; (2) "Humanity" has been the core axis in the process of service enhancement, and the continuous development of science and technology has brought about more diversified experiences, rich products, and better service quality. The continuous development of technology has brought more diversified experiences, rich commodities, and a compounded field; (3) The innovative process of combining the concept of sustainable development through emotional space, emotional experiences, and qualia products can expand the influence of the specialty outlets, continue to build the image of "good neighbors in the community," and provide a new way of thinking about the development of "human nature" and "technology".

Keywords: Service Experience Development Process · Creative Lifestyle Industry · Quaila Management · 7-Eleven

1 Introduction

The modern design movement originates from the 19th-century Arts and Crafts Movement in England, emphasizing the idea that technology should be human-centered. The entire development process of design has been pursuing harmony between "technology" and "humanity." The "design philosophy" and "life advocacy" proposed by Bauhaus have had a profound impact on modern design (Sun et al., 2019). With the rise of the service economy, design methods are increasingly positioned as valuable means to achieve service innovation (Andreassen et al., 2016). Approaches such as experience-centered service design recognize insights into customer experiences (Zomerdijk & Voss, 2010) and support the development of new services (Bitner et al., 2008), thus facilitating service innovation. Furthermore, from the perspective of experience design, the core of experiential marketing lies not in emphasizing the performance and benefits of products but in creating unique experiences for customers (Pine II & Gilmore, 1999).

In Taiwan, characteristic cases represented by convenience stores offer diverse services, continuously innovate, and meet the public's needs for dining, living, mobility, and entertainment. Even foreigners have fallen in love with Taiwan's convenience stores, which proves the richness of their services and reflects that service innovation is a key factor in retail enterprise differentiation and increased relative attractiveness (Lee et al., 2022). Taiwanese convenience stores not only engage in cross-industry integration, multi-channel operations, and online-offline integration but also realize a large number of convenient life services through dense networks.

The contemporary society's emphasis on the cultural and creative industries validates Rungtai Lin viewpoint proposed in 2011: "Culture" is not only a way of life but also a "lifestyle taste" nurtured by a group of people's "life advocacy" and formed into a "lifestyle" through extensive recognition. By satisfying consumers' life advocacies through "products," "services," "activities," and "environments," constructing unique lifestyle tastes, and forming consumer-identified lifestyles. Therefore, this study aims to explore the development of modern convenience store environments and service changes, examining the changes and lost details brought about by the integration of "technology" while pursuing "humanity."

This study will investigate the following two research questions: How have the service experiences and innovations provided by characteristic stores, using 7-Eleven as a case study, evolved from the consumer's perspective? What are the commonalities between related characteristic environments and sustainable development, and how do they provide new ways of thinking for the development of "humanity" and "technology"?

2 Literature Review

2.1 Exploring Experience Design with Qualia Integration

Qualia are the unique sensations and emotional touches that constitute human feelings, leaving unforgettable memories throughout one's life. This pursuit of "happiness" is manifested in tangible objects through sensations, experiences, and memories, eliciting wonder and emotional responses (Idei, 2002). Incorporating Qualia into experiences opens up a whole new process, where consumers are no longer satisfied with singular product choices, services, or simple needs. Qualia, through experiences that touch the depths of consumers' emotions, based on the meaning of life and emotional satisfaction, becomes a delightful sensation that strikes a chord and triggers emotions. The inclusion of Qualia in experiences allows consumers to evoke inner joy and happiness, and the emotional impact they receive may vary due to individual differences.

Experience design aims to create unforgettable memories for consumers through clever design behaviors, with the goal of enhancing the overall consumer experience. This evolution process has developed from agricultural economies, industrial economies, service economies to the present "Experience Economy" era. With the advancement of technology since the millennium, the concept of experience economy has been widely applied in commercial activities and deeply integrated into consumers' daily lives (Pine II & Gilmore, 1999). As a convenience store integrated with local characteristics (local industry), how it shapes the connotation of experience design, allowing consumers to feel the consumption value that balances leisure and entertainment, is mainly achieved by leveraging the nation's traditional culture and lifestyle to shape and foster differentiation. This corresponds to Taiwan's high regard for the cultural and creative industries, as well as the intensive development of convenience stores against the backdrop of urbanization and high population density (Chang et al., 2015).

2.2 Aesthetic Characteristics of Qualia Products

"Qualia goods" include intangible services and tangible products, based on quality as an extension of the five qualia forces, constructing the added value and uniqueness of the products and services, and being able to evoke fond memories of the experience at the time when seeing the related goods in the future (Lin, 2014).

Thanks to the pre-experience emphasis on the emotional aspects of consumer behavior and the creation of experiential values that emphasize context-specific aspects of non-functional features of products and services (Cleff et al., 2014), the essence of the product is extracted and applied to tangible, physical and interactive experiences. Qualia goods not only fulfill functional needs but also stimulate sensory and emotional experiences compared to single products (Zarantonello & Schmitt, 2010).

Regarding the aesthetic aspect of Qualia Products, because aesthetically pleasing products can bring joy and psychological satisfaction, they need to have storytelling elements, align with aesthetic economy, and be suitable for everyday lifestyles (Fig. 1). Good Qualia Products typically have a compelling story or concept. For example, 7-Eleven's themed stores such as "Starlux Airlines Store," "Future Store," and "Snoopy Research Institute Store" all have concrete brand stories and concepts. The specialty products sold therein, through subtle storytelling, enrich consumers' experiences, making Qualia Products emotionally compelling.

2.3 Spatial Design and Development Patterns of Emotional Space

The creative lifestyle industry pursues a sense of "Qualia" and "touching joy." Among them, "high-quality aesthetics" not only involves the creative use of "space" and "products" but also emphasizes the environment space as a perceived and sensed "affair." In such spaces, consumers not only release emotions but also perceive the creative aesthetics emanating from the space atmosphere, thereby creating and enjoying the beauty of spatial design. For example, nostalgic spaces evoke emotional resonance between consumers and the atmosphere due to their historical or unique characteristics. To shape

Fig. 1. Three Elements of Creative Industry Services (Lin, 2014)

consumers' unique impressions of the atmosphere and the unique charm of consumption, the overall environment and atmosphere of the consumption place directly affect consumers' experiential senses (Kunkel & Berry, 1968).

Pine II and Gilmore (2007) point out that experiencing in appealing places is the most effective way to stimulate demand, whether it's materials, goods, services, or other experiences. Therefore, after experiencing Emotional Experience and Qualia Products, it is necessary to create a service environment experience with unique attractiveness to influence consumers' cognitive behavior and aesthetic experience processes.

Through the conceptual model of developing Qualia Business proposed by relevant scholars (Fig. 2), the importance of "core stories" in touching consumers is emphasized. 7-Eleven's themed stores all have thematic or narrative elements, and this clear storytelling theme is transformed into an Emotional Space through physical space creation, providing emotional experiences (Chien et al., 2017).

2.4 Service Innovation and Sustainability in the Consumer Arena

Economist Schumpeter proposed the concept of "Innovation" in 1934, considering innovation as the primary force driving economic growth. Innovation can drive the renewal of industry technology through creative destruction, while in related consumer fields, it can provide high-quality life experiences through creative lifestyle design. This innovative concept can develop into an integrated service innovation design business model. Additionally, innovation is crucial for the competitiveness of retailers (Feng et al., 2020).

Scholars Johnson, Menor, and Chase (2000) categorized innovation services into two main types: "Radical Innovation" and "Incremental Innovation." Taking 7-Eleven as an example, its service innovation mainly involves expanding existing service lines (increasing service content) and improving the characteristics of existing services (innovative service quality). The major changes are in style, influencing consumers' perceptions, emotions, and attitudes through visual changes, thus primarily belonging to "Incremental Innovation." In the process of service innovation, driven by economic development,

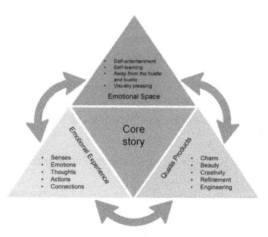

Fig. 2. A Conceptual Model for Development of Qualia Business (Lin, 2014)

businesses are compelled to continuously pursue unique, unforgettable experiences for customers. This trend aligns with the development trajectory of 7-Eleven (Fig. 3).

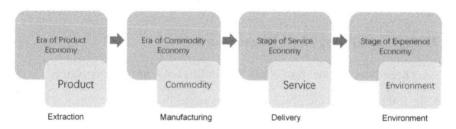

Fig. 3. Evolution of the economy (Pine & Gilmore, 1999)

3 Research Methods

This study is based on the "case study method," selecting specific cases and utilizing three Qualia dimensions-"Emotional Space", "Emotional Experience," and "Qualia Products,"-as well as relevant theories of service innovation as the theoretical framework. The research explores four dimensions of creative lifestyle industry design: core knowledge, deep experience, high-quality aesthetics, and lifestyle advocacy, ultimately drawing conclusions. Depending on the adopted perspective and analytical level, service innovation can be conceptualized in various ways, and the degree of innovation can be judged by customers or companies (Lin, 2019). By adopting the perspective of the experience, this study references relevant scholars' theories (Lin, 2016) and views service innovation through the judgment of experiencers.

In terms of data collection, this study mainly utilizes two methods: secondary data compilation (official data, relevant books) and on-site experiential data collection at themed stores. Those with rich experiential processes and clear category differences are selected as sources of information. Since there are many specialty stores throughout Taiwan, we chose the ones with rich experiences and significant differences in categories for reference. Through the analysis of factors in case comparisons (Table 1), this study provides new insights and directions for relevant research from the perspectives of Qualia design and service innovation, The elements of evaluation were revised and innovated mainly with reference to the paper written by Chang (2016).

Table 1. Comparison between standardized stores and special stores

Emotional Experience	Qualia Products	Emotional Space
In-store Feelings	Product Texture	In-store Decor
Service Offerings	Product Visuals	Space Design
Service Attitude	Product Content	Store Features
Additional Services	Product Story	In-store Display
Distinctive Value	Product Features	Themed Activities

4 Research Findings and Discussion

4.1 A Three-Part Exploration of the Sensibilities of Convenience Stores

Looking at the development process of 7-Eleven, as the market for chain retail stores has matured, there have been too many stores with a single style, necessitating aesthetic innovation. For example, in 2019, the introduction of "branded collaborative stores" aimed to create distinctive consumer spaces. Unlike the previous uniformity in materials, colors, and decorations, themed stores now present different styles (IP), distinguishing them from previous store scales. As a result, various themed concept stores, branded collaborative stores, future image stores, etc., have all become relevant spaces for shaping brand image and enriching creative levels (space design, IP decoration, selfie spots). Having a "core story, unique IP" consumer space reflects the existence of "Emotional Space."

Brands convey stories to consumers through visual elements, which is an effective way to enhance emotional connections with the brand or product. This approach also leaves a vivid and positive impression in consumers' minds, thereby increasing their identification and support for the brand or product (Dhote & Kumarl, 2019). In the context of experiential environments, consumers differentiate brands through firsthand experiences and visual observations. Through the experience of specialty stores, especially for consumers who understand the relevant "core stories" such as Kanna Hara, Snoopy, and Line Friends, they can immerse themselves in these stories. This experience, whether through tangible or intangible objects, helps evoke consumers' past memories, trigger emotions, deepen emotional depth, and create moving experiences.

Through the selection of design elements, materials, shapes, and colors, products can effectively convey brand values, potential stories, and underlying philosophies. Purely practical functions are no longer sufficient to meet the needs of some consumers. Therefore, emphasizing the Qualia needs and conveying the aesthetic, significance, and connotations of products through "core stories" becomes crucial. Understanding the Qualia aspects of 7-Eleven's specialty stores through Table 2 specifically explains what these three aspects are.

Table 2. Three-Part Summary of the Sense of 7-Eleven Specialty Outlets

Emotional Experience	Qualia Products	Emotional Space
Immersive experience of these IP stories helps evoke past memories	Selling IP merchandise, exclusive tables and chairs, exclusive CITY beverage cups	Theme concept stores, brand collaboration stores, and future image stores shape the brand image and enrich the level of creativity

4.2 Analysis of the Development of the "Consumer-Centered" Service Innovation Design Phase

Since the 1980s, Taiwan has introduced foreign hypermarkets and convenience stores. These emerging convenience stores have flourished with low prices and a diverse range of products, with the 24-h operation being a characteristic that traditional grocery stores find hard to match. This has led to the gradual elimination of traditional grocery stores, as people increasingly prefer to shop at convenience stores.

This study proposes the stages of service innovation development from Function to Feeling through the three levels of sensation (Fig. 4). Throughout this process spanning more than seventy years, "human nature" has always been present, conveying emotional experiences to consumers through continuous introduction of "culture" and "creativity" into services. In terms of "technology," it can be seen from the later stages of development that it has become increasingly diverse and humanized. A rich variety of products and features shift the focus from consumption scenes to innovation, with elements such as brand collaborations, intelligent experiences, and composite spaces showcasing diverse characteristics.

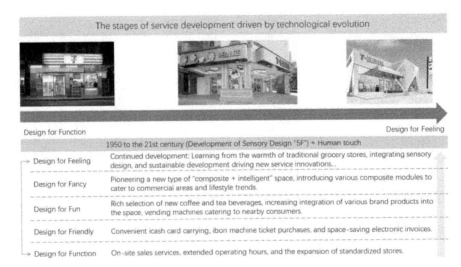

Fig. 4. Different Stages of Service Innovation in Convenience Supermarkets (Organized by this Study)

4.3 Comprehensive Analysis

As a convenience store, 7-Eleven has successfully integrated Taiwanese cultural elements, such as the iconic OPEN mascot created in Taiwan, offering delicious freshly brewed coffee, and the emblematic 7-Eleven clerks. These have become unique features of 7-Eleven in Taiwan, and through these elements, it has gained immense popularity among the new generation of consumers. This not only reflects the role of "technology" in the development process but also highlights the indispensable nature of "humanity." The added value of technology makes the consumer experience more convenient, while integrating more technological elements into the traditional convenience store foundation further enhances the technological sense and innovation of the service process (Table 3).

Table 3. Comparison table of standardized stores and special stores

Qualia aspect	Items	Standardized 7–11 stores	Specialized 7–11 stores
Emotional Experience	In-store Feelings	Technological innovation, Convenience	Technological innovation, Convenience, Novelty
	Service Offerings	On-site sales services, Fee collection services, In-store pickup for online shopping, Printing, Ticket booking, etc	On-site sales services, Fee collection services, In-store pickup for online shopping, Printing, Ticket booking, etc

(continued)

Table 3. (*continued*)

Qualia aspect	Items	Standardized 7–11 stores	Specialized 7–11 stores
	Service Attitude	Friendly	Friendly
	Additional Services	Package collection, Delivery	Introducing co-branded products
	Distinctive Value	24-h operation, convenient and fast	24-h operation, convenient and fast, with added co-branded thematic features
Qualia Products	Product Texture	Modern product style	Having aesthetic characteristics
	Product Visuals	Diverse presentation	Featuring IP elements
	Product Content	Abundant products, fresh groceries	Cultural and creative products, fresh groceries
	Product Story	Integration of corporate characteristics	Integration of brand stories
	Product Features	Various and diverse	Various and diverse
Emotional Space	In-store Decor	Simple and clear	In line with co-branded characteristics
	Space Design	Spacious and bright	Spacious and bright, with rich details
	Store Features	Creating a cross-channel lifestyle circle	Specially decorated in collaboration with IPs
	In-store Display	Abundant products, but neatly displayed	Featuring co-branded counters and limited edition foods
	Themed Activities	Different themes each period	Varies according to the IP

Source: Tabulated from this study

5 Conclusions and Recommendations

Throughout the development of 7-Eleven, each stage has showcased unique characteristics. "Humanity" has always been the main focus, while "technology" continues to innovate in societal development. 7-Eleven's specialty stores are based on captivating consumers with "core stories" and integrate Taiwan's industrial foundation and cultural life into "high-quality aesthetics." These specialty stores provide consumers with firsthand experiences of moving stories and creative designs, thereby strengthening consumer support and identification with the brand. The entire process realizes a service innovation of "shaping differences and building identities."

The core of "deep experience" lies in touching consumers' hearts through "service" and "activities." Thanks to special décor designs, the exterior walls of the stores can exhibit themed features, attracting consumers to experience them. In terms of service, the interaction with clerks and the purchasing experience play a role, while activities are reflected in the selection process of specialty products, evoking resonance with consumers through IP stories. Whether it's brand experience or the aesthetics of emotion, both can lead consumers into deeper emotional experiences.

In the realm of "high-quality aesthetics," 7-Eleven has since 2019 focused more on opening "brand-themed joint stores" and "future stores." This new "composite + smart" store model has gradually attracted attention from different groups, providing a unique experience in 7-Eleven's specialty stores. Through changes in the venue and a diverse selection of products, 7-Eleven's service process has been innovative, forming a distinct differentiation.

In terms of "lifestyle advocacy," with the development of society and the advancement of technology, people's consumption habits have changed from chatting in grocery stores to selecting goods in convenience stores, reflecting a lifestyle shift. Specialty stores incorporate moving stories (core stories), creativity, and innovative designs (technologization) into commercial spaces, expanding from a single brand to joint brands, enhancing the impact of services through emotional design. This combination of "humanity" and "technology" embodies the concept of service innovation design centered on the "consumer."

Finally, in terms of sustainable development, as the United Nations Sustainable Development Goals (SDGs) receive increasing attention, convenience stores also actively respond to this trend by launching the "cross-channel ecosystem" to provide consumers with a more convenient lifestyle. This also reflects the humanized and intimate design of the 21st century. The commonality between related spaces and sustainable development can be presented through Emotional Space, Emotional Experience, and Qualia Products, serving as a new way of humanized and intimate design. The expanded influence of specialty stores will continue to play the image of a "good neighbor in the community," further providing new ways of thinking for the development of "humanity" and "technology."

References

Andreassen, T.W., Kristensson, P., Lervik-Olsen, L., Parasuraman, A., McCollKennedy, J.R., Edvardsson, B.: Linking service design to value creation and service research. J. Serv. Manag. **27**(1), 21–29 (2016)

Bitner, M.J., Ostrom, A.L., Morgan, F.N.: Service blueprinting: a practical technique for service innovation. Calif. Manage. Rev. **50**(3), 66–94 (2008)

Chang, J.T.: Transforming Image of Taiwanese Traditional Grocery Store into Cultural Creative Product Design. National Taiwan University of Arts, Taiwan, M.S. Thesis, Department of Craft Design (2016)

Chang, S.H., Ma, J.P., Chen, S.J., Lin, P.-H.: Exploration of experiencing design model according to majoee tea factory of Yunnan. J. Nat. Taiwan Coll. Arts **97**, 103–134 (2009)

Chien, C.W., Chen, S.J., Lin, R.T.: The research on transforming the intangible qualia experience to tangible products. J. Nat. Taiwan Coll. Arts **101**, 57–77 (2017)

Cleff, T., Lin, I.C., Walter, N.: Can you feel it?-The effect of brand experience on brand equity. J. Brand Manag. **11**(2), 7–27 (2014)

Feng, C., Ma, R., Jiang, L.: The impact of service Innovation on firm performance: a meta-analysis. J. Serv. Manag. **32**(3), 289–314 (2020)

Idei, N.: Quantum leaps. Taipei: Businessweekly (2003)

Johnson, S.P., Menor, L.J., Roth, A.V., Chase, R.B.: A critical evaluation of the new service development process: integrating service innovation and service design. Sage Publications, Inc, Thousand Oaks, CA (2000)

Kunkel, J.H., Berry, L.L.: A behavior conception of retail image. J. Mark. **32**, 21–27 (1968)

Lee, W.L., Liu, C.H., Tseng, T.W.: The multiple effects of service innovation and quality on transitional and electronic word-of-mouth in predicting customer behaviour. J. Retail. Consum. Serv. 64 (2022)

Lin, C.Y.: How does perceived retail service innovativeness affect retail patronage intentions? Creativity Innovation Manage. **28**(4), 519–532 (2019)

Lin, C.Y.: Perceived convenience retailer innovativeness: how does it affect consumers? Manag. Decis. **54**(4), 946–964 (2016)

Lin, R.T.: Cultivate oneself, put the family in order, govern the state, and pacify the world - Cultural creativity in the world. New Taipei City: Graduate School of Creative Industry Design of National Taiwan University of Arts (2014)

Lin, R.T.: The essence and research of cultural and creative industry. J. Des. **16**(4), i–iv (2011)

Pine, B.J., II., Gilmore, J.H.: Authenticity: What Consumers Really Want. Harvard Business School Press, Massachusetts (2007)

Pine, B.J., II., Gilmore, J.H.: The Experience Economy: Work Is Theatre & Every Business a Stage. Harvard Business School Press, Boston (1999)

Schumpeter, J.A.: The Theory of Economics Development. Harvard University Press, Boston (1934)

Sun, Y., Lin, S., Sun, M.: The evaluation of the classic design in contemporary perspective: reflection on Bauhaus Hundred years of Prosperity. J. Des. **24**(3), 49–72 (2019)

Dhote, T., Kumar, V.: Long-duration storytelling: study of factors influencing retention ability of brands. J. Creative Commun. **14**(1), 31–53 (2019)

Zarantonello, L., Schmitt, B.H.: Using the brand experience scale to profile consumers and predict consumer behaviour. J. Brand Manag. **17**(7), 532–540 (2010)

Zomerdijk, L.G., Voss, C.A.: Service design for experience-centric services. J. Serv. Res. **13**(1), 67e82 (2010)

The Impact of Reusing Old Houses on Community Sustainable Development: A Case Study of Longquan Street in Taipei City

Chang-Wei Chang[1,2(✉)], Ying-Shueh Shih[1], and Yi-Fu Hsu[1]

[1] Graduate School of Creative Industry Design, National Taiwan University of Arts, Taipei, Taiwan
ansondata@gmail.com
[2] Department of Marketing and Logistics Management, Taipei City University of Science and Technology, Taipei, Taiwan

Abstract. From the Japanese colonial period (1895–1945) to the era of the Nationalist government's relocation to Taiwan (1949–1987), Taipei City has become a crucial center for Taiwan's historical, economic, and cultural development. Since the late 1980s, the city has undergone rapid economic growth and urban construction. In recent years, it has embraced policies focused on cultural sustainability, regeneration, and regional revitalization, leading to an urban structure where tradition and modernity intertwine. The strategy of repurposing old houses for community space reuse, appreciated for its cultural context and historical continuity, has become increasingly popular and now represents a distinctive cultural feature of both the city and its communities. Under urban renewal initiatives, utilizing vacant spaces for community building has emerged as a critical issue. This study centers on the bidirectional communication between groups occupying old houses and community residents, examining the role of these buildings in community development and their impact on sustainable development. The research begins by analyzing policies, community demographics, transportation geography, commercial structures, and cultural characteristics to determine the community's functional needs and service directions. It then evaluates how the operational models of these groups and community networks can form effective symbiotic relationships, enhancing the sustainable development value of old houses. Through case studies with similar cultural backgrounds, this study assesses the implementation process and outcomes of repurposing old houses in various regions, evaluating their potential to boost community economy, social cohesion, and cultural sustainability. The findings suggest that plans aligned with community needs can effectively stimulate economic revitalization and strengthen residents' sense of cohesion and belonging. For future urban planning policies on old house reuse, this study provides empirical evidence and recommends that policymakers thoroughly consider community characteristics and resident needs, effectively set commercial and profit conditions for occupying groups, encourage innovative services and experiences, and promote goals of multi-win and sustainable community development.

Keywords: Urban Renewal · Old House Reuse · Community Co-Creation · Sustainable Development · Group Symbiosis

1 Introduction

1.1 Research Background

Since 2023, the Taipei City Government has been promoting the "Renovation and Maintenance 2.0 Project," with "openness" and "sharing" as its core principles. This initiative aims to revitalize old houses and idle spaces, thereby altering the public's traditional perception of urban renewal (Taipei City Urban Renewal Office, 2023). As the capital of Taiwan, Taipei plays a pivotal role in demonstrating how urban renewal and community revitalization policies can enhance residents' quality of life and promote sustainable development. These policies, which involve protecting and reusing old buildings, not only preserve the city's cultural heritage but also inject new vitality and economic opportunities into the community. The implementation of these policies also includes improving basic public infrastructure and providing diverse activity spaces. The goal is to enhance the quality of the local environment to meet modern needs and ensure alignment with international development trends and local demands, further achieving sustainability objectives (Taipei City Government, 2022). In this policy context, the focus of urban and regional development is not just on the scale of urban construction but also on how to improve community inclusivity, diversity, and quality of life, strengthen environmental protection, and enhance overall community vitality. These strategies collectively shape a quality living foundation and reinforce the Taipei City Government's commitment to sustainable development in terms of "environmental resource cyclical symbiosis, shared social safety progress, and smart economic and technological growth." This study takes the reuse of old houses at Longquan Street Dormitories, operated by the Central News Agency in Taipei, as a case study to examine their specific impacts on community sustainable development, aiming to provide deeper insights and empirical support for future related policies.

1.2 Research Motivation

With the global push for urban renewal and the United Nations Sustainable Development Goals (SDGs), the model of repurposing old houses for community co-creation has emerged as a significant trend in contemporary urban development. The motivation for this research stems from exploring the specific impacts of this reuse strategy on community sustainable development. The old house reuse strategy focuses not only on cultural preservation and architectural repurposing but also emphasizes leveraging the strengths of buildings and community resources through innovative reuse models to serve community practical functions more effectively (Guo, 2011). However, despite the implementation of related policies and programs, the actual effectiveness of old house reuse in promoting community sustainable development and its impact remain relatively underexplored areas. Therefore, this study aims to conduct an in-depth case analysis to investigate the actual performance of the old house reuse strategy in fostering community co-creation and development, as well as the actual expectations and feedback of community residents. The objective of the research is to provide empirical evidence and references for urban renewal and community revitalization strategies for Taipei City and other cities, thereby more precisely shaping urban development policies and promoting the economic, social, and cultural sustainable development of communities.

1.3 Research Objectives

The primary purpose of this research is to explore the specific impacts of the repurposing of old houses at Longquan Street Dormitories, operated by the Central News Agency in Taipei City, on local community sustainable development. With growing global emphasis on sustainable lifestyles and environmental responsibility, the reuse of old buildings has become one of the crucial strategies for achieving sustainable urban development. Particularly in Taipei City, the repurposing project at Longquan Street not only involves the reuse of buildings but also includes multiple aspects of community revitalization and economic regeneration. This study aims to analyze how the operational strategies proposed by occupying groups effectively align with community development needs, thereby enhancing community cohesion and economic vitality. It will explore how the management of emotional spaces can facilitate a symbiotic relationship between residents' needs and the occupying teams, supporting sustainable community development. Furthermore, this research will investigate the role and significance of old house reuse in urban regeneration, and how it can foster broader community participation and development through "community co-creation." Through a case study of the Longquan Street Dormitories by the Central News Agency in Taipei, this research will provide empirical data to advise the formulation and implementation of future urban policies, ensuring that strategies for community revitalization and economic regeneration not only meet technical and environmental standards but also fulfill the specific needs and expectations of community members.

1.4 Research Questions

This study utilizes multiple data sources, including government policy documents, field participation observations, and semi-structured in-depth interviews, to gain a comprehensive understanding of the old house reuse projects in Taipei City. The subjects include local residents and commercial occupant groups, aiming to explore the actual impact and development potential of old house reuse from the perspectives of community needs and the occupant groups, thereby promoting more inclusive and sustainable urban development strategies. The research focuses on the following two core questions:(1) Community demographic composition, transportation geography, economic structure, and cultural characteristics: This part investigates how these factors affect the community's functional needs and expectations for sustainable development services. The goal is to identify the unique needs of the community to better tailor the old house reuse projects to effectively serve the local community. (2)Operational models of occupant groups and their interaction with community networks: This analysis examines how occupant groups establish effective symbiotic relationships with the community through their operational strategies and discusses how these relationships promote the sustainable development and reuse of old buildings. This analysis will provide insights on how commercial activities can enhance community cohesion and economic vitality.

Through the research methods mentioned above, this study aims to provide deep insights to assist local governments and occupant teams in understanding the unique needs and expectations of the community, thus designing more effective old house reuse strategies. The findings will offer empirical support for future urban renewal and community revitalization policies, contributing to the promotion of sustainable development goals in Taipei City and beyond.

2 Literature Review

2.1 Urban Regeneration Through Old House Reuse

Marx noted in "Das Kapital" that all commodities possess a use value, which is a basic attribute that satisfies or resolves specific human needs. In a capitalist market economy, the production of goods primarily aims to achieve exchange value, thus maximizing profit, rather than merely fulfilling people's use needs (Marx, 1867). In the process of urban development and regeneration, an excessive pursuit of the exchange value of buildings or real estate often overlooks the living needs and community networks of residents. Therefore, promoting urban development and regeneration requires not only environmental renewal but also the protection and strengthening of residents' public life and community relations.

Healthy urban development should preserve the city's inherent diversity and complexity, focusing on the vitality created by human activities (Jacobs, 1964; Fornal & Myna, 2017) and considering the cultural values and social network needs of community residents (Gans, 1962). Urban regeneration models are categorized into "urban construction" and "old house reuse," with the latter emphasizing transforming the exchange value of buildings into a use-value-centered urban development approach, highlighting the importance of use value when formulating urban renewal policies and conducting community revitalization economic activities. According to the urban development stage theory (Van Den Berg,, 1987), the urbanization process includes four stages: Urbanization, Suburbanization, Disurbanization, and Reurbanization. Currently, Taipei City is experiencing stages of disurbanization and reurbanization, facing challenges such as outward migration, aging population, and declining birth rates, leading to community space decay and "urban shrinkage" in downtown areas (Berry et al., 1963).

Taiwan's urban regeneration projects must address key issues including population aging, urban voids, development of livable communities, and transformation needs of the community economy. This includes promoting economic resilience, immigrant economy, grassroots economic reconstruction, inclusive economy, and emerging fields such as innovative technology education and cultural and creative industries (Aulia, 2016). The old house reuse strategy has been strengthening the preservation, revitalization, and reuse of old houses in Taipei City since 2013, initiated by the municipal government to assist in the repair and reuse of old buildings. Through public tendering and selection processes, combined with private creativity and funding, a "Cultural Movement of Old Houses" matchmaking mechanism has been formed, not only enhancing the efficiency of cultural asset restoration but also strengthening the willingness of private teams to participate in the preservation and revitalization of cultural assets (Lin, 2013).

Moreover, the Cultural Movement of Old Houses, through various cultural activities and projects like "On-site Inspections," invites public participation, increasing awareness of the historical and cultural value of old houses, thereby strengthening community cohesion and residents' sense of belonging (Cultural Movement of Old Houses Project

Team, 2021). The strategy of old house reuse demonstrates a unique role and crucial mission in urban regeneration.

2.2 Cognitive Framework for Sustainable Development

The concept of sustainable development traces its origins to the 1987 report "Our Common Future" by the Brundtland Commission, which defined sustainable development as "meeting the needs of the present without compromising the ability of future generations to meet their own needs." This theory emphasizes the balance among economic, environmental, and social factors and has been widely accepted by the international community as a guiding principle for development policies (World Commission on Environment and Development, 1987). In the context of urban regeneration, this means integrating the reuse of old buildings with modern needs rather than pursuing short-term economic benefits.

Old houses carry a rich cultural history and are part of the shared lifestyles and collective memory of a community. The creativity involved in reusing old houses needs not only local community recognition but also to appeal to and resonate with visitors and consumers, thereby fostering sustainable community economic development. For example, the Dihua Street (Dadaocheng) district in Taipei City, a significant urban regeneration case of old house reuse, demonstrates how to balance urban construction with strategies for reusing old buildings. The district is considered an important historical and cultural area by the government and has undergone urban regeneration with a historical and cultural focus (Lin, 2021), successfully blending traditional culture with contemporary styles and becoming an attractive urban destination as the cultural and creative industries cluster there (Yin, 2016).

Sustainable development in community building should be approached from the perspectives of needs and experiences, establishing engagement and involvement through shared cultural history, memories, and profound atmospheres and experiences, thus promoting active community and economic activities. Therefore, the strategy of reusing old houses should return to a cultural mindset, starting anew from cultural inheritance and maintenance (Lin, 2014).

From the external, internal, and emotional attributes of cultural design (Hsu, 2004) and the three levels of emotional design: visceral, behavioral, and reflective (Norman, 2014), the reuse of old houses should serve as a medium for communication between occupant groups and community residents. This model, through conveying feelings and cognitions, achieves mutual support of form and content. This research will establish a Community Resident Cognitive Communication Model (Fig. 1), which through the experience of old houses, gradually evolves on three levels to promote interactive activities among community residents, incorporating new technologies and knowledge, allowing residents to re-experience the value and significance of reusing old buildings, thereby enhancing participation and the attractiveness of activities (Taylor & Francis, 2020).

As shown in Fig. 1, the communication process is divided into three stages. From top to bottom, it begins with the planning process, extending the new use of old buildings to promote local markets. Then, community classrooms push forward technical research, leading to the sharing of results that combine multiple technologies. Creative courses drive collective learning, fostering a unique emotional connection within the community and achieving distinctive creations.From left to right, the external (instinctive) aspect of architectural development involves transforming existing buildings into classrooms and developing workshops. The internal (behavioral) aspect involves repurposing spaces, creating and accumulating the expressive meaning and emotional significance that architecture brings to the community. As it delves into the emotional (reflective) aspect, the market gradually showcases the co-creation brought about by sharing. This co-creation is not just the creation of art or products but also an essential intersection and connection between the community and its residents, demonstrating its value and impact. It culminates in the realization of important community development goals that are recognized and communicated by the community residents.

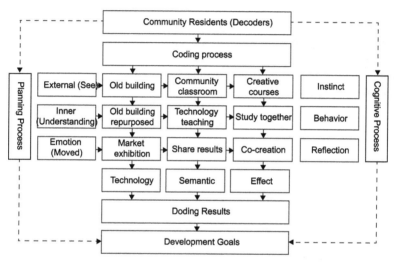

Fig. 1. Community Resident Cognitive Communication Model (Modified from Lin, 2005).

2.3 The Impact of Emotional Fields on Community Engagement and Interaction

An emotional field refers to the concept of connecting people to a place through sensory experiences, playing a critical role in community revitalization. According to research from the "Journal of Environmental Psychology," emotional fields can enhance community members' emotional connections to public spaces and stimulate civic participation (Brown & Green, 2021). For instance, elements such as music, art installations, and community gardens can effectively stimulate residents' sensory experiences, thereby fostering communication and collaboration within the community.

The primary goal of the old house reuse programs is to create an enriched community atmosphere, utilizing the "core knowledge" of the creative living industry to provide "deep experiences" and subsequently demonstrate the "high-quality aesthetics" of creative community living (Lin, 2011). This experience-centered field service and value not only balance the needs of residents and occupant groups but also create a unique community lifestyle. Transparent and continuous communication strategies help build trust and consensus, promoting a symbiotic model among groups, effectively aligning the intentions of occupant groups with the needs of the community, and thereby enhancing mutual understanding and cooperation (Johnson, 2022).

Starting from the cultural perspective of community residents, the core lies in the added value creation of cultural creativity, which not only drives the development of the cultural and creative industries but is also essential for creating community value and livable conditions (Lin & Lin, 2009; Lin & Wang 2008). This study utilizes the research framework of the creative living industry as the structure for constructing the old house reuse framework. The creative living industry research framework (Lin, 2010b) outlines the horizontal dimensions of community resident participation levels and the vertical dimensions of field relevance. Through appropriate cultural and creative experiences, it is possible to establish a willingness among residents to participate in experiences, gradually shifting from a coexistence stage to a co-creation stage, showcasing the community-building value of old house reuse (Fig. 2). The process moves from mere sensory reception (coexistence) to active participation in cultural history and interaction (sharing, co-learning), and ultimately to a collective co-creation experience (co-creation).

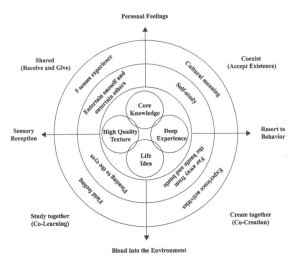

Fig. 2. Creative Living Industry Research Framework (Lin, 2010b).

2.4 Current Development Status of Longquan Street in Taipei City

Longquan Street in Taipei City's Da'an District is adjacent to National Taiwan Normal University, Shida Night Market, and the MRT Taipower Building Station. It boasts strong living and economic functions, enriched with cultural heritage and a unique historical background. According to the "Da'an District Cultural Map" (Taipei City Department of Cultural Affairs, 2020), Longquan Street and its surrounding areas were once early commercial and residential mixed zones in Taipei City, preserving many buildings from the Japanese colonial period and the early post-war era, including the historically significant former site of the Central News Agency. These old buildings not only witness the transformation of Taipei City but also carry local memories and historical sentiments.

According to the "Gufengli Community Development Report" (2023), Gufengli in Da'an District experienced significant demographic changes between 2012 and 2022, with a population decrease of 8.27% and notable changes in age structure, particularly a 50% increase in the elderly population. Additionally, Da'an District has the lowest birth rate in Taipei City, and its negative population growth rate is six times that of Taiwan overall. In age distribution, the population under 15 years old decreased by about 5%, those aged 15–64 decreased by 20%, and those over 65 increased by 50%. These statistics highlight the changing needs of residents, demanding corresponding adjustments to community services and infrastructure to better accommodate the increasing elderly population and the needs of the younger generation.

Starting from the needs for livable community development and economic transformation, the Taipei City Department of Cultural Affairs has been promoting policies on cultural memory, revitalization of buildings, and utilization since 2021, aiming to achieve economic goals in the cultural industry in response to the strategy requirements of reusing old houses. Proper old house reuse strategies not only protect historical buildings but also effectively enhance the regional economy, reinvigorate the area, attract new investments and collaborations, create job opportunities, and thereby achieve sustainable development (Mommaas, 2004; Taipei City Department of Cultural Affairs, 2021).

3 Research Methods

3.1 Case Study and Subject Description

This study selects the Central News Agency Dormitories located on Longquan Street in Taipei's Da'an District as the research case. This location was chosen as a government target case for community interaction and sustainable management, receiving government subsidies and guidance. It is a key focus of the Taipei City Government Urban Renewal Office for 2024. Before its demolition for urban renewal, the state-owned Central News Agency Dormitories were opened for use by occupying teams, serving as a model and experimental site.

3.2 Research Methodology Description

This study employs qualitative methods, combining policy analysis, field participation observation, and semi-structured in-depth interviews. Firstly, it analyzes policy documents, project reports, community meeting records, and related research data from the Taipei City Government to gain a comprehensive understanding of the area's developmental history and current status. Secondly, field participation involves directly observing the implementation process of the old house reuse project, recording changes in the physical environment and their impact on the community atmosphere. Finally, through semi-structured in-depth interviews, the study deeply analyzes the needs and expectations of community residents and occupying groups to ensure the comprehensiveness and depth of the research.

This study utilizes Grounded Theory, employing open coding to analyze respondents' experiences in participating in community activities and their interactions with other community residents. The research explores how these interactions foster community cohesion and economic vitality. Subsequently, it further examines the relationship between community cohesion and economic vitality to evaluate the impact of the adaptive reuse of old buildings on promoting community economic benefits. Based on interviews, the study integrates the needs analysis of community members and incoming groups to propose corresponding community needs plans and policy recommendation models.

3.3 Research Framework and Semi-structured Interview Structure and Questions

This study's framework is based on the theories of community engagement and economic revitalization, aiming to thoroughly explore the specific impacts of old house reuse projects on sustainable community development.

Table 1. Interview topics from four aspects.

Theme	Questions for interviewing community residents	Questions for in-person group interviews
Exterior	What is the environmental impact of the existence of old houses?	Creative ideas brought about by the existence of the old house?
Need	What are the needs of community residents?	What are the needs of resident groups?
Economy	What kind of experiences will be provided that residents will participate in?	What experience and services are provided to residents?
Sustainable	What kind of services are provided that residents will return for return visits?	What kind of business services can operate sustainably?

The research focuses on the following core aspects, planned for deep exploration during interviews: (1) the concept of interior and exterior restoration plans for old houses and their impact on the community; (2) the interaction needs and expectations between community residents and occupant groups; (3) how the planning and needs of community building are realized through interactions between both parties; (4) the specific economic benefits and challenges brought about by old house reuse (Table 1). The research methodology includes qualitative analysis, particularly semi-structured in-depth interviews, to gain detailed insights into the perspectives and expectations of various stakeholders. The interview structure is designed as follows to collect in-depth insights from participants of diverse backgrounds: Resident interviewees: one male and one female aged between 60 and 65 who have participated in community-building activities, capable of providing insights into past and present needs changes. Occupant group interviewees: one male and one female aged between 50 and 55 who have been involved in several community-building projects in Taipei City, capable of providing expert strategies and operational experience (Table 2).

Table 2. Qualifications for selection of interviewees.

Interview Subjects	Position	Number of Interviews
Residents	Market Management	Participate in community creation and creation
	Limin Services	Coordinate with the actual needs of community residents
Representative	Settled in Merchants	Participated in social construction in the past five years and has experience in serving the community

The purpose of the interviews is to collect data to analyze specific community needs and development strategies, and based on the data gathered, to propose practical recommendations and policy adjustment plans. This process will help researchers design a sustainable development model that meets the current and future needs of the community.

4 Research Findings and Discussion

This study examined the impact of "old house reuse on sustainable community development" using the case of the Central News Agency Dormitories on Longquan Street in Taipei City. It focused on changes in the economic, social, and cultural dimensions of the community and provided a detailed analysis of the actual effects following the implementation of such projects. Below is a detailed description of the research findings:

4.1 Exploring Community Composition and Functional Needs and Service Directions

This research highlights the widespread distribution of the elderly population within the community, thereby discussing effective implementation of the old house reuse projects. Through community interviews, it was found that residents look forward to diverse activities such as specialized learning, craft creation, market exchanges, therapeutic living, and material recycling to revitalize the community economy. Moreover, utilizing historic buildings not only promotes commercial activities but also meets the community's needs for preserving local historical and cultural contexts. Further observations indicated that activities such as cultural performances and handicraft workshops enhance social connections and cohesion within the community. These activities not only facilitate communication among neighbors but also strengthen the sense of community belonging. Residents' participation in and impact on community development have increased, highlighting the importance of community involvement in the decision-making process.

Old house reuse projects play a key role in the protection and revitalization of cultural assets. By restoring old buildings, the community's historical memory and cultural identity are preserved while also injecting new functionalities, making them a part of residents' daily lives. Additionally, these projects offer a sustainable development model that promotes environmental, economic, and social harmony.

It is recommended that policymakers and community planners draw lessons from the Longquan Street Dormitories old house reuse case to explore how to balance commercial development with cultural preservation needs and how to more effectively encourage broad community participation and shared outcomes. This study combines policy analysis, interview data, and the dual cognitive communication between community residents and occupying teams, along with the creative living industry integration model, to propose a community sustainable management framework involving sharing, co-learning, and co-creation (Fig. 3).

Fig. 3. Community Sustainable Management Framework (Illustrated by this study).

4.2 Operational Models of Occupant Groups and the Formation of Effective Symbiotic Relationships with the Community Network

This study examines the operational models of occupant groups establishing effective symbiotic relationships with the community, a key factor in implementing old house reuse strategies. Beyond meeting the basic needs of community residents, occupant groups must maintain continuous interaction with the community. The research identified common issues such as service overlaps, singular directions, and conflicts in professional skills among the teams, which require reevaluation and adjustment.

Through semi-structured interviews, this study revealed complex interactions among the government, multiple occupant groups, and residents. To meet the diverse needs of the community, the operational models of the occupant groups should encompass and integrate the varied demands of residents, thereby creating diversified interaction and service opportunities. For instance, community demands for childcare, pet care, technology learning, and performance activities reflect the diversity of service scopes. Moreover, when organizing learning and cultural activities, occupant groups should achieve more effective community engagement through coherent planning and strategies. This not only helps to enhance the cultural identity and knowledge level of community members but also maintains the continuity and richness of community culture. Holding regular coordination meetings to ensure that activities and services meet the residents' needs and expectations is crucial for reducing overlaps and enhancing residents' willingness to participate.

In summary, through diversified operational models and active community interaction, occupant groups can effectively promote ongoing community participation and learning, facilitating the co-creation of environmental and economic opportunities, thus strengthening community cohesion and residents' sense of belonging. This model of sustainable community management not only improves the overall quality of life in the community but also provides empirical support for the successful implementation of old house reuse strategies.

5 Conclusions and Recommendations

This study thoroughly explores the multifaceted impact of the old house reuse at the Central News Agency Dormitories on Longquan Street in Taipei City on sustainable community development, demonstrating its multifunctional and positive effects on economic, social, and cultural levels. The findings reveal that old house reuse not only strengthens the revitalization of the community economy but also promotes the preservation and transmission of community culture, and enhances residents' sense of belonging. To continue advancing these positive effects, it is recommended that policymakers strengthen support for old house reuse projects and optimize related policies to better meet the needs and expectations of community residents. Additionally, more community participation, cross-governmental, and private sector collaboration should be encouraged to ensure the successful implementation and long-term sustainability of old house reuse programs.

References

Ancient Style: BPM. https://bpm.com.tw/block/63000030-013/ (2023). [In Chinese, semantic translation]

Ancient Style Community Development Report: Analysis of population and community structure in Ancient Style. Da'an District Community Development Association (2023)

Aulia, D.N.: A framework for exploring livable community in residential environment. Case study: public housing in Medan, Indonesia. Procedia-social and behavioral sciences **234**, 336–343 (2016). [In Chinese, semantic translation]

Berry, B.J., Simmons, J.W., Tennant, R.J.: Urban population densities: structure and change. Geogr. Rev. 389–405 (1963)

Brown, S., Green, T.: Sensory fields and community activation: the role of aesthetic experiences in urban spaces. J. Environ. Psychol. **71**, 101–112 (2021)

City Property Activation Introduction: Taipei City Property Activation Integration Platform. https://propertyspace.gov.taipei/cp.aspx?n=008B81344940F1C4 (2023). [In Chinese, semantic translation]

Fornal, R., Myna, A.: The development of residential areas on the example of Łuków: reurbanization and suburbanization. Barometr Regionalny **55**(2), 45–53 (2017). https://doi.org/10.56583/br.436

Gans, H.J.: The Urban Villagers. Free Press, New York (1962)

Guo, L.L.: Research on the Spirit of Place Reuse. Beijing Jiaotong University, Beijing (2011). https://d.wanfangdata.com.cn/thesis/Y1962901

Jacobs, J.: Death and Life of Great American Cities. Pelican, London (1964)

Johnson, L.: Communicating regeneration: narrative approaches in urban redevelopment projects. J. Commun. Dev. **53**(3), 234–251 (2022)

Lee, P.X.: Social Value Study of Historical House Reuse. National Cheng Kung University, Tainan City (2018)

Lin, R.T.: Exploring sensory experience design from a service innovation perspective. Des. Stud. **14**(S), 13–31 (2011). [In Chinese, semantic translation]

Lin, R.T.: Establishing peace and order: cultural creativity unifying the world. Yucheng, New Taipei City (2014). [In Chinese, semantic translation]

Lin, R.T., Li, X.M.: Poetic beauty – Experiences shared through the beauty of immortal clouds. National Taiwan University of Arts, New Taipei City (2015). [In Chinese, semantic translation]

Lin, R.T., Liu, P.H.: Integrating culture and aesthetics to promote emerging industries of cultural creativity design. J. Arts **5**(2), 81–105 (2009). [In Chinese, semantic translation]

Lin, R.T., Wang, M.X.: Exploring the current situation and vision of Taiwan's design industry development. J. Arts **4**(1), 49–69 (2008). [In Chinese, semantic translation]

Lin, R.T., Liu, P.H., Li, Y.J., Su, J.H., Zhang, S.H.: Sensory fields and sensory products in creative living industries – a case study of The One South Garden. In: Proceedings of the 2010 Taiwan Sensory Society Academic Symposium, pp. 61–66. Donghai University, Taichung (2010). [In Chinese, semantic translation]

Lin, W.I., Zhang, J.R.: Governance shaping, practice, and its specificity in the preservation special zone of Dihua Street. City and Plann. **48**(3), 311–346 (2021). [In Chinese, semantic translation]

Lin, X.C.: Hao Long-bin: "Old House Cultural Movement" is a Win-Win Situation. MyGoNews. https://www.mygonews.com/news/detail/news_id/107285 (2013). [In Chinese, semantic translation]

Marx, K.: Capital Vol.3: A Critique of Political Economy. PENGUIN GROUP (USA) INC (1867)

Ministry of Culture: Urban and Rural Construction - Cultural Life Circle Construction Project. https://www.moc.gov.tw/cp.aspx?n=139 (2021)

Mommaas, H.: Cultural clusters and the post-industrial city: towards the remapping of urban cultural policy. Urban Stud. **41**(3), 507–532 (2004)

Old House Cultural Movement Project Team: 2021 Old House Opening Tour Guide. Taipei City Department of Cultural Affairs. https://oldhouse.taipei/home/zh-tw/handbook/696885 (2021). [In Chinese, semantic translation]

Old House Cultural Movement Project Team: 2022 Old House Time Tunnel I Tour Guide. Taipei City Department of Cultural Affairs. https://oldhouse.taipei/home/zh-tw/handbook/696885 (2022). [In Chinese, semantic translation]

Old House Cultural Movement Project Team: 2023 Taipei Old House Adventure Team I Tour Guide. Taipei City Department of Cultural Affairs. https://oldhouse.taipei/home/zh-tw/handbook/2186480 (2023). [In Chinese, semantic translation]

Su, Y.W.: Place attachment and space transformation. Dayeh University, Changhua County (2013). [In Chinese, semantic translation]

Taipei City Department of Cultural Affairs: Da'an District Cultural Map. https://www.culture.gov.tw/ (2020). [In Chinese, semantic translation]

Taipei City Department of Cultural Affairs: Cultural Asset Activation and Reuse Policy Report. https://www.culture.gov.tw/ (2021). [In Chinese, semantic translation]

Taipei City Government: The Road to Sustainable Development (2022). https://sdg.gov.taipei/page_en/eng_area2/21

Taipei Urban Renewal Office: Taipei City Reconstruction and Maintenance 2.0 Project Pilot Program. https://uro.gov.taipei/News_Content.aspx?n=971AC16B4D3BACA1&sms=25BE17A71805E3E3&s=0C24D892BCAB1A86 (2023). [In Chinese, semantic translation]

Taylor, A., Francis, B.: Enhancing community engagement in urban regeneration through augmented reality experiences. J. Urban Technol. **27**(2), 3–20 (2020)

Van Den Berg, L.: Urban systems in a dynamic society. Gower Pub Co, UK (1987). https://repub.eur.nl/pub/104236/

Wang, H.X., Weng, Q.L., Zheng, Y.P., Zhang, Z.J.: Emotional design: why some designs make you fall in love at first sight. Yuan-Liou, Taiwan (2011). [In Chinese, semantic translation]

Wu, T.Y., Hsu, C.H., Lin, R.: The study of Taiwan aboriginal culture on product design (2004). [In Chinese, semantic translation]

Wolff, M., Wiechmann, T.: Urban growth and decline: Europe's shrinking cities in a comparative perspective 1990–2010. Eur. Urban Reg. Stud. **25**(2), 122–139 (2018)

World Commission on Environment and Development: Our Common Future. Oxford University Press, Oxford (1987)

Yin, B.N.: Creative district, food culture and urban regeneration: the culinary landscape and cultural transformation of Dadaocheng Dihua street in Taipei City. J. Cult. Resour. **10**, 029–066 (2016). [In Chinese, semantic translation]

Generative Artificial Intelligence to Enhance the Sustainability of Traditional Crafts: The Case of Ceramic Teapots

Yi-Fu Hsu[1](✉), Chang-Wei Chang[1,2], and Chih-Long Lin[1]

[1] Graduate School of Creative Industry Design, National Taiwan University of Arts, Taipei, Taiwan
light617617@gmail.com

[2] Department of Marketing and Logistics Management, Taipei City University of Science and Technology, Taipei, Taiwan

Abstract. This study aims to investigate the application of Generative Artificial Intelligence (GAI) in enhancing the sustainability of traditional craft products, using ceramic teapots as a case study. Through an analysis of the Qualia attributes and cultural connotations of ceramic cultural and creative products, a cognitive space was constructed using Multi-Dimensional Scaling (MDS) to explore the similarity and distribution of characteristics among the artworks. The results indicate that artworks generated by GAI received higher scores in public evaluations, demonstrating greater acceptance. Through multiple regression analysis, it was found that GAI effectively extracts cultural elements and transforms them into design elements, creating ceramic cultural and creative products with Qualia. This study provides valuable insights into the status and trends of traditional craft products in contemporary society, which is crucial for their sustainability.

Keywords: Generative Artificial Intelligence · Ceramic Craft · Cultural Products · Stable Diffusion · Qualia Commodities

1 Introduction

Since Taiwan embarked on its journey towards a "knowledge economy," there has been a growing focus on cultural and creative industries, including design, with various craft industries now considered part of this sector. The approach to craft product design has shifted from merely functional considerations to prioritizing human-centric, cultural, and aesthetic experiences (Lin and Lin 2009). This evolution has broadened the scope of cultural products, moving beyond mere functional satisfaction to encompass deeper internal experiences and emotions. Yen et al. (2014) emphasized the importance of Qualia attributes and cultural connotations in cultural products. Consequently, sustainability issues have gained prominence, especially in the production and design of traditional crafts, where cultural connotations and Qualia attributes play a significant role.

The rapid advancement of technology, particularly the emergence of Generative Artificial Intelligence (GAI), has significantly impacted traditional craft industries, including traditional ceramic craftsmanship. Traditionally, these crafts faced challenges in technological advancement and digitization, leading to issues such as a lack of innovative design patterns and difficulties in technical inheritance, resulting in aging talent and cultural continuity concerns. However, GAI presents a new opportunity for traditional ceramic craftsmanship by intervening in design processes, potentially creating sustainable momentum for local crafts.

Preserving cultural heritage is crucial for the enduring welfare of human civilization, especially concerning traditional crafts. However, contemporary traditional ceramic craftsmanship is facing decline, a challenge that GAI technology can address. GAI not only promises to enhance traditional ceramic craft design but also has the potential to bridge technological and digital gaps within the craft domain, thus promoting sustainable progression.

This study aims to explore how GAI technology can enhance the sustainability of traditional ceramic craftsmanship products, focusing on ceramic teapots as a case for analysis. It seeks to investigate audience perceptions of Qualia attributes in cultural products integrated with GAI, preferences for human-machine collaboration in cultural and creative products, and potential avenues for GAI collaboration in ceramic craft product design. Through these inquiries, the study aims to shed light on the transformative potential of GAI in reshaping traditional craftsmanship and fostering sustainable progress amidst contemporary challenges.

2 Literature Review

2.1 Generative Artificial Intelligence (GAI)

The release of ChatGPT by OpenAI in November 2022 marked the beginning of rapid advancements in various forms of Generative Artificial Intelligence (GAI), such as Bing, Midjourney, Stable Diffusion, DALL-E 3, and Sora. GAI has become a widely discussed topic, surprising the public with its impressive capabilities and significant impact. The development of GAI is an irreversible trend, prompting humanity to consider how best to integrate GAI into various industries to enhance competitiveness.

GAI technology has evolved to generate not only high-quality artistic works but also intelligent content based on user input text. This innovative human-machine collaboration model offers new opportunities for creators and the cultural and creative industries. Integrating culture with AI in the creative industry is a powerful driving force and a necessary condition for the development of these industries (Freedman 2023; Liu and Song 2022; Andrejevic 2020).

In Taiwan, where over 80% of cultural and creative industries are small-scale enterprises (Taiwan Creative Content Agency 2024), the utilization of GAI, including technologies like Stable Diffusion, has gained traction. Stable Diffusion, known for producing coherent and contextually relevant text, has sparked interest in research experiments. In the visual arts, practitioners are exploring and leveraging Text-to-Image (Txt2Img) technology within GAI, both domestically and internationally (Oksanen et al. 2023). This human-machine collaboration model offers advantages such as reducing personnel costs, bridging the technological and digital gap, and improving efficiency. However, it also presents challenges. For instance, how can cultural elements be transformed into cultural products through GAI? Can the public perceive cultural connotations in these products?

In summary, while GAI holds great potential for enhancing the cultural and creative industries, it requires careful integration and consideration of cultural elements to ensure its effectiveness and public acceptance.

2.2 The Concept of Qualia

The concept of Qualia was introduced by the Small and Medium Enterprise Administration of the Ministry of Economic Affairs in Taiwan in 2013. It encompasses attributes such as Attractiveness, Beauty, Creativity, Delicacy, and Engineering, as defined by the Taiwan Design Center (Yen et al. 2014). Yen et al. (2014) argued that cultural significance is one of the factors that attract consumers, emphasizing the importance of not overlooking cultural significance as one of the attributes of Qualia. They advocated that cultural significance is a crucial element in enhancing attractiveness. Considering the literal meaning of "cultural products," these products should inherently integrate cultural elements and generate cultural creative designs through modern forms. Therefore, how to integrate cultural elements through GAI to innovate ceramic craft product designs becomes a significant focus of this study.

Recent research has further expanded on the concept of Qualia and its impact on consumer behavior. Niu (2019) discusses how Qualia, when applied to tangible objects, can be concretely presented through products or services, touching the inner needs of consumers' hearts. This underscores the relevance of Qualia in creating emotional connections with consumers through product design.

Moreover, Li et al. (2024) explore factors influencing consumers' purchase intentions for Gejia batik, highlighting that consumer demands extend beyond mere functionality to include aesthetic and emotional attributes such as beauty, creativity, and delicacy. They introduce the theory of product personality, which posits that products possess both physiological qualities (e.g., design, functionality) and psychological qualities (e.g., attractiveness, creativity). This duality is crucial for meeting consumers' comprehensive needs.

Yen and Hsu (2017) investigate college student perceptions of cultural elements in fashion design, reinforcing the idea that products satisfying both physiological and psychological qualities can meet consumers' functional and emotional needs. This aligns with the notion that integrating cultural elements into product design can enhance both the attractiveness and the cultural significance of the product.

In light of these findings, this study aims to explore how cultural elements can be integrated through Generative AI (GAI) to innovate ceramic craft product designs, leveraging the comprehensive framework of Qualia to assess and enhance both the aesthetic and functional attributes of these designs. By doing so, the study seeks to provide a nuanced understanding of how these factors influence consumer preferences and purchase intentions, thereby contributing to the broader field of design and consumer behavior research.

2.3 Cultural Connotation

Culture, the cornerstone of society, is shaped by community-specific values, norms, and communication styles (Lin 2011), and reflects both compatibility and changeability (Lorusso 2015). In traditional craft industries, cultural elements are crucial, influencing technique transmission and development while preserving cultural values. For example, ceramic craftsmanship represents significant cultural heritage, and preserving these crafts is essential for both industry and heritage. Generative AI technology can enhance the sustainability of traditional craft products, but understanding the audience's cultural background is key (Hofstede 1984; Nöth 1990). This understanding enables the design of products that resonate with the audience, promoting cross-cultural communication and understanding. Thus, a deep knowledge of the target audience's cultural background and values is necessary for effective and appropriate design, ensuring the sustainable development of craft products.

2.4 Cognitive Models Between Cultural Products and the Public

Norman (2002) proposed three cognitive models—design models, user models, and system models—that are closely related to cognition. Creators use design models to generate and implement product creativity based on personal life experiences. User models focus on consumers' perceptions of product form, enabling them to experience and understand the products and explore their meanings and stories to evoke resonance. System models pertain to the form, semantics, and effects of products, serving as platforms for interaction with consumers. These three models interact to form cognitive patterns between creators and consumers, requiring consideration of multiple levels such as technology, semantics, and effects to effectively convey stories and connotations (Lin et al. 2017). Additionally, Norman (2004) proposed three levels of emotional design: visceral, behavioral, and reflective, to deconstruct design thinking within products. Integrating Norman's two concepts not only provides a basis for transforming cultural elements into design elements in this study but also becomes a dimension for evaluating audience cognition.

Jakobson (1987) proposed a communication model emphasizing six factors and six functions in the communication process: sender, receiver, context, message, contact, and code. These factors and functions interact to form encoding and decoding communication activities. This perspective complements the cognitive models proposed by Norman and aids in analyzing the cognitive process between narrative stories and the public.

This study builds on these theoretical foundations to establish evaluation criteria. Based on this, Fig. 1's cognitive model is formed, depicting interactions among creators, viewers, and products. This model helps to understand how creators convey connotations through products and how viewers decode and understand these connotations. Further research and application of this model will enhance the efficiency and quality of cultural product dissemination, promoting cultural exchange and understanding.

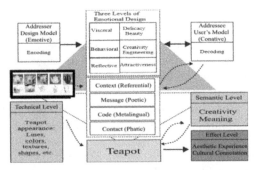

Fig. 1. Cognitive model of creators and audience (Adapted and Redrawn from Lin, R. et al., 2017, as synthesized in this study)

3 Research Methods

3.1 Methodology

This study employs a mixed-methods approach, incorporating elements of case study analysis, literature review, and questionnaire surveys. Analysis techniques include Multi-dimensional Scaling (MDS) and regression analysis. The collected data from the questionnaires undergo synthesis and analysis to reveal its underlying significance.

The research process unfolds in three distinct stages. Initially, a comprehensive literature review and relevant discussions are undertaken to establish the theoretical framework of the study. Subsequently, an experimental design is crafted, and questionnaires are distributed to gather data for multidimensional analysis, focusing on perceptual characteristics, cultural product connotations, and preference evaluations. The total number of participants in this study is 203 individuals. Finally, the findings of the analysis are juxtaposed with the literature review to elucidate the research outcomes and conclusions.

In participant selection, the study imposes a minimum age requirement of 18 years, considering the investigation into purchase willingness. Questionnaires are disseminated among university students from diverse disciplines, members of pottery clubs, and individuals aged 18 and above. To ensure a broad spectrum of opinions, the Likert 5 Point Scale is adopted as the questionnaire completion format.

3.2 Stimuli

Table 1 shows the two groups of stimuli for this experiment. Group A consists of cultural ceramic products from the National Palace Museum's boutique, with the primary focus being on "ceramic teapots." The selection criteria for Group A are based on three standards: the formation of decorations using cultural elements (color, pattern, line), shape transformation (turning specific shapes into product appearances), and symbolic representation (using a symbol to express abstract meanings). Group B comprises a series of teapots generated through artificial intelligence, incorporating design elements derived from Group A's analysis along with additional elements from the creators. These teapots are distinguished by the English letters A and B, respectively, and are interspersed in the questionnaire survey to mitigate respondent fatigue.

Table 1. Stimuli

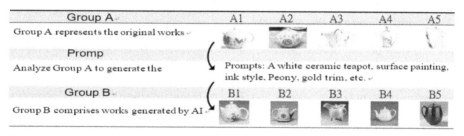

(Source of the Image: National Palace Museum Collection, 2023. Tabulated for this Study)

3.3 Evaluation Framework

Based on the theories and cognitive models discussed, Fig. 2 presents the evaluation framework used in this study to assess stimuli through dimensions of external form, meaning, and Qualia attributes within a cultural context. The framework employs a modified questionnaire from Yen et al (2014), consisting of 18 items that combine Qualia attributes with preference evaluations.

The modified questionnaire incorporates one preference dimension and five Qualia attribute dimensions, with each dimension comprising three items, totaling 18 items. These Qualia attributes are categorized into the following five dimensions.

- Attractiveness: Evaluates emotional appeal and cultural significance.
- Beauty: Assesses aesthetic qualities such as form and color.
- Creativity: Measures originality and innovation.
- Delicacy: Refers to craftsmanship and precision.
- Engineering: Evaluates functionality and durability.

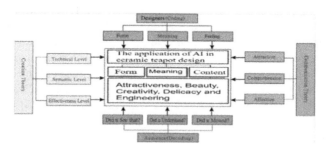

Fig. 2. Evaluation Framework (synthesized in this study)

This comprehensive framework allows for an in-depth assessment of both aesthetic and functional aspects, providing valuable insights into how these factors impact consumer preferences and purchase intentions. The integration of preference evaluations with Qualia attributes ensures a holistic approach, capturing the multifaceted nature of consumer perception and appreciation in a cultural context.

4 Research Findings and Discussion

4.1 Cognitive Space of the Artworks

This study adopts the Multi-Dimensional Scaling (MDS) framework to construct the cognitive space of the tested artworks. Based on the data obtained from the questionnaire survey, the 10 research objects are positioned in the four quadrants constructed by the intersecting hori-zontal and vertical axes. Besides evaluating the participants' perception of the similarity between different artworks, this cognitive space also delineates the cultural product and Qualia attribute characteristics of these 10 ceramic teapots (Fig. 3). The adoption of this analytical method serves two main purposes: firstly, the cognitive space formed by multidimensional scaling helps pinpoint the positions of the five Qualia attributes to assess their interrelationships; secondly, based on the positions of the 10 teapots in the cognitive space, the distinctive features, attributes, and common-alities of the artworks can be determined.

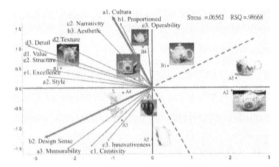

Fig. 3. Cognitive Space of the Artworks (Organized by this Study)

The study conducts multidimensional scaling for the 10 ceramic teapots and 15 Qualia attributes. Using data obtained from the questionnaire survey, SPSS statistical software is employed to generate the cognitive space, yielding a Stress value of 0.06562 and an RSQ coefficient of 0.98668. With the Stress value below 0.1 and the RSQ coefficient nearing 1, it indicates a good fit and descriptive adequacy of the cognitive space fitted based on the questionnaire survey data. From the cognitive space, it can be observed that artworks A1, A2, A5, and B1 are situated in the right half position, with A1 and A2 located near the end of the horizontal axis, forming a cluster indicative of their similar attributes. Artworks B2, B3, B4, B5, A3, and A4, on the other hand, are distributed in the left half, forming another cluster.

Further delving into the distribution of Qualia attributes in the cognitive space, Fig. 3 exhibits three clusters. The first cluster comprises attributes close in nature, namely a1(Cultura), b1(Proportioned), b3(Aesthetic), c2(Narrativity), and e3(Operability). The second cluster includes attributes such as a2(Style), d1(Value), d2(Texture), d3(Detail), e1(Excellence), and e2(Structure), which share similarities. Lastly, the third cluster consists of attributes like a3(Memorability), b2(Design Sense), c1(Creativity), and c3(Innovativeness), indicating their proximity.

Through multiple regression analysis calculating the axes formed by each attribute on the cognitive map, it can be observed that the correlation angle between "a1–b2" is close to 90 degrees. Therefore, the 90° angle formed by "a1 Cultura" and "b2 Design Sense" serves as the primary dimension, explaining the relationship between the vertical and horizontal axes of the cognitive space constructed by "a1–b2". The top end of the attribute axis of "a1 Cultura" is occupied by artwork B3, consistent with its data. Artwork A5, however, is situated at the end of this axis, indicating the lowest value among the attributes, which is also in line with the data obtained from Table 2.

Table 2. Cultural Attributes Evaluation Results

Code	B3	B1	B2	A4	A3	B5	A1	B4	A2	A5
Work										
Mean	4.14	3.36	3.28	3.22	3.16	3.03	3.01	3.00	2.97	2.45
Turn	1	2	3	4	5	6	7	8	9	10

(Source of the Image: National Palace Museum Collection, 2023. Tabulated for this Study.)

Artwork B3 occupies the forefront of the attribute axis of "b2 Design Sense," with the highest data in this attribute. Conversely, artworks A1 and A2 are located at the end of this attribute axis, indicating a significant disparity in design sense attributes compared to other artworks, which is also corroborated by the data in Table 3.

Table 3. Creativity Attributes Evaluation Results

Code	B3	A3	A5	A4	B2	B4	B5	B1	A2	A1
Work										
Mean	3.73	3.55	3.43	3.33	3.33	3.22	3.21	2.93	2.45	2.44
Turn	1	2	3	4	5	6	7	8	9	10

(Source of the Image: National Palace Museum Collection, 2023. Tabulated for this Study.)

Furthermore, from Fig. 3, it can be observed that artwork A1 and B1 are situated in the first quadrant, artworks B2, B3, and B4 are in the second quadrant, artworks A3, A4, A5, and B5 are located in the third quadrant, and only artwork A2 is in the fourth quadrant. Based on the presented results, out of the 10 artworks, 6 are positioned in the second and third quadrants. Corresponding to their Qualia attribute axes, it is evident that most of these artworks possess emotional characteristics and are relatively more favored in this survey. Among them, artwork B3 stands out prominently in terms of overall Qualia attributes and preference, leading to a noticeable distance from all artworks in the left half. Furthermore, among the artworks located in the first and second quadrants, artworks A1 and A2 are notably distant from the left half, exhibiting a slightly conservative and outdated style, making them less favored overall in this survey.

This study utilizes Multi-Dimensional Scaling (MDS) to delineate the cognitive space of ceramic teapots, analyzing their Qualia attributes and cultural significance. The findings illustrate the nuanced interplay between traditional and AI-generated designs, revealing insights into public perceptions and preferences. To build on this, it is crucial to establish a clear definition of ceramic culture, craftsmanship, and creative products.

Ceramic culture encompasses the historical and artistic traditions that shape ceramic artistry, while craftsmanship refers to the skills and techniques employed in creating these artifacts. Creative products represent the innovative application of these traditions and skills to contemporary design challenges. Addressing sustainability in this context involves not only preserving these traditional elements but also integrating them with modern technological advancements.

4.2 Qualia Attribute Evaluation Results

From Table 4, it can be observed that artworks generated by artificial intelligence, except for those generated based on the visual elements of artwork A1, all fall within the second and third quadrants. This indicates a high level of acceptance among the public for artworks generated by artificial intelligence. Furthermore, from the performance of various qualia attributes, it is evident that artworks generated by artificial intelligence, particularly B3, exhibit significantly higher values compared to other artworks. Additionally, most artworks generated by artificial intelligence score above 3 in the comprehensive qualia attributes, demonstrating public recognition of the qualia attributes designed by artificial intelligence. In other words, this also implies that utilizing artificial intelligence as a tool interface to extract and transform cultural elements into design elements can create ceramic cultural and creative product designs with qualia.

Table 4. Qualia Attribute Evaluation Results

Code	B3	A4	A3	B2	B4	B5	B1	A5	A2	A1
Work										
Mean	3.70	3.26	3.24	3.22	3.19	3.08	3.01	2.99	2.57	2.56
Turn	1	2	3	4	5	6	7	8	9	10

(Source of the Image: National Palace Museum Collection, 2023. Tabulated for this Study.)

4.3 Preference and Purchase Intention Analysis

The sustainability of traditional craftsmanship is closely intertwined with consumer support, with consumer preferences emerging as potential influencing factors for purchasing decisions. Therefore, this study evaluates audience preferences based on data collected from the questionnaire survey and arranged by mean values (Table 5). The results show that the top four rankings are all occupied by artworks generated by artificial intelligence, with B4 and B3 being the only works to receive ratings exceeding 3 in preference evaluations. Notably, B3 not only scores the highest in both comprehensive Qualia attributes and cultural attributes but also ranks second in preference, indicating a strong appeal to the audience.

Table 5. Preference Assessment

Code Work	B4	B3	B2	B5	A4	B1	A3	A5	A1	A2
Mean	3.29	3.20	2.9	2.78	2.75	2.74	2.52	2.18	2.17	1.92
Turn	1	2	3	4	5	6	7	8	9	10

(Source of the Image: National Palace Museum Collection, 2023. Tabulated for this Study.)

Given these findings, we chose to conduct a regression analysis on the "purchase intention" for both A3 and B3. The selection of A3 and B3 for this analysis was driven by several factors. B3 emerged as the highest-ranking artwork in terms of comprehensive Qualia attributes and cultural relevance, significantly surpassing other stimuli. Its prominence in both Qualia and preference metrics warranted an in-depth exploration. Consequently, A3, which serves as a counterpart to B3, was selected for comparison to understand how the preference for B3 influences purchase intention relative to a comparable non-AI-generated artwork.

Our regression analysis reveals that factors such as "liking this artwork" and "intending to recommend it to friends" significantly influence "purchase intention" (Table 6). The significant influence of these factors suggests that audience preferences for these artworks indeed impact their purchase intentions. This analysis provides valuable insights into how the distinct attributes of AI-generated artworks, as exemplified by B3, can affect consumer behavior and decision-making processes.

Table 6. Analysis of Purchase Intentions

		Regression Analysis of Purchase Intention					
		Unstandardized Coefficients		Standardized Coefficients			
Code	ttribute Name	B	Standard Error	Standardized Coefficients	T	Significance
	Constant	−0.065	0.132		−0.491	.624
R Square for A3 = 0.842	liking this artwork	0.272	0.063	0.277	4.316	0.000***
	intending to recommend it to friends	0.613	0.062	0.613	9.960	0.000***
R Square for B3 = 0.750	Constant	−0.066	0.223		−0.295	0.768

(continued)

Table 6. (*continued*)

Code	ttribute Name	Regression Analysis of Purchase Intention					
		Unstandardized Coefficients		Standardized Coefficients		
		B	Standard Error	Standardized Coefficients	T	Significance
	liking this artwork	0.568	0.072	0.509	7.893	0.000***
	intending to recommend it to friends	0.492	0.065	0.491	7.599	0.000***

* $p < 0.05$. ** $p < 0.01$. *** $p < 0.001$.
Source: Tabulated from this study

5 Conclusions and Recommendations

5.1 Conclusion

With the advancement of society, the demand for products has evolved beyond mere functionality to encompass a focus on qualia attributes and cultural connotations. This study explores the application of Generative Artificial Intelligence (GAI) in crafting ceramic teapot cultural and creative products. By analyzing the qualia attributes and cultural connotations of Palace Museum fine products, converting them into design element keywords, and utilizing GAI to generate product prototypes, we conducted comprehensive evaluations of qualia attributes, cultural connotations, and preference ratings.

This study also emphasizes the potential of Generative Artificial Intelligence (GAI) to enhance the sustainability of traditional ceramic craftsmanship. By integrating traditional ceramic techniques into modern design processes facilitated by human-machine collaboration, it promotes the preservation and development of traditional ceramic craftsmanship and culture, ensuring its sustainability and viability in contemporary markets. Furthermore, our research findings indicate that products developed through human-machine collaboration generally garner favor and support from audiences. This fusion of technology and culture fosters innovation and adaptation in traditional craftsmanship to address evolving consumer preferences and environmental issues.

5.2 Recommendations

This study highlights the potential application of Generative Artificial Intelligence (GAI) in designing ceramic cultural and creative products, while also acknowledging the associated limitations and risks. Future research efforts could focus on both further exploring GAI's application in this domain to enhance its design capabilities and accuracy, and cautioning against excessive reliance on GAI in product design processes to avoid stifling creativity and promoting product homogenization.

References

Andrejevic, M.: Automated Media. Routledge, London (2020)
Freedman, E.: Artificial Intelligence and Playable Media. Routledge, London (2023)
Hofstede, G.: Culture's consequences: international differences in work-related values. Sage (1984)
Jakobson, R.: Language in Literature. Harvard University Press, Cambridge (1987)
Li, X., Romainoor, N.H., Sun, Z.: Factors in consumers' purchase intention for Gejia batik. Heliyon **10**(1), e23085 (2024). https://doi.org/10.1016/j.teler.2023.100113
Lin, R.T., Lin, P.H.: A study of integrating culture and aesthetics to promote cultural and creative industries. Taiwan J. Arts **85**, 81–105 (2009). [in Chinese, semantic translation]
Lin, R.T.: From service innovation to qualia product design. J. Des. Sci. **14**, 13–31 (2011). [in Chinese, semantic translation]
Lin, R., Qian, F., Wu, J., Fang, W., Jin, Y.: A pilot study of communication matrix for evaluating artworks. In: Proceedings of the 9th International Conference, CCD, Held as Part of the 19th HCI International Conference, HCII 2017, Vancouver, BC, Canada, 9–11 July 2017, pp. 356–368 (2017)
Liu, Y., Song, P.: Creating sustainable cultural industries: the perspective of artificial intelligence and global value chain. Hindawi J. Environ. Public Health **11**, 1–11 (2022)
Lorusso, A.M.: Cultural Semiotics: For a Cultural Perspective in Semiotics. Palgrave Macmillan, London (2015)
Niu, H.-J.: Qualia: touching the inner needs of consumers' hearts. Australas. Mark. J. **27**(1), 41–51 (2019)
Norman, D.A.: The Design of Everyday Things. Basic Books, New York (2002)
Norman, D.A.: Emotional Design: Why We Love (or Hate) Everyday Things. Basic Books, New York (2004)
Nöth, W.: Handbook of Semiotics. Indiana University Press, Bloomington (1990)
Oksanen, A., et al.: Artificial intelligence in fine arts: a systematic review of empirical research. Comput. Human Behav. Artif. Humans **1**(2), 100004 (2023)
Taiwan Creative Content Agency. 2022–2023 Taiwan Cultural & Creative Industries Annual Report. Taipei, Taiwan Creative Content Agency (2024). [in Chinese, semantic translation]
Yen, H.Y., Lin, P.H., Lin, R.T.: Qualia characteristics of cultural and creative products. J. Kansei **2**(1), 34–61 (2014). [in Chinese, semantic translation]
Yen, H.-Y., Hsu, C.-I.: College student perceptions about the incorporation of cultural elements in fashion design. Fash. Text. **4**, 1–16 (2017)

Innovative Frontiers in Visual Arts: AI's Role in Interdisciplinary Collaboration

Jen-Feng Chen[1(✉)], Yun-Song Chu[2], and Po-Hsien Lin[1]

[1] Graduate School of Creative Industry Design, National Taiwan University of Arts, Taipei, Taiwan
jenfeng0328@gmail.com

[2] Department of Chinese Music, National Taiwan University of Arts, Taipei, Taiwan

Abstract. This study investigates the application of Artificial Intelligence (AI) in the field of visual arts, focusing particularly on its impact on musical experiences and the perception of music across different educational backgrounds. The findings reveal that AI significantly enhances efficiency in animation production and music visual design, reducing the learning curve for technical skills, which allows non-professionals to participate in the creative process. However, survey results indicate that AI visual designs do not significantly improve audience understanding of music and may even negatively affect their emotional responses and immersion in some instances. Additionally, the study finds that participants with a design and arts background exhibit superior comprehension and emotional responses to music compared to those without such a background. Overall, this research highlights the dual role of AI in interdisciplinary artistic creation: on one hand, AI brings efficiency and new possibilities to artistic creation; on the other hand, it underscores the need for cautious use of AI in artistic expression to preserve the emotional depth and artistic value of the works.

Keywords: Artificial Intelligence · Visual Arts · Musical Experience · Interdisciplinary Artistic Creation

1 Introduction

In today's era of rapid digitalization and technological innovation, the field of arts has also welcomed a new revolution. Particularly in the integration of music and visual arts, this has brought audiences a new sensory experience that transcends traditional arts. The combination of music and visual arts stimulates both visual and auditory senses of the audience, not only creating a multidimensional sensory environment but also achieving deep emotional resonance and cognitive interaction [1–3]. This multimodal art form leverages the crucial role of visual information in the musical experience and its impact on auditory processes, enriching the expressiveness of artworks and deepening the audience's understanding and feelings towards art [4–6]. From multisensory stimulation to emotional resonance, and then to narrative storytelling and symbolic meaning, the fusion of music and visuals makes music appreciation a unique, complex, and creative psychological activity [7]. Modern technology, especially the widespread use of digital information technology and the internet, plays a significant role in the creation and distribution of art, providing artists and designers with a broader creative space, and also bringing audiences a richer and more diverse artistic experience [8, 9]. Interactive elements in modern artworks, such as how the audience's behavior can influence music or visual effects, further enhance the audience's immersion and participation [10, 11], driving the popularization of music and the revolution of multidimensional sensory experiences.

Technological innovations in fields like AR (Augmented Reality) and VR (Virtual Reality) enable artists to create entirely new music and visual fusion experiences. For instance, "The Singing Gallery" project uses virtual reality technology to construct an interactive art gallery, offering a new multidimensional auditory experience [12]. Similarly, virtual concerts and augmented reality art exhibitions, through innovative technological means, have expanded the boundaries of artistic expression, providing a more immersive viewing environment [13]. Interactive art installations, using sensor technology, create unique visual and auditory experiences, emphasizing the participation and individual experiences of the audience [14]. These technologies not only provide immersive sensory experiences but also resonate with the audience on an intellectual and emotional level, bringing them entirely new sensory experiences and emotional interactions. The advancements in technology in the 21st century, especially the rise of generative AI, have profoundly changed the ways of artistic creation, reshaping the interaction between art and the audience. AI provides new tools and methods for creation, enhancing the personalization and interactivity of artworks, allowing artists and designers to focus more on creativity and innovation while enriching the artistic experiences of audiences [15].

This study aims to deeply explore the impact of AI-assisted visual design in the "Zhuolu Fantasy" on music performances. Through expert interviews and surveys, this research will thoroughly analyze the impact of AI on the visual design process and how this new form of art affects the audience's musical experience and emotional responses. The purpose of the study is to evaluate the role of AI in artistic creation and to explore how to better integrate AI technology with artistic creation to enhance the expressiveness of artworks and the immersion of the audience. This in-depth analysis will provide valuable insights for future artistic creation and interdisciplinary innovation.

2 Literature Review

2.1 Development of AI Technology

The evolution of Artificial Intelligence (AI) technology has brought revolutionary changes across technology, art, and business sectors. In the late 20th century, AI primarily focused on pattern recognition and data generation, with landmark technologies such as Hopfield networks and Boltzmann machines. However, these were less effective at handling complex problems [16].Entering the 21st century, AI began embracing the era of deep learning, especially with the emergence of Generative Adversarial Networks (GANs) in 2014. GANs consist of two main components: a generator that creates realistic data samples and a discriminator that attempts to differentiate between real and generated data, significantly enhancing the quality of generated data [17]. Additionally, the integration of Convolutional Neural Networks (CNNs) and Recurrent Neural Networks (RNNs) has led to significant advancements in AI's capabilities for processing images, videos, and sequential data [18]. In recent years, the development of Natural Language Processing (NLP) has been noteworthy. Advanced models like GPT-3 and BERT have achieved significant success in understanding and generating natural language, excelling in various tasks [19, 20]. Another important advancement is the integration of deep learning with reinforcement learning, enhancing AI's decision-making and self-adjusting learning capabilities [21]. Overall, AI technology has shifted from simple pattern recognition to complex data processing and decision-making. As technology continues to advance and innovate, AI still holds vast potential for future development and application.

In terms of applications, AI has shown outstanding performance in fields such as healthcare, autonomous driving, finance, manufacturing, and retail. For instance, the AlphaFold program has made significant breakthroughs in protein folding predictions, while Tesla and Waymo have demonstrated highly automated driving [22–24]. In the creative industry, AI's application has also shown potential in gaming, advertising, and artistic creation [25, 26].As AI technology matures, we are witnessing its revolutionary applications in creative design. AI not only has the potential in automating basic tasks but also plays a key role in fostering innovative thinking and interdisciplinary collaboration. However, research on effective collaboration between AI and designers to drive innovation and enhance audience experience is still relatively limited. Therefore, exploring the collaborative relationship between AI and designers, and its impact on the creative design field, has become an important research direction. Through research and practice, we can better understand the interaction between AI and human creativity, and explore AI's potential in promoting socio-cultural development and economic growth, thereby paving the way for creating richer, more interactive, and personalized audience experiences.

2.2 AI in Visual Design

Artificial Intelligence (AI) is a major driver of economic growth and social progress in the current global context [27], particularly in areas such as image classification, sentiment analysis, and speech understanding [28]. Communication between AI and designers has become more streamlined, making it possible to create from scratch [29, 30]. Recently, AI models based on the Diffusion Model have attracted widespread attention for their excellence in image generation quality and diversity. These models have been applied in tasks such as image generation, super-resolution, restoration, editing, and image-to-image transformation [31]. These tools have enhanced the quality of creation and have made design more diverse [32, 33]. In specific applications of visual design, Adobe Sensei's automated image editing features demonstrate AI's role in simplifying designers' workflows [34]. Personalized design experiences represent another significant application of AI in visual design. For example, Spotify uses AI to generate personalized playlist covers for each user, providing a more personalized and enriched user experience [35]. This personalized visual design not only enhances the overall user experience but also showcases AI's capability in designing personalized solutions. Using AI for character animation is a significant advancement in the animation field. For instance,

Disney Research Studios has developed AI tools that can automatically generate complex character animations. By analyzing motion capture data from real actors, AI can create realistic character movements, greatly enhancing the efficiency of animation production. AI technology can assist animators in making more effective decisions when designing scene layouts. By analyzing a large volume of animation clips, AI can suggest how animators should arrange scenes to achieve the best visual effects [36, 37].

In summary, AI technology is demonstrating its vast potential and impact in the field of visual design. From improving design processes and enhancing creation efficiency to enriching personalized experiences, AI not only provides powerful tools for designers but also expands the possibilities of design, leading new trends in the design industry. As the technology continues to evolve, AI is expected to play an even more crucial role in the field of visual design in the future.

2.3 AI in Interdisciplinary Applications of Design and Art

In today's increasingly digital world, Artificial Intelligence (AI) has become not just a technological tool but a significant driver of innovation in the fields of design and art. AI's interdisciplinary applications between design and art reveal its new role in cross-disciplinary integration, breaking traditional boundaries and creating new possibilities for creation.

In the field of visual design, AI enables designers to interact through natural language, creating unprecedented visual artworks [38], and also plays a crucial role in fashion design by predicting trends and assisting designers in creating personalized clothing [39, 40]. Furthermore, AI is key in interior design and architecture, analyzing user needs to provide customized design solutions [41], and simulating the environmental impact of different design scenarios to help designers make more sustainable and efficient decisions [42]. In the arts, AI's applications range from music to visual arts, dance, and filmmaking. For example, research by Andrew Starkey and others explores AI's application in transforming live painting into music [43]. AI also plays a crucial role in music video production and interactive musical experiences, analyzing musical elements to create coordinated visual effects [44]. AI can analyze the rhythm and style of music to assist choreographers in creating innovative dance arrangements and providing new creative inspiration [45, 46]. In film scoring, AI automatically generates or adjusts music based on the emotional and narrative style of the film to enhance visual storytelling [47, 48]. Additionally, AI adjusts music based on game scenarios and player interactions, adding new dimensions to the gaming experience and opening new creative spaces for interactive music creation [49]. Despite its great potential in artistic creation, AI also faces challenges such as maintaining originality and managing copyright issues [50]. Looking forward, AI is poised to play a more significant role in artistic creation and audience experience, driving new developments in the arts [51, 52]. In summary, the diverse applications of AI demonstrate its important role in promoting the integration of different art forms and technologies, providing designers and artists with rich creative methods and offering audiences new experiences.

3 Research Methods

This study employs a mixed-method approach, combining expert interviews and a survey to explore the impact of AI-assisted visual design on musical experiences. Initially, through expert interviews, we compared the differences between AI-assisted and traditional design processes to gain industry insights into the application of new technologies. Subsequently, we conducted a survey with 126 participants to assess their reactions to the 'Zhuolu Fantasy Symphony', a music piece infused with visual art. The survey was structured around three dimensions: comprehension, emotional response, and musical experience, aiming to comprehensively evaluate the participants' reactions to music pieces with and without AI-assisted visual design. All discussions were recorded in text form and a detailed comparison and analysis of the results was conducted.

3.1 Introduction to the AI Visual Creation Process of 'Zhuolu Fantasy Symphony'

The collaboration between AI and traditional design processes shows significant differences. In traditional processes, designers rely on drawing skills and spend a considerable amount of time from concept to completion. In contrast, AI acts as an executor, quickly generating high-quality images, while designers set the stylistic direction and choose appropriate visuals to meet the musicians' needs, significantly reducing design time and allowing designers to focus more on creativity and communication. The AI-assisted visual creation process includes several key steps: content planning, text generation, image generation, image screening, image confirmation, and image completion, involving AI tools such as Perplexity, Chat GPT, Midjourney, and HitPaw.

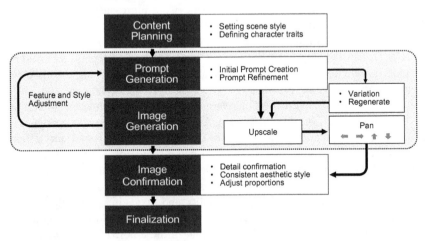

Fig. 1. AI-Assisted Design Process Diagram (drawn for this study).

The specific process is as follows:

1) Content Planning: Using Perplexity Ask to understand the background and characters of 'Zhuolu Fantasy Symphony' and write a creative base on the dialogue results.

2) Text Generation: Input the story into ChatGPT for in-depth discussions to precisely define each character's image and scene descriptions, ensuring story coherence.
3) Image Generation: A core step where images are generated using Midjourney based on text descriptions. Adjustments to prompts are made to optimize visual presentation.
4) Image Confirmation: Screen AI-generated images using HitPaw AI to enlarge and adjust them, ensuring consistency in style and composition.

The entire process took one week, generating about 7,000 images, from which 27 were chosen to combine with music for a performance at Taiwan National University of the Arts in October 2023. The AI collaboration process is illustrated in Fig. 1.

3.2 Expert Interviews

In exploring the application and impact of Artificial Intelligence (AI) in animation production, the AI-assisted visual design in the "Battle of Zhuolu" case is particularly noteworthy. This example demonstrates AI's exceptional ability to rapidly generate images, providing significant benefits to the time-consuming and costly process of animation production. Expert interviews reveal that the main advantages of AI in animation visual design are improving work efficiency and lowering the threshold for learning specialized technical skills. Traditionally, newcomers to the industry must learn multiple drawing software programs, each with varying learning curves; however, with AI tools, even non-professionals can quickly engage in professional domains and realize creative concepts.

The animation design process is generally divided into three stages: pre-production, production, and post-production. AI technology plays a transformative role, especially in speeding up the creative process and enhancing communication between designers and directors during the pre-production stage. This phase usually requires significant time for character design, scene layout, and the creation and revision of the overall visual style. AI's image generation tools, with their rapid generation capabilities and diverse styles, can transform textual descriptions into visual images, which are crucial for presenting preliminary concepts and facilitating quick iterations. For example, in the "Battle of Zhuolu" case, traditional drawing methods would take a week, but AI can complete the task in just 2–3 h, saving a substantial amount of time. This efficient method of generating visual design images not only shortens the time required for hand drawing and digital drawing but also enables creative teams to quickly explore and compare different design options, enhancing communication efficiency and making the creative decision-making process more dynamic and interactive. In terms of story creation, AI writing tools like ChatGPT can assist in data collection and organization, stimulate innovative thinking, help perfect script structures, mimic different writing styles, and improve text quality, thereby saving a considerable amount of writing time. The integration of such technology not only boosts overall creative efficiency but also fosters close collaboration among teams, making the entire animation production process smoother and more efficient. In summary, the application of AI technology in the early stages of animation production not only accelerates the creative process but also significantly enhances the creative quality of the work and team collaboration efficiency, while also reducing the learning curve for technical skills, allowing designers to focus more on creative ideation and artistic expression.

3.3 Questionnaire Design

Music appreciation is a comprehensive artistic experience that relies on both auditory and visual elements to enhance the audience's perception. This study explores how AI-assisted visual design affects music experience, particularly its impact on the audience's emotional response, cognitive understanding, and immersion. With the widespread application of AI technology in the creative industries, its role in music visual design has attracted attention from both academic and practical fields. This study evaluates the role of AI in music appreciation through a questionnaire survey, investigating how this new technology affects audience perception and emotional experience. The questionnaire uses a five-point scale to assess understanding of the music's content and theme (Level A), emotional response elicited by the music (Level B), and the immersion and time perception produced by the music (Level C). Additionally, overall preference for the music piece is assessed (Level D). Details of the questionnaire are shown in Table 1.

Table 1. Survey Questions.

Level		questions
a Understanding	a1	I understand this music's content or theme
	a2	I feel this music's cultural or historical ties
	a3	I grasp the symbols or metaphors in this music
b Emotional Response	b1	This music has touched my emotions
	b2	This music has led me through a change in feelings
	b3	This music has inspired my imagination or sparked insights
c Music Experience	c1	While listening to this music, I feel immersed in it
	c2	While listening to this music, I perceive time differently
	c3	I find this music to be highly expressive
	d	I really like this piece of music

Participants in this study were aged between 16 to 22 years and were divided into two main categories based on their educational background: those with a design background and those without a design background. These participants were then randomly assigned to two groups: a control group that experienced the music solely through audio, and an experimental group that experienced the same music with AI-assisted visual design. This design aims to reveal the potential impact of visual elements on music experience through the questionnaire test, thereby eliminating interference from other variables and ensuring the reliability of the research results.

The results of this study will provide insights for visual artists and music producers, demonstrating the potential of AI in music visual design and revealing how AI-designed visual elements can alter or enhance the audience's music perception and emotional experience. These findings will help to promote the integration of music and visual arts, opening new possibilities for future artistic expressions and audience interaction. Through this interdisciplinary research approach, we can gain a more comprehensive understanding of the trends in multisensory music experiences in the current technological environment (Table 2).

Table 2. Impact of AI Visual Design on Musical Understanding, Emotional Response, and Experience.

Items	Variable	n	M	SD	t
b2. This music has led me through a change in feelings	WV	61	3.54	1.177	2.171*
	WoV	65	3.11	1.062	
b3. This music has inspired my imagination or sparked insights	WV	61	3.74	1.063	1.996*
	WoV	65	3.37	1.009	
c3. I find this music to be highly expressive	WV	61	4.34	.750	2.575*
	WoV	65	3.92	1.050	
Level b	WV	61	3.55	.941	2.284*
	WoV	65	3.16	.963	
Level c	WV	61	3.80	.678	2.381*
	WoV	65	3.47	.843	

*$p < 0.05$

Note: WV refers to 'With Visuals', indicating the group that experienced AI-assisted visual design. WoV refers to 'Without Visuals', indicating the group that experienced only the audio component without visual design

4 Results and Discussions

4.1 The Impact of AI Visual Design on Understanding, Emotional Response, and Music Experience

AI visual design, as an emerging auxiliary tool, aims to enhance listeners' music experiences through visual elements. This study, using independent sample T-tests, explored the specific impacts of AI visual design on music understanding, emotional response, and music experience, assessing AI's actual role in music appreciation.

In terms of understanding, the comparison between Group A (without visual design) and Group B (with AI visual design) revealed that Group B scored slightly higher on average in understanding the music content. However, the difference was not statistically significant, indicating that AI visual design did not significantly enhance cognitive understanding of music (Table 3).

Regarding emotional responses, Group A scored higher in experiencing emotional changes and inspiring imagination than Group B, reflecting a richer and deeper emotional experience without AI design. Specific scores were 3.54 and 3.74 for Group A, compared

Table 3. T-test Results Comparing Design Background and No Design Background Groups on Responses of the Questionnaire.

Items	Variable	n	M	SD	t
a1. I understand this music's content or theme	DB	46	3.83	.739	2.350*
	NDB	80	3.48	.914	
a2. I feel this music's cultural or historical ties	DB	46	3.80	.833	2.468*
	NDB	80	3.39	1.037	
a3. I grasp the symbols or metaphors in this music	DB	46	3.96	.918	2.252*
	NDB	80	3.56	.992	
b1. This music has touched my emotions	DB	46	3.54	.982	3.023**
	NDB	80	2.96	1.130	
b2. This music has led me through a change in feelings	DB	46	3.85	.894	4.532***
	NDB	80	3.01	1.153	
b3. This music has inspired my imagination or sparked insights	DB	46	3.85	.918	2.603*
	NDB	80	3.38	1.084	
c1. While listening to this music, I feel immersed in it	DB	46	3.74	.905	2.194*
	NDB	80	3.35	1.045	
c2. While listening to this music, I perceive time differently	DB	46	3.54	.862	2.499*
	NDB	80	3.11	1.043	
c3. I find this music to be highly expressive	DB	46	3.76	.923	2.705**
	NDB	80	3.29	.983	
Level a	DB	46	3.86	.694	2.812**
	NDB	80	3.48	.825	
Level b	DB	46	3.75	.838	3.834***
	NDB	80	3.12	.968	
Level c	DB	46	3.86	.726	2.381*
	NDB	80	3.50	.788	

* $p < 0.05$, ** $p < 0.01$, *** $p < 0.001$

Note: DB (Design Background) refers to participants with a background in design. NDB (No Design Background) refers to participants without a background in design

to 3.11 and 3.37 for Group B, with corresponding t-values of 2.171 and 1.996 and p-values less than 0.05, indicating significant differences in emotional experience.

In the aspect of music experience, including immersion and expressiveness, Group A without AI visual design scored significantly higher than Group B with AI design. This further confirmed that the music experience without AI design was perceived more intensely, suggesting that in pure music listening, listeners might immerse more deeply in the emotional and artistic expression of the music. Specifically, Group A scored 4.34

in music expressiveness, compared to 3.92 for Group B, with a t-value of 2.575 and a p-value less than 0.05.

In summary, our analysis indicates that listeners scored higher in understanding, emotional responses, and music experience without AI visual design compared to with AI design. This suggests that when designing AI-assisted visual elements, deeper consideration should be given to how they integrate with the music content to promote understanding. While AI's efficiency and stylistic diversity bring advantages, careful integration into creative fields is necessary to maintain artistic depth. The study shows that human-machine collaboration in the integration of auditory and artistic elements has its limitations and requires careful consideration of coordination with music. Future research should explore interdisciplinary approaches to deepen understanding of the interaction between AI and human creativity, fostering innovative thinking and collaboration.

4.2 Impact of Different Educational Backgrounds on Music Perception with and Without AI Visual Design

This study explores how participants from different educational backgrounds perceive music with and without the aid of AI visual design. The findings indicate that participants with a background in design and art consistently scored higher across all evaluation dimensions than those without a design background, regardless of the presence of AI visual design.

In terms of music understanding, participants with a design background scored an average of 3.83 on understanding the content or theme of the music (a1), significantly higher than the 3.48 scored by those from a non-design background, with a t-value of 2.350*. Similarly, in feeling the connection between the music and its cultural or historical context (a2), design-oriented participants scored higher (average 3.80 compared to 3.39), with a t-value of 2.468*. This highlights the advantage design-background participants have in understanding musical content. The same trend is observed in understanding the symbolic meanings or metaphors in the music (a3), with scores of 3.96 for design background versus 3.56 for non-design, and a t-value of 2.252*. Regarding emotional responses, participants from a design background exhibited a higher sensitivity and intensity in their responses, scoring higher in how the music touched their emotions (b1), led them through a change in feelings (b2), and inspired imagination or insights (b3), with corresponding t-values of 3.023**, 4.532***, and 2.603* respectively. For music experience, although there was no significant difference in the strong expressiveness of music (c3), participants with a design background scored significantly higher on feeling immersed in the music (c1) and perceiving a different flow of time while listening (c2), with t-values of 2.194* and 2.499*, indicating a more nuanced and complex musical experience. Overall, participants with a design and arts background scored higher on all levels of assessment, especially in emotional response (Level B) and understanding (Level A), with significant differences noted (t-values of 3.834* and 2.812** respectively).

In summary, participants with a design and arts background outperformed those from non-design backgrounds in understanding music, emotional responses, and overall music experience. These differences reflect the higher sensitivity of design-background participants to artistic elements and their profound understanding of musical structure and emotional expression. Music producers and AI designers should consider the audience's background to create music experiences that better meet their needs, promoting the integration of music with visual arts and catering to a broader audience.

5 Conclusion

This study, through expert interviews and survey analysis, explored the impact of AI on the musical experience within visual arts and examined how different educational backgrounds affect musical perception. The results indicate that AI visual design introduced novel sensory stimuli; however, it did not significantly enhance cognitive understanding of the music across all participant groups. Notably, participants with design backgrounds were more likely to benefit from the visual design in terms of comprehension and emotional response, suggesting that the effectiveness of AI-assisted visual elements may vary depending on the audience's prior experience and background.

Expert interviews revealed that AI significantly improved efficiency in visual design, especially during the pre-production phase, drastically reducing the time needed for design and implementation. AI also lowered the barriers to learning professional techniques, enabling non-professionals to quickly engage in creative processes. This highlights AI's role as a creative accelerator and its potential in promoting interdisciplinary collaboration and innovation.

Survey findings suggest that while AI visual design provided new sensory stimuli, it did not significantly enhance listeners' understanding of the music and, in some cases, even distracted from emotional engagement, reducing the music's emotional appeal. This highlights the need for a delicate balance between technological application and artistic expression when integrating AI into visual and musical creation.

Furthermore, the study found significant differences in music perception among participants with different educational backgrounds, with those having a background in design and arts performing better in understanding and emotional response. This underscores the importance of considering the listener's background in music creation and performance to better cater to diverse audience needs.

In summary, this research provides valuable insights into AI's dual role in art creation: enhancing efficiency in creative processes and opening new possibilities while also showing the need for cautious use of AI to maintain the emotional depth of the artwork. Future research should consider more rigorous designs, such as using multiple control groups or larger sample sizes, to better isolate AI's impact. Additionally, it is important to explore how discrepancies between visual and auditory information, as seen in the McGurk effect, might influence the perception of music and emotional engagement. This could provide deeper insights into the interplay between different sensory modalities in the context of AI-assisted visual design, further enhancing the application of AI in art creation without compromising artistic expression, particularly in interdisciplinary collaboration.

References

1. Nijs, L., Coussement, P., Moens, B., Amelinck, D., Lesaffre, M., Leman, M.: Interacting with the Music Paint machine: relating the constructs of flow experience and presence. Interact. Comput. **24**(4), 237–250 (2012). https://doi.org/10.1016/j.intcom.2012.05.002
2. Ou-Yang, W.S.: Visualization of music: an aesthetic study of the combination of music and animation. In: Proceedings of the 2nd Annual International Conference on Social Science and Contemporary Humanity Development, Article 24 (2015). https://doi.org/10.2991/sschd-16.2016.24
3. Tiihonen, M., Brattico, E., Maksimainen, J., Wikgren, J., Saarikallio, S.: Constituents of music and visual-art related pleasure – a critical integrative literature review. Front. Psychol. **8**, 1218 (2017). https://doi.org/10.3389/fpsyg.2017.01218
4. Byun, J., Loh, C.S.: Audial engagement: effects of game sound on learner engagement in digital game-based learning environments. Comput. Hum. Behav. **46**, 129–138 (2015). https://doi.org/10.1016/j.chb.2014.12.052
5. Chibashvili, A., Kharchenko, P., Savchuk, I., Sydorenko, V., Bezuhla, R.: Practices of visual arts in the music of the twentieth and early twenty-first centuries. Studia Universitatis Babes-Bolyai Musica **67**(Special Issue 1), 79–101 (2022)
6. Schutz, M.: Seeing music? what musicians need to know about vision. Empir. Musicol. Rev. **3**(3), 83–108 (2008). https://doi.org/10.18061/1811/34098
7. Xu, Z., Xu, Q.: Students' psychological state, creative development, and music appreciation: the influence of different musical act modes (exemplified by a video clip, an audio recording, and a video concert). J. Psycholinguist. Res. **52**, 3001–3017 (2023). https://doi.org/10.1007/s10936-023-10035-8
8. Dixon, S., Smith, B.: Digital Performance: A History of New Media in Theater, Dance, Performance Art, and Installation. The MIT Press, Cambridge (2007)
9. Walmsley, B.: From arts marketing to audience enrichment: how digital engagement can deepen and democratize artistic exchange with audiences. Poetics **58**, 66–78 (2016). https://doi.org/10.1016/j.poetic.2016.07.001
10. Avraam, I.: Visuals within Classical music concerts - how can we attract young audience? Master's thesis, Utrecht University (2018). https://studenttheses.uu.nl/handle/20.500.12932/31176
11. Shirzadian, N., Redi, J.A., Röggla, T., Panza, A., Nack, F., Cesar, P.: Immersion and togetherness: how live visualization of audience engagement can enhance music events. In: Cheok, A.D., Inami, M., Romão, T. (eds.) ACE 2017. LNCS, vol. 10714, pp. 488–507. Springer, Cham (2017). https://doi.org/10.1007/978-3-319-76270-8_34
12. Bramah, O., Cheng, X., Morreale, F.: the singing gallery: combining music and static visual artworks. In: Proceedings of the 16th International Audio Mostly Conference (AM 2021), pp 117–120 (2021). https://doi.org/10.1145/3478384.3478396
13. Çelik, G.: A new field in music production: metaverse concerts. Ege Univ. Fac. Commun. Media Commun. Res. Peer-Rev. E-J. **12**, 4–24 (2023). https://doi.org/10.56075/egemiadergisi.1230583
14. Vi, C.T., Ablart, D., Gatti, E., Velasco, C., Obrist, M.: Not just seeing, but also feeling art: mid-air haptic experiences integrated in a multisensory art exhibition. Int. J. Hum. Comput. Stud. **108**, 1–14 (2017). https://doi.org/10.1016/j.ijhcs.2017.06.004
15. Verganti, R., Vendraminelli, L., Iansiti, M.: Innovation and design in the age of artificial intelligence. J. Prod. Innov. Manag. **37**(3), 212–227 (2020). https://doi.org/10.1111/jpim.12523
16. Ackley, D.H., Hinton, G.E., Sejnowski, T.J.: A learning algorithm for Boltzmann machines. Cogn. Sci. **9**(1), 147–169 (1985)

17. Goodfellow, I.J., et al.: Generative adversarial nets. In: Proceedings of the 27th International Conference on Neural Information Processing Systems, vol. 2, pp 2672–2680 (2014)
18. Sharma, V., Gupta, M., Kumar, A., Mishra, D.: Video processing using deep learning techniques: a systematic literature review. IEEE Access **9**, 139489–139507 (2021). https://doi.org/10.1109/ACCESS.2021.3118541
19. Li, J., Tang, T., Zhao, W.X., Nie, J.Y., Wen, J.R.: Pretrained language models for text generation: a survey (2022). arXiv preprint arXiv:2201.05273. https://doi.org/10.48550/arXiv.2201.05273
20. Yang, B., Luo, X., Sun, K., Luo, M.Y.: Recent progress on text summarisation based on BERT and GPT. In: Jin, Z., Jiang, Y., Buchmann, R.A., Bi, Y., Ghiran, A.-M., Ma, W. (eds.) Knowledge Science, Engineering and Management: 16th International Conference, KSEM 2023, Guangzhou, China, August 16–18, 2023, Proceedings, Part IV, pp. 225–241. Springer Nature Switzerland, Cham (2023). https://doi.org/10.1007/978-3-031-40292-0_19
21. Pawan, S.J., Rajan, J.: Capsule networks for image classification: a review. Neurocomputing **509**, 102–120 (2022). https://doi.org/10.1016/j.neucom.2022.08.073
22. Perrakis, A., Sixma, T.K.: AI revolutions in biology: the joys and perils of AlphaFold. EMBO Rep. **22**, e54046 (2021). https://doi.org/10.15252/embr.202154046
23. Shakhatreh, H., et al.: Unmanned aerial vehicles (UAVs): a survey on civil applications and key research challenges. IEEE Access **7**, 48572–48634 (2019). https://doi.org/10.1109/ACCESS.2019.2909530
24. Tian, Y., Pei, K., Jana, S., Ray, B.: DeepTest: automated testing of deep-neural-network-driven autonomous cars. In: Proceedings of the 40th International Conference on Software Engineering (ICSE 2018), pp 303–314 (2018). https://doi.org/10.1145/3180155.3180220
25. Anantrasirichai, N., Bull, D.: Artificial intelligence in the creative industries: a review. Artif. Intell. Rev. **55**(4), 589–656 (2022). https://doi.org/10.1007/s10462-021-10039-7
26. Lyu, Y., Wang, X., Lin, R., Wu, J.: Communication in human–AI co-creation: perceptual analysis of paintings generated by text-to-image system. Appl. Sci. **12**(22), 11312 (2022). https://doi.org/10.3390/app122211312
27. Hall, D.W., Pesenti, J.: Growing the artificial intelligence industry in the UK (2018). https://assets.publishing.service.gov.uk/government/uploads/system/uploads/attachment_data/file/652097/Growing_the_artificial_intelligence_industry_in_the_UK.pdf
28. Samek, W., Wiegand, T., Müller, K.R.: Explainable artificial intelligence: understanding, visualizing and interpreting deep learning models (2017). arXiv preprint arXiv:1708.08296. https://doi.org/10.48550/arXiv.1708.08296
29. Bontrager, P., Lin, W., Togelius, J., Risi, S.: Deep interactive evolution. In: Liapis, A., Cardalda, J.J.R., Ekárt, A. (eds.) Computational Intelligence in Music, Sound, Art and Design, pp. 267–282. Springer, Cham (2018). https://doi.org/10.1007/978-3-319-77583-8_18
30. Lin, L., Yang, B.B.: From perception to creation: a forefront discussion on image frequency generation methods. Acta Optica Sinica **43**(15) (2023). https://doi.org/10.3788/AOS2023758
31. Rodriguez-Ruiz, V., Garnacho, D., Moratal, D.: Creating comic-style images using deep learning-based text-to-image models to make science, technology, engineering, and mathematics (STEM) disciplines accessible to the young generations. In: INTED2023 Proceedings, pp 4947–4954 (2023)
32. Shi, Y., Gao, T., Jiao, X., Cao, N.: Understanding design collaboration between designers and artificial intelligence: a systematic literature review. In: Proceedings of the ACM on Human-Computer Interaction, vol. 7(CSCW2), Article 368 (2023). https://doi.org/10.1145/3610217
33. Nada, O.A.E., Dawood, M.E.T.: Digital twin: methodologies for modeling the work environment during the design and development processes. Inter. Des. J. **12**, 225–242 (2022). https://doi.org/10.21608/IDJ.2022.260602

34. Singh, K.D., Duo, Y.X.: Future design: an analysis of the impact of AI on designers' workflow and skill sets. In: Vasant, P. (ed.) Intelligent Computing and Optimization: ICO 2023. Lecture Notes in Networks and Systems, vol. 852. Springer, Cham (2023). https://doi.org/10.1007/978-3-031-50330-6_23
35. Björklund, G., Bohlin, M., Olander, E., Jansson, J., Walter, C.E., Au-Yong-Oliveira, M.: An exploratory study on the Spotify recommender system. In: Rocha, A., Adeli, H., Dzemyda, G., Moreira, F. (eds.) Information Systems and Technologies: WorldCIST 2022, Volume 2, pp. 366–378. Springer International Publishing, Cham (2022). https://doi.org/10.1007/978-3-031-04819-7_36
36. Disney Research Studios. Pose and Skeleton-Aware Neural IK for Pose and Motion Editing (2023). Accessed 28 Jan 2024. https://studios.disneyresearch.com/2023/12/11/pose-and-skeleton-aware-neural-ik-for-pose-and-motion-editing/
37. Zhang, R.: Automatic generation of real-time animation game learning levels based on artificial intelligence assistant. Sci. Program. **2022**, 1–19 (2022). https://doi.org/10.1155/2022/1557302
38. Manovich, L.: Computer vision, human senses, and language of art. AI & Soc. **36**(4), 1145–1152 (2021). https://doi.org/10.1007/s00146-020-01094-9
39. Jeon, Y., Jin, S., Shih, P.C., Han, K.: FashionQ: an AI-Driven creativity support tool for facilitating ideation in fashion design. In: Proceedings of the 2021 CHI Conference on Human Factors in Computing Systems (CHI 2021), Article 576 (2021). https://doi.org/10.1145/3411764.3445093
40. Liang, Y., Lee, S.H., Workman, J.E.: Implementation of artificial intelligence in fashion: are consumers ready? Cloth. Text. Res. J. **38**(1), 3–18 (2020). https://doi.org/10.1177/0887302X19873437
41. Kán, P., Kurtic, A., Radwan, M., Rodríguez, J.M.L.: Automatic interior design in augmented reality based on hierarchical tree of procedural rules. Electronics **10**(3), 245 (2021). https://doi.org/10.3390/electronics10030245
42. Almusaed, A., Yitmen, I.: Architectural reply for smart building design concepts based on artificial intelligence simulation models and digital twins. Sustainability **15**, 4955 (2023). https://doi.org/10.3390/su15064955
43. Starkey, A., Steenhauer, K., Caven, J.: Painting music: using artificial intelligence to create music from live painted drawings. Draw. Res. Theory Pract. **5**(2), 209–224 (2020). https://doi.org/10.1386/drtp_00033_1
44. Deruty, E., Grachten, M., Lattner, S., Nistal, J., Aouameur, C.: On the development and practice of AI technology for contemporary popular music production. Trans. Int. Soc. Music Inf. Retr. **5**(1), 35–49 (2022). https://doi.org/10.5334/tismir.100
45. Zhou, L.: Research on the application of dance movement skill analysis in teaching in the context of artificial intelligence in universities. Appl. Math. Nonlinear Sci. **9**(1) (2024). https://doi.org/10.2478/amns.2023.1.00482
46. Zulić, H.: How AI can change/improve/influence music composition, performance and education: three case studies. INSAM J. Contemp. Music Art Technol. **2**, 100–114 (2019)
47. Amato, G., et al.: AI in the media and creative industries (2019). arXiv preprint arXiv:1905.04175. https://doi.org/10.48550/arXiv.1905.04175
48. Cunningham, S., Ridley, H., Weinel, J., et al.: Supervised machine learning for audio emotion recognition. Pers. Ubiquit. Comput. **25**, 637–650 (2021). https://doi.org/10.1007/s00779-020-01389-0
49. Hutchings, P.E., McCormack, J.: Adaptive music composition for games. IEEE Trans. Games **12**(3), 270–280 (2020). https://doi.org/10.1109/TG.2019.2921979
50. Selvadurai, N., Matulionyte, R.: Reconsidering creativity: copyright protection for works generated using artificial intelligence. J. Intellect. Property Law Pract. **15**(7), 536–543 (2020). https://doi.org/10.1093/jiplp/jpaa062

51. Cetinic, E., She, J.: Understanding and creating art with AI: review and outlook. ACM Trans. Multimedia Comput. Commun. Appl. **18**(2), 66 (2022). https://doi.org/10.1145/3475799
52. McCormack, J., Hutchings, P., Gifford, T., Yee-King, M., Llano, M.T., D'Inverno, M.: Design considerations for real-time collaboration with creative artificial intelligence. Organ. Sound **25**(1), 41–52 (2020). https://doi.org/10.1017/S1355771819000451

Impact of the Audience's Aesthetic Perceptions on the Traditional Dance-Drama: Eternal Love Across the Magpie Bridge

Tze-Fei Huang(✉)

Graduate Student, Graduate School of Creative Industry Design,
National Taiwan University of Arts, Taipei, Taiwan
s8820219@gmail.com

Abstract. The main purpose of this study is to analyze the differences in aesthetic perceptions of the classic Taiwanese ethnic dance drama "Eternal Love Across the Magpie Bridge." This study applies Aristotle's "Six Elements of Drama" as evaluation criteria, and uses questionnaire surveys to conduct a comprehensive comparative analysis of the differences in the perceptions of the messages conveyed by dance among respondents with different backgrounds. Statistical analysis is also used to further examine these differences based on academic background and gender. The results indicate that in the evaluation elements of "Traditional Dance-drama-Eternal Love across the Magpie Bridge," the dancers' skills, melody of the music, and profound themes are all related to the audience's preferences for the work. Furthermore, respondents' perceptions of the work did not differ based on different backgrounds or age groups. This demonstrates that as a paradigmatic classic dance drama, "Eternal Love Across the Magpie Bridge" does not vary according to changes in space and time.

Keywords: Aesthetic · Perception · Six Elements of Drama

1 Introduction

Art encompasses music, visual arts, literature, drama, dance, and film and serves as a tangible manifestation of the human inner creative impulse. These diverse forms of expression not only constitute a vital component of culture and provide economic benefits and employment opportunities but also enrich people's lives with joy and meaning. Dance has been intricately intertwined with various aspects of life throughout history, playing roles in religious and ceremonial practices [1, 2], storytelling, animal mimicry, rainmaking, war dances [3], and the performing arts. Aesthetic perception research is a broad field that explores how individuals perceive, evaluate, and comprehend beauty in various forms such as artwork, literature, music, film, and architecture. These studies often focus on understanding audiences' aesthetic experiences, preferences, and evaluations while examining their influencing factors. The primary objective of this study is to analyze the dance drama "Eternal Love Across the Magpie Bridge", utilizing Aristotle's Six Elements of Tragedy as the evaluative criteria, questionnaire surveys, and quantitative analysis methods.

Dance drama is a fusion of multiple art forms that demands meticulous professional artistic cultivation, from choreography to stage performance. This requires precise mastery and judgment of rhythm and visual aesthetics and utilizing the body to convey nonverbal forms of aesthetic artistic pursuits. Consequently, a work's success is gauged by its widespread recognition and acceptance, ability to convey direct primal emotions through the artistry of the body, and capacity to engender responses from and dialogues with the world.

"Eternal Love Across the Magpie Bridge" is a large-scale dance drama that premiered at the Taipei Arts Festival in 1984, when it was one of the few ethnic dance dramas of its time. Composed by Zheng Sisen and choreographed by Xu Huimei, the drama is based on the story of the Cowherd and the Weaver Girl and divided into the prologue, the first act "Dance of the Fairies," the second act "The Story of the Cowherd and the Weaver Girl," the third act "The Magpie Dance," and the epilogue. Zheng Sisen integrated unique tones and musical forms from various instruments using narrative forms to advance the plot. He used solemn, gentle, lively, and plaintive music to depict the layers of twists and turns in the situations the characters face and the emotions they experience. "Eternal Love Across the Magpie Bridge" also pioneered Taiwanese ethnic dance dramas and large-scale productions. In addition to Xu Huimei, composer Zheng Sisen specifically composed this dance drama, which had life accompaniment from the Municipal Chinese Orchestra. Stage lighting was designed by Nie Guangyan, costume design by Huo Rongling, and still photography by Wang Xin, all of whom were leading figures in the artistic community at the time, making the performance acclaimed as a collaboration among artistic elites. The drama showcases Xu Huimei's choreography, which inherits the characteristics of classical ethnic dance while incorporating the delicacy of Western classical and modern dance. The passing of Zheng Sisen and Xu Huimei has made this dance drama, created jointly by these two renowned artists, particularly precious. In 2023, Xu Limei reconstructed "Eternal Love Across the Magpie Bridge," with Zhu Yunsong rearranging the music and Zheng Libin, the former conductor of the Municipal Chinese Orchestra, conducting. This production utilizes new stage technologies to combine music, dance, theater, and multimedia to reconstruct this classic masterpiece.

As the performing arts continue to flourish, attending such events has become a leisure activity for many consumers. Participation in cultural and artistic activities, as well as related exhibitions, has also become a part of daily life. Thus, this study aims to analyze aesthetic differences from the perspective of performing arts spectators and other professionals' viewing motivations. The research uses data collected through questionnaire surveys to understand viewers' evaluations, reactions, and emotional experiences after watching "Eternal Love Across the Magpie Bridge," which can help provide a comprehensive understanding of viewers' aesthetic perceptions. Overall, this study analyzes audiences' viewing experiences and responses from the perspective of Aristotle's Six Elements of Tragedy, including plot, characters, thought, diction, melody, and spectacle, by integrating viewers' cultural backgrounds, educational levels, and personal preferences, as well as the characteristics of the drama itself. This in-depth reflection aims to explore the factors that influence viewers' aesthetic perceptions of the ethnic dance drama" Eternal Love Across the Magpie Bridge. "This will help develop a better understanding of how audiences perceive and evaluate this art form and how artistic performances can be improved to meet audiences' needs and expectations.

2 Literature Review

2.1 Aesthetic Perception

Human aesthetic ability far exceeds specific biological factors, encompassing a wide range of objects (e.g., paintings, poems, architecture, jewelry, sonatas, and dance) and perceptual features (e.g., color, form, balance, rhythm, and harmony). Moreover, what is considered "beautiful" depends on both the surrounding culture and environment, and artistic practices can vary significantly across time and space. Despite the longstanding biased belief that these variable factors are completely independent of human biology, the fact remains that all human cultures exhibit some form of artistic practices, seemingly unmatched by any nonhuman species in terms of flexibility and diversity.

To be successful, aesthetics must include a rich and thorough understanding of human aesthetic culture, its preferences, and the sense of order and beauty on which it is based. The visual and sensorimotor areas of the brain may play a role in automatic aesthetic responses to dance [4]. To some extent, these findings explain why dance exhibits a cultural universality.

While aesthetics are manifested in cultural traditions that vary drastically worldwide, the overall universality of art across cultures and time suggests that aesthetic expression and appreciation are deep-seated biological traits. Cultural environments determine language, social behaviors, and aesthetic preferences. Cognitive processing is also shaped by experiences that individuals acquire throughout their lives. An artist's cultural background and personal experiences are crucial for the creation of artwork. However, cultural and individual factors also play significant roles in how humans observe artwork. As cultural environments and individual interactions with them vary in society, the cultural experiences of artists and audiences are always intertwined. Emotional processing is often considered an important component of aesthetic perception [5–9]. Emotions can play a role in both creating and viewing artwork. Artists convey their emotions to viewers through their artwork, which can induce basic emotions such as joy, sadness, disgust, fear, and anxiety through their attributes and content. Furthermore, audiences have emotional experiences with viewing dance performances. This includes viewers' emotional responses to different dance styles and themes and how their emotions are influenced by movements, music, and visual elements in dance, with individual backgrounds, education, and dance experience affecting aesthetic perception. Thus, different viewers may have varying levels of knowledge and understanding of dance performances. Visual perception of dance movements and understanding of bodily movements, including how viewers perceive and evaluate dancers' techniques, coordination, and performance; music and sound play a role in dance performances; and viewers' perceptions of music affect their aesthetic evaluation of dance, are all important factors. Overall, viewers' cultural backgrounds and social factors such as gender, age, and religious beliefs influence their aesthetic perceptions of different styles of dance.

2.2 Aristotle's Six Elements of Tragedy

Aristotle's Poetics, written in the fourth century BC, is the first treatise on the art of drama and continues to be widely used and respected as a theory of dramatic criticism over 2000 years. In Poetics, Aristotle (384–322 BC) defines "tragedy" as an art form that imitates serious and complete, actions of a certain magnitude. Its language is adorned with various artistic embellishments, and several forms of expression can be found in different parts of a drama. Drama is presented in the form of action, not narrative, and it affects people's emotions through the proper purification of pity and fear. Aristotle also elaborated on the formation of tragedy from both historical and local perspectives. Tragedy itself is viewed as a complex whole and not as a simple individual. Specifically, Aristotle identified six elements in drama: plot, characters, thought, diction, melody, and spectacle. Aristotle believed that plot is the root and soul of tragedy. The purpose of tragedy is to imitate a serious and complete action of a certain magnitude. Regarding plot, seriousness is the difference between tragedy and comedy, in that it must imitate serious rather than lighthearted humorous events. Tragedy is not only that of great people but also that of ordinary people. Sufficient similarities can be found between the forms of tragedy and dramatic dance, indicating that the above concepts are not only applicable to tragedy but also to successful dance dramas. To better suit dance dramas, this study adjusts Aristotle's Six Elements of Tragedy into six dimensions for measurement items: plot, characters, diction, thought, spectacle, and melody.

2.3 Artistic Creation and Audience

Viewing a performance "is a social activity of sharing homogeneous experiences and feelings while simultaneously acknowledging these experiences and feelings, should be understood as a process of manufacturing or doing something, rather than simply consuming it.

Baumol and Bowen emphasize the importance of audience research from two perspectives in understanding art in terms of economics. First, from a policy perspective, if art is a public good that has a positive impact on society, then participating in the performing arts as an audience member should contribute to individual welfare, thereby determining the characteristics of the audience and whether the government should support performing arts. Second, from a managerial perspective, audience research is the basis for making rational decisions and formulating investment policies through market predictions.

Artists express their creativity through the process of encoding, while viewers understand artwork through the process of decoding. According to communication theory, in the process of communication from the artist (sender) and creating an artwork for the audience (receiver) [10, 11], the content of successful artwork should satisfy three levels: technical, semantic, and effect [12]. First, at the technical level, the artwork should be vividly conveyed to the audience. Second, at the semantic level, the audience should be allowed to understand the meaning of the artwork. Third, at the effect level, the audience should be able to understand the artwork and move toward it emotionally.

This study provides an analysis based on a literature review, combined with communication studies, cognitive theory, and Aristotle's Six Elements of Tragedy (i.e., plot, characters, diction, thought, spectacle, and melody). Accepting artwork is a reciprocal relationship that can affect people's senses, emotions, perceptions, tastes, and how they experience themselves and the world. No artwork provides a universal experience, and perceptions will differ with each viewer. The involvement of both the audience and the dancers can help provide a better understanding of the resonance between their experiences and those of the audience. The audience's attention plays an important role in aesthetic [13] and artistic experiences by combining all categories of art [14]. According to aesthetic theory, the degree of absorption of a work, regardless of its artistic category and whether it is classical or contemporary, is a primary factor of aesthetic experience and resonance with the work [15].

Fig. 1. Dance of the Fairies.

3 Research Methods

This study used a questionnaire survey method based on the research objectives, with the audience of "Eternal Love across the Magpie Bridge" as the participants, divided into four major groups: dance professionals, music professionals, arts and design professionals, and other fields. The study was conducted using an online questionnaire with 233 valid questionnaires collected. The sample included 46 male and 189 female participants, of which 70 were under 19 years old, 44 were aged 20–29 years, 37 were aged 40–49 years, and 65 were aged 50 and above. Regarding professional backgrounds, 106 participants were in the dance field, 12 were in the music field, 13 were in the arts and design field, and 104 were in other fields. The evaluation was conducted using a 5-point Likert scale (1 = strongly disagree to 5 = strongly agree) provided in the questionnaire survey. In terms of technical, semantic, and effect levels, the evaluated factors were characters, diction, plot, thought, spectacle, and melody. In addition, dance techniques, traditional Chinese music techniques, cultural backgrounds, and preferences were included. First, descriptive statistical analysis was used to evaluate the audience's ratings of the performance in each dimension. Then, multiple regression analysis was conducted using Aristotle's Six Elements of Tragedy and Preference as the basis for audience members' aesthetic perceptions. Finally, chi-square tests were used to explore the relationship between the participants' gender and their favorite scenes (Fig. 1).

3.1 Research Design

The participants were audience members of a live performance of "Eternal Love across the Magpie Bridge," who filled out the questionnaire online after watching the performance. A questionnaire survey method was used for performing data collection and analysis, making comparisons, and summarizing the results to confirm the validity of the research.

The questionnaire was designed based on the literature theory, focusing on the relationships between different variables according to the research objectives and questions. This study utilized Aristotle's Six Elements of Tragedy, including characters, diction, plot, thought, spectacle, and melody, to analyze and statistically summarize audience members' perceptions of dance techniques and traditional Chinese music techniques throughout the performance as well as their cultural backgrounds. Additionally, the study investigated participants' degree of preference by asking them to indicate their preference level on a scale that served as the basis for the study's recommendations and conclusions.

3.2 Questionnaire Content

This study used an assessment matrix of aesthetic perception with 20 evaluation items rated using a 5-point Likert scale. Participants completed the questionnaire based on their aesthetic perceptions of "Eternal Love across the Magpie Bridge," evaluating it through the three levels of communication and based on the six elements of tragedy, as shown in Table 1. Questions 1–6 pertain to the "technical level," evaluating factors such as the choreography's ability to showcase the dancers' exquisite technique, the dancers' ability to use expressions and body language effectively, the clear rhythm and pace of the performance, the clear presentation of the main theme, the exquisite quality of costumes and props, and the professional quality of the sound effects. Questions 7–12 focus on the "semantic level," evaluating factors such as the appropriate design of different characters' movement techniques, the ability of dancers' expressions and body language to convey rich emotions, the complete structure of the performance in presenting a clear narrative, the conveyance of educational implications in the main theme, the ability of costumes and props to enhance the context and highlight character personalities, and the selection of music in capturing the content and rhythm of the dance.

Questions 13–18 relate to the "effect level," evaluating factors such as the ability of character development to drive the emotional and atmospheric elements of the stage, the

Table 1. Measurement Questions.

The three levels of communication	The six elements of tragedy	Evaluation Items
Technical Level	Character	1. Choreography's ability to showcase the dancers' exquisite technique
	Diction	2. The dancers' ability to use expressions and body language effectively
	Plot	3. The clear rhythm and pace of the perfor mance
	Thought	4. The clear presentation of the main theme
	Spectacle	5. The exquisite quality of costumes and props
	Melody	6. The professional quality of the sound effects
Semantic Level	Character	7. The appropriate design of different characters' movement techniques
	Diction	8. The ability of dancers' expressions and body lan guage to convey rich emotions
	Plot	9. The complete structure of the performance in presenting a clear narrative
	Thought	10. The conveyance of educational implications in the main theme
	Spectacle	11. The ability of costumes and props to enhance the context and highlight character personalities
	Melody	12. The selection of music in capturing the content and rhythm of the dance
Effect Level	Character	13. The ability of character development to drive the emotional and atmospheric elements of the stage

(*continued*)

Table 1. (*continued*)

The three levels of communication	The six elements of tragedy	Evaluation Items
	Diction	14. The ability of dancers' expressions and body language to influence the audience's emotions
	Plot	15. The clever arrangement of content to engage the audience
	Thought	16. The wonderful cultural connotations expressed through dance
	Spectacle	17. The creation of a good atmosphere through overall stage design
	Melody	18. The skillful integration of music and stage atmosphere
Overall Evaluation		preferences for their favorite act
		the exquisite technique displayed by the dancers
		the skillful technique of the Chinese music ensemble
		the profound cultural connotations depicted
		overall preferences for the performance

ability of dancers' expressions and body language to influence the audience's emotions, the clever arrangement of content to engage the audience, the wonderful cultural connotations expressed through dance, the creation of a good atmosphere through overall stage design, and the skillful integration of music and stage atmosphere. Questions 19–23 reflect respondents' preferences for their favorite act, the exquisite technique displayed by the dancers, the skillful technique of the Chinese music ensemble, the profound cultural connotations depicted, and audiences' overall preferences for the performance, with questions focusing on creativity intensity and preference levels.

4 Results and Discussions

4.1 Descriptive Statistics

This study used the descriptive analysis feature of SPSS to calculate and analyze sample data from the questionnaire. Statistical analyses included deriving the means and standard deviations of each scale from the survey data. The statistical analysis of each variable dimension is presented below to provide a clearer understanding of the analysis of each dimension in the survey responses, as shown in Table 2, which presents the average scores and standard deviations for each dimension. A higher average score indicated a higher level of preference for the respective items.

Table 2. The mean and standard deviation of three assessment items.

Variables	Item	Mea	Standard Deviation
The six elements of tragedy	Character	4.65	.570
	Diction	4.70	.589
	Plot	4.61	.603
	Thought	4.59	.597
	Spectacle	4.66	.568
	Melody	4.71	.541
The three levels of communication	Technical Level	4.65	.534
	Semantic Level	4.64	.548
	Effect Level	4.67	.558
Overall Evaluation	Dance Technique	4.64	.676
	Chinese Music Technique	4.79	.530
	Cultural Connotations	4.65	.647
	Preference Levels	4.58	.685

$^*p < .05.$ $^{**}p < .01.$ $^{***}p < .001.$

Regarding the six elements of tragedy (i.e., plot, characters, diction, thought, spectacle, and melody), the analysis showed that melody had the highest average score of 4.71, with a standard deviation of 0.541, indicating that the performance of the music in this work received a good evaluation from the audience.

Regarding the three levels of communication (technical, semantic, and effect), the analysis showed that the effect level had the highest average score of 4.67, with a standard deviation of 0.558, indicating that this aspect of the performance received a good evaluation from the audience.

Regarding the overall evaluation (dance technique, Chinese music technique, cultural connotations, and preference levels), Chinese music technique was found to have the highest average score of 4.79, with a standard deviation of 0.530, indicating that this aspect of the performance received a good evaluation from the audience.

Overall, the audience provided quite high evaluations of the melody, effect level, and Chinese music technique, reflecting the excellent performance of the work in these aspects. However, these aspects should continue to receive focus in future productions while opportunities for improvement in other dimensions should also be explored to further enhance audience preferences.

4.2 Multiple Linear Regression

The multiple linear regression results as shown in Table 3 were used to predict the relationship between the dependent variable (preference level) and multiple independent variables (plot, characters, diction, thought, spectacle, and melody). The correlation coefficients between the predictor variables and preference levels were .717, .637, .714, .685, .696, and .763, respectively, all showing significance ($p < .001$). Among these, the predictor variables for characters and melody had significant t-values of 2.423 and 5.903, respectively, whereas the predictor variables for diction and thought had non-significant t-values of $-.075$ and $-.323$, respectively.

Overall, the correlation coefficient R between the predictor and dependent variables was .800, with the independent variables explaining the variance in audience preferences at .640. The relative F-value was 66.873 and statistically significant ($p < .001$).

In summary, the multiple regression model showed that the Character and Melody aspects statistically significantly predicted audience preferences, whereas the other aspects did not reach statistical significance. The entire model demonstrated high explanatory power ($R^2 = 0.640$), indicating that the six elements of tragedy have a strong ability to explain audience preferences.

Table 3. Multiple regression analysis predicting preference level using the six elements of tragedy.

Dependent variable	Independent variable	B	r	β	t
Preference Level	Character	.269	.717***	.223	2.423*
	Diction	−.007	.637***	−.006	−.075
	Plot	.186	.714***	.163	1.789
	Thought	−.035	.685***	−.030	−.323
	Spectacle	.100	.696***	.083	1.085
	Melody	.548	.763***	.433	5.903***
	R = .800		$R^2 = .640$	F = 66.873***	

*$p < 0.05$ **$p < 0.01$ ***$p < 0.001$

Using technical, semantic, and expressive factors to predict audience preference for a work, the multiple regression analysis in Table 4 shows that the correlation coefficients for predicting audience preference for the work from the three aspects are technical at .775, semantic at .713, and expressive at .736, all reaching significance ($p < .001$). The t-value for the technical variable was 6.476 ($p < .001$), indicating significance. However, the semantic variable has a non-significant t-value of $-.253$.

Overall, the correlation coefficient R between the predictor and dependent variables was .789, with the technical, semantic, and expressive variables explaining the variance in audience preferences at .623. The relative F-value was 126.303 and statistically significant ($p < .001$).

In summary, the multiple regression model showed that the Technical and Expressive aspects statistically significantly predicted audience preferences, whereas the semantic aspect did not reach statistical significance. The entire model demonstrated high explanatory power (R2 = 0.623), indicating that Technical, Semantic, and Expressive aspects have a strong ability to explain audience preferences.

Table 4. Multiple regression analysis predicting preference levels using the three levels of communication: technical, semantic, and effect.

Dependent variable	Independent variable	B	r	β	t
Preference Level	Technical Level	.698	.775***	.544	6.476***
	Semantic Level	−.033	.713***	−.026	−.253
	Effect Level	.368	.736***	.300	3.017
	R = .789a		R^2 = .623	F = 126.303***	

*p < 0.05 **p < 0.01 ***p < 0.001

4.3 Chi-Square Test

As Table 5 shows, the 233 participants included 46 men and 187 women. Gender was used as the independent variable and preference for the three performance scenes was used as the dependent variable. The results showed significant gender differences in performance scene preferences. Of the male participants, 39.1% preferred the third scene (The Magpie Dance), while 66.3% of the female participants preferred the second scene (The Cowherd and the Weaving Maid), with a chi-square value of 15.6, indicating a statistically significant difference in favorite scenes based on gender.

After analysis and comparison, the audience's evaluations of the six aspects of performance preference were found to be high. No significant differences were found in preference levels based on professional background. These findings provide an overall evaluation of the audience's preference for the performance and an understanding of the different aspects. Melody and characters played important roles in determining preferences, and the technical and expressive aspects also played statistically significant predictive roles for audience preferences. Gender analysis further revealed differences in audience preferences. These results provide valuable insights for further understanding audience preferences.

Table 5. Chi-square test of gender on favorite act

Independent variable			Item			x^2
			First Scene	Second Scene	Third Scene	(df)
Gender	women	f	22	124	41	15.613***
		%	11.8	66.3	21.9	(2)
	men	f	12	16	18	
		%	26.1	34.8	39.1	
	Total	f	34	140	59	233
		%	14.6	60.1	25.3	100.0%

***p <.001

5 Conclusion

This study applied Aristotle's Poetics to analyze the six elements of tragedy: characters, diction, plot, thought, spectacle, and melody. The results reflect the aesthetic cognition of general audiences toward ethnic dance dramas, providing a theoretical basis for the aesthetic judgment of dance art and future aesthetic assessments of dance dramas. This study's aims were achieved by analyzing the technical, semantic, and expressive aspects using communication theory, which aims to achieve three functions: expression, representation, and communication (Fang, 2012). Expression means that ideas can be expressed through dance, representation reproduces the thoughts and emotions of the creator through the work, and communication is the result of expression and representation, transmitting emotions to the audience to achieve emotional resonance. People (dancers), events (performance content), and things (props and music) transform an artist's artistic expression into objects, allowing the audience to have aesthetic, meaningful, and emotional experiences. Through the performance of the six elements (i.e., characters, diction, plot, thought, spectacle, and melody) and the analysis of technical, semantic, and expressive aspects, the audience was found to have different preferences for the three levels. However, no significant differences were observed among the evaluation items across participants with different backgrounds.

A descriptive statistical analysis revealed that the performance of deep-seated storylines in ethnic dance creation gained audience support and approval in terms of preferences. From a technical perspective, which involves the skill of performers, the presentation of dance aesthetics through the dancers' skills was a key factor in their preferences. Regarding character, diction, plot, thought, spectacle, and music, after the multiple regression analysis, all six evaluation items showed significant predictive effects for audience performance preferences. The results of the regression model showed that music and characters had a significant positive correlation with preference, whereas plot had a slightly positive correlation with preference. Characters (role) ranked second in importance among the six elements of tragedy, ensuring that the outcome of the plot aligned with its development and showcasing the characters' personalities. Characters and plots are closely linked with actions arising from the characters and character traits being displayed through dance movements. Aristotle believed that "the orchestra should be regarded as part of the whole performance and play a constructive role in the performance." This work of ethnic dance creation, conveys deep-seated storylines in terms of people, events, and things, garnering support and approval from the audience. The audiences' preferences for the technical aspect involving a performer's skills, and the presentation of dance aesthetics through the performer's skills were key factors affecting preferences. A chi-square analysis of gender revealed further differences in audience preferences. These results provide valuable insights for further understanding audience preferences. Both male and female audiences showed a preference for characters, indicating no significant difference in preference between the genders. Overall, character was the most preferred of the six elements in terms of gender and professional background.

The findings of this study demonstrate that the creative choreography of "Eternal Love Across the Magpie Bridge" conveys rich emotional expressions through dancers' role-playing. In addition to showcasing their exquisite dance skills, dancers' adept use of facial expressions and body language stimulated the audience's emotions and created an atmosphere. The adapted emotions and plots of traditional stories are integrated into the roles in ethnic dance performances. The cross-boundary integration of traditional dance drama elements and performance techniques presents a diverse cultural and artistic outlook. Through the integration of Eastern and Western, as well as ancient and modern, elements, and the use of multimedia, the value of art and its greater possibilities are highlighted. These methods, which break traditional patterns, expand the audience's horizons and imagination within the change and constancy in dance dramas. However, the uniqueness and artistic nature of cultural thinking are the highlights of sustainable development. The millennia-old tale of the love story between the Cowherd and Weaver Girl's love story has long been a cross-temporal national emblem in Chinese society. The sentiment of "through ages, our love remains, like the Qixi Rain year by year in the mortal world," is timeless. This sentiment, spanning from ancient times to the present, is deeply nostalgic.

References

1. Lewis, J.: A cross-cultural perspective on the significance of music and dance to culture and society: insight from BaYaka Pygmies (2013). https://doi.org/10.7551/mitpress/9780262018104.003.0002
2. Merker, B.H., Madison, G.S., Eckerdal, P.: On the role and origin of isochrony in human rhythmic entrainment. Cortex **45**(1), 4–17 (2009). https://doi.org/10.1016/j.cortex.2008.06.011
3. Massin, M.: Expérience esthétique et art contemporain. Presses universitaires de Rennes (2013)
4. Calvo-Merino, B., Glaser, D.E., Grèzes, J., Passingham, R.E., Haggard, P.: Action observation and acquired motor skills: an fMRI study with expert dancers. Cereb. Cortex **15**, 1243–1249 (2005). https://doi.org/10.1093/cercor/bhi007
5. Chatterjee, A., Vartanian, O.: Neuroaesthetics. Trends Cogn. Sci. **18**, 370–375 (2014). https://doi.org/10.1016/j.tics.2014.03.003
6. Jacobsen, T.: Individual and group modelling of aesthetic judgment strategies. Br. J. Psychol. **95**, 41–56 (2004). https://doi.org/10.1348/000712604322779451

7. Leder, H., Nadal, M.: Ten years of a model of aesthetic appreciation and aesthetic judgments: the aesthetic episode - developments and challenges in empirical aesthetics. Br. J. Psychol. **105**, 443–464 (2014). https://doi.org/10.1111/bjop.12084
8. Leder, H., Belke, B., Oeberst, A., Augustin, D.: A model of aesthetic appreciation and aesthetic judgments. Br. J. Psychol. **95**, 489–508 (2004). https://doi.org/10.1348/0007126042369811
9. Silva, P.J.: Human emotions and aesthetic experience: an overview of empirical aesthetics. In: Shimamura, A.P., Palmer, S.E. (eds.) Aesthetic Science. Connecting Minds, Brains, and Experience, pp. 250–275. Oxford University Press, Oxford (2014)
10. Baumol, W.J., Bowen, W.G.: Performing Arts-the Economic Dilemma: a Study of Problems Common to Theater, Opera, Music and Dance. MIT Press, Cambridge (1993)
11. Lin, R., Qian, F., Wu, J., Fang, WT., Jin, Y.: A pilot study of communication matrix for evaluating artworks. In: Rau, P.L. (ed.) Cross-Cultural Design, CCD 2017, LNCS, vol. 10281, pp. 356–368. Springer, Cham (2017). https://doi.org/10.1007/978-3-319-57931-3_29
12. Schaeffer, J.M.: L'EXPÉRIENCE ESTHÉTIQUE. Gallimard (2017)
13. Dewey, J.: Art as experience. In: The Richness of Art Education, pp. 33–48. Brill (2008)
14. Fang, W., Gao, Y., Zeng, Z., Lin, B.: A study on audience perception of aesthetic experience in dance performance. J. Design **23**, 23–46 (2018)

Research on Perceptual Differences in Corporate Identity Systems

Jhih-Ling Jiang(✉) and Rungtai Lin

Graduate School of Creative Industry Design, National Taiwan University of Arts,
New Taipei City 220307, Taiwan
`jausten@ntua.edu.tw`

Abstract. Perception refers to the process by which individuals receive external stimuli through their sensory systems and process and interpret these stimuli in the brain. Perceptual differences can establish a Visual Identity (VI) that highlights brand spirit and creates market visual differentiation. The Corporate Identity System (CIS) plays an important role in building brand image, encompassing product design, brand positioning, and overall packaging. The visual identity (VI) within the CIS is the first impression that guides consumers to recognize the company. Consequently, the concepts of MI (Mind Identity) and BI (Behavior Identity) in modern CIS have become less emphasized, with a growing focus on CVIS (Corporate Visual Identity System).

Research has shown that CIS plays a crucial role in establishing brand image, forming market competitive advantages, enhancing customer experience, reinforcing social responsibility, and sustaining corporate culture. However, whether the CIS concept can be positively conveyed as intended by the designer depends on the audience's perception. The effectiveness of the CIS concept should consider the perceptual differences of the audience. This study utilized a Likert 5-point scale for evaluation, where participants selected their preferred case based on their subjective feelings. The results indicated that audience perceptions vary, leading to differences in CIS experiences. Classic CIS brands are likely to be favored because they resonate with the audience's life experiences, suggesting that classic CIS brands have a sustainable and representative significance.

Keywords: Corporate Identity System · Brand Image · Perceptual Differences

1 Introduction

The Corporate Identity System (CIS) is a structured visual design that expresses a company's intangible business philosophy through tangible media that interact with the public. These media include products, advertisements, business cards, buildings, stationery, and employee attitudes. The ultimate goal is to create a unified corporate image for consumers, fostering confidence and brand loyalty. This concept originated from Germany's AEG during World War I and gained prominence post-World War II. With the international economy thriving, companies like Coca-Cola, IBM, and MAZDA adopted CIS to communicate their corporate image amid increasing diversification and globalization of business operations.

CIS, also known as the Corporate Image Design System, is a comprehensive system used by enterprises to highlight their uniqueness and differentiate their products or image in a competitive market (Melewar and Jenkin 2002). This system aims to effectively convey the business philosophy and cultural spirit of the enterprise to relevant stakeholders through well-crafted visual and cultural elements, ensuring recognition and identification with the enterprise. Therefore, CIS combines modern design concepts and corporate management theories to portray the company's personality, highlighting its corporate spirit. This leads to consumer recognition and leaves a lasting impression, ultimately achieving promotional goals (Balmer 2008; Harwood 2011; Lee 2008; Li 2003; Deng 2002; Lin 1994, 1997, 2018; Lin and Chang 1993).

Given that CIS is a means for companies to introduce or market themselves to consumers, it raises questions about whether consumers can perceive the carefully designed CIS as intended. Do the characteristics that designers want audiences to experience receive positive feedback as expected? Or do different biases emerge during the perception process? These are the goals this study aims to explore.

2 Literature Review

2.1 Corporate Identity Systems and Perceptual Differences

The Corporate Identity System (CIS) aims to strengthen the audience's impression of a company, thereby establishing recognition and ultimately achieving promotional objectives. Therefore, CIS can be seen as the company's business card in product marketing, allowing the audience to understand and identify with the company through CIS. However, the perception that consumers form based on a company's CIS may not always align perfectly with the company's intended image. Perception involves the process by which individuals receive external stimuli through sensory systems (such as vision, hearing, touch, etc.) and process and interpret these stimuli in the brain. This process enables us to understand and recognize the world around us (Schacter et al. 2011). Since the perceptual process is an individual's sensory experience and understanding, the same thing can be interpreted differently by different people. Therefore, the objective information conveyed by CIS may be perceived differently by the audience due to varying backgrounds such as gender, age, education, experience, and other factors.

2.2 A Comparative Case Study

This study selected three cases to explore perception differences in CIS: IBM, Formosa Plastics Group, and Startup Island TAIWAN. These cases were chosen for their distinctive characteristics: IBM represents "innovation, technology, and intelligence"; Formosa Plastics Group is noted for its localization and foresight; and Startup Island TAIWAN is the youngest, embodying contemporary spirit. It is important to note that CIS is a complex and diverse system, with people generally being most familiar with corporate logos as the initial representation of a company's image and culture. Therefore, the logos of these three companies or institutions were chosen as the subjects for this study. Below are brief introductions to the three subjects (information compiled from relevant official websites):

1. IBM

IBM's logo, designed by Paul Rand, was inspired by the American flag. This simple logo consists of a rectangle formed by blue and white stripes combined with the letters "IBM," showcasing a modern and sleek style. Paul Rand's design has gained widespread recognition in the global market, becoming a classic example of brand design. His pursuit of perfection and attention to detail have made IBM's CIS a model of modern brand design. This simple yet stylish logo symbolizes IBM's corporate spirit and values, emphasizing innovation, customer satisfaction, and global expansion. In summary, Paul Rand's design created a unique brand image for IBM, making it a pioneer in modern corporate logos.

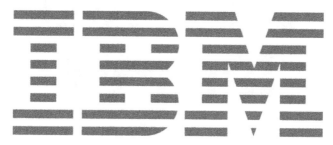

Fig. 1. IBM.

2. Startup Island TAIWAN

"Startup Island TAIWAN" is a national startup brand advocated by the National Development Council, symbolizing Taiwan's entrepreneurial energy and innovative potential. The brand elements include imagery of mountains and seas, representing the island's geography and the resilience of its entrepreneurs. The butterfly symbolizes Taiwan's diverse culture and innovative energy, reflecting the spread of creative power globally. Taiwan's rich cultural history and diverse ecology make it an ideal place for entrepreneurs. The brand's logo features an island image with mountains reflected on the ocean, combined with the infinity symbol and DNA symbol, representing the limitless potential of Taiwan's startup energy and the entrepreneurial gene inherent in every Taiwanese person. The overall design cleverly integrates the characteristics and values of Taiwan's startup culture, creating a unique visual symbol. Through government and private sector collaboration, "Startup Island TAIWAN" aims to propel the nation's startup power globally, assisting Taiwanese startups in entering the international stage and becoming a significant player in the global startup ecosystem.

3. Formosa Plastics Group

In the 1970s, during its reorganization and expansion, Formosa Plastics Group recognized the critical importance of enhancing its corporate image and thus introduced a new Corporate Identity System (CIS). The Ming Chi Industrial Design Center was responsible for designing this system, which included elements such as the logo, typography, colors, and advertisements. This CIS garnered widespread attention at the time and is still considered a successful case in design history. A CIS is a set of brand image elements designed uniformly to shape a company's brand image. A successful CIS not only effectively shapes the corporate image but also increases brand

Fig. 2. Startup Island TAIWAN.

visibility and recognition. Formosa Plastics Group was the first company in Taiwan to launch a corporate identity image, designed by the Ming Chi Industrial Design Center under the leadership of Professor Kuo Shu-Hsiung. Their professional and exquisite design created a unique and successful brand image for Formosa Plastics Group, establishing it as a pioneer in corporate identity design at the time (Figs. 1, 2 and 3).

Fig. 3. Formosa Plastics Group.

2.3 Corporate Identity Systems

These elements intertwine to form the overall structure of the CIS. A CIS not only helps a company establish a unique image but also enhances brand recognition and loyalty in a competitive market, strengthens the company's social image, and further promotes sales and profitability. Therefore, companies should prioritize and properly construct their identity systems to shape their corporate image and ensure sustainable development (Fig. 4).

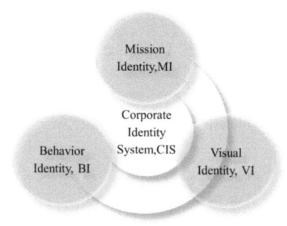

Fig. 4. Corporate Identity System Design Framework. (As Illustrated in this study)

3 Research Methods

3.1 Research Hypotheses

CIS design often carries the personal style of the designer, leading to different perceptions among respondents based on their field of study. This research hypothesizes that respondents from humanities fields such as art and design will evaluate CIS more critically and that their assessments will differ significantly from those of respondents from other fields. Additionally, most people assess a company's image primarily through the quality of its most prominent symbol (the logo), making the CIS of well-known companies more likely to be well-received. On the other hand, for newer companies or brands, even if their design concepts align closely with current trends, they may not be as easily appreciated due to lower recognition, and some people may not even be aware of them.

3.2 Respondents

The questionnaire was created and distributed through Google Forms, with 161 valid responses collected from participants in online communities (Table 1). Research suggests that the sample size should be 3–5 times or 5–10 times the number of items on the largest subscale being tested (Comrey 1988; Tinsley and Tinsley 1987; Wu Ming-Long and Tu Jin-Tang 2011). Therefore, the number of valid questionnaires meets the sample size requirement. Upon inspection, all questionnaires were deemed valid.

Table 1. Respondent Profile Table.

Aspect	Category	Number of Respondents	Ratio
Gender	Male	79	49.07%
	Female	82	50.93%
Age	Under 20 years old	23	14.29%
	21–40 years old	62	38.51%
	41–65 years old	59	36.65%
	Over 65 years old	17	10.56%
Highest education level	High School/Vocational High School	26	16.15%
	Bachelor's degree	54	33.54%
	Master's degree	44	27.33%
	Doctoral degree	37	22.98%
Field of expertise	Education	13	8.07%
	Humanities & Arts	53	32.92%
	Social Sciences, Business, and Law	17	10.56%
	Science	2	1.24%
	Engineering, Manufacturing, and Construction	27	16.77%
	Agricultural Science	2	1.24%
	Medicine, Health, and Social Welfare	8	4.97%
	Service	5	3.11%
	Other	34	21.12%

3.3 Questionnaire Design

Based on literature review, theoretical framework, and research hypotheses, the questionnaire design consists of three parts. The first part includes basic demographic information of the participants, such as gender, age, highest education level, and field of expertise. The classification of participants' fields of expertise is based on the nine disciplines established by the Ministry of Education, aiming to gain insights from respondents across different fields.

The second part involves the evaluation of three cases, each accompanied by a brief description. Nine questions are rated using a Likert 5-point scale, where 5 indicates "Strongly Agree" and 1 indicates "Strongly Disagree" (see Table 2).

The third part entails a comprehensive judgment: participants select their preferred case based on their subjective preferences.

Table 2. Nine questions in the second part of the questionnaire (illustrated with Case One as an example).

	Case One: IBM					
					Case Introduction (brief)	
Contents of the assessment	Criteria for assessment	Strongly agree	Agree	Uncertain	Disagree	Strongly disagree
VI	Is the selection of A1 appropriate for this CIS visual type?	☐	☐	☐	☐	☐
	Is the selection of A2 appropriate for this CIS visual element?	☐	☐	☐	☐	☐
	Is the consistency of this CIS design high?	☐	☐	☐	☐	☐
BI	Are the behavioral objectives of this CIS easy to understand?	☐	☐	☐	☐	☐
	Is the selection of market positioning for this CIS appropriate?	☐	☐	☐	☐	☐
	Is this CIS easily recognizable in its brand image?	☐	☐	☐	☐	☐
MI	Is the assessment of this CIS corporate philosophy easy?	☐	☐	☐	☐	☐
	Are the criteria and characteristics of this CIS clear and understandable?	☐	☐	☐	☐	☐
	Does this CIS have competitive advantages?	☐	☐	☐	☐	☐

4 Research Findings and Discussion

4.1 Overall Assessment

To assess the reliability of the questionnaire, Cronbach's α was used to examine reliability, and factor analysis was employed to assess validity. The analysis revealed that the reliability coefficients ranged from 0.912 to 0.954, all exceeding 0.9 ($p < .001$).

The analysis of Corrected Item-Total Correlation (CITC) values indicated that all CITC values corresponding to analysis items were above 0.4, indicating good inter-item relationships and high reliability levels among the analysis items. The data collected by the questionnaire are suitable for further analysis.

Table 3 presents the average, standard deviation, and ranking of scores for 161 respondents' evaluations of three cases (C1 to C3) on nine questions (A1 to A9). Table 4 shows the average and ranking of respondents' evaluations of three dimensions (visual identification, behavioral identification, and philosophical identification) in CIS design. Table 5 displays respondents' preferences for the three cases.

Table 3. Mean, Standard Deviation, and Ranking of Evaluations by Participants on Three Cases.

		C1	C2	C3
VI	Is the selection of visual type A1 appropriate for this CIS?	**4.043 (0.839)**	3.559 (0.928)	3.720 (1.038)
		C1 > C3 > C2		
	Is the selection of visual elements A2 appropriate for this CIS?	**3.975 (0.873)**	3.596 (0.938)	3.839 (0.987)
		C1 > C3 > C2		
	Is the consistency of the design of this CIS high?	**4.248 (0.742)**	3.429 (0.920)	3.938 (0.992)
		C1 > C3 > C2		
BI	Are the behavioral objectives of this CIS easy to understand?	**3.789 (0.958)**	3.329 (1.065)	3.776 (1.090)
		C1 > C3 > C2		
	Is the selection of market positioning for this CIS appropriate?	**3.950 (0.765)**	3.335 (0.866)	3.801 (0.961)
		C1 > C3 > C2		
	Is this CIS easily recognizable in its brand image?	**4.149 (0.875)**	3.323 (1.034)	3.876 (1.100)
		C1 > C3 > C2		
MI	Is the assessment of the corporate philosophy of this CIS easy?	**3.814 (0.917)**	3.280 (1.014)	3.795 (0.988)
		C1 > C3 > C2		
	Are the criteria and characteristics of this CIS clear and understandable?	**3.988 (0.887)**	3.230 (1.097)	3.832 (0.983)
		C1 > C3 > C2		
	Does this CIS have competitive advantages?	**3.957 (0.869)**	3.180 (0.941)	3.696 (1.084)
		C1 > C3 > C2		

N = 161; C1: IBM, C2: Startup Island TAIWAN, C3: Formosa Plastics Group.

To further interpret whether there are perceptual differences among participants of different genders, ages, educational backgrounds, and fields of expertise, the study found no significant differences in the evaluations of the three cases among participants from different backgrounds. However, there were certain cognitive differences in different aspects among participants of different genders, ages, and educational backgrounds. It should be noted that these perceptual differences among participants in different aspects are not the main focus of this study. The main focus of this study is whether this model

Table 4. Participants' evaluations of three aspects (nine questions) in CIS design: average values and ranking.

	VI (A1-A3)			BI (A4-A6)			MI (A7-A9)		
	A1	A2	**A3**	A4	A5	**A6**	A7	A8	A9
C1	4.04	3.98	**4.25**	3.79	3.95	**4.15**	3.81	3.99	3.96
	A3>A6>A1>A8>A2>A9>A5>A7>A4								
	A1	**A2**	A3	A4	A5	A6	A7	A8	A9
C2	3.56	**3.6**	3.43	3.33	3.34	3.32	3.28	3.23	3.18
	A2>A1>A3>A5>A4>A6>A7>A8>A9								
	A1	A2	**A3**	A4	A5	**A6**	A7	A8	A9
C3	3.72	3.84	**3.94**	3.78	3.80	**3.88**	3.80	3.83	3.70
	A3>A6>A2>A8>A5>A7>A4>A1>A9								

N=161; C1: IBM, C2: Startup Island TAIWAN, C3: Formosa Plastics Group.

Table 5. Participants' preferences for the three cases.

Cases	Number of people	Proportion
C1: IBM	84	52.17%
C2: Startup Island TAIWAN	35	21.74%
C3: Formosa Plastics Group	42	26.09%

N = 161; C1: IBM, C2: Startup Island TAIWAN, C3: Formosa Plastics Group.

can link the perceptions of the participants with the intentions of the designers, serving as a basis for future research.

4.2 Analysis and Discussion

Based on the trends presented in Table 4, it can be observed that among the nine questions, the highest scores in the evaluations of the three cases are for A3, A2, and A3, all of which belong to the visual identification category. This indicates that participants are more concerned with and attentive to the visual effects of CIS. The lower scores are concentrated in the areas of philosophical identification and behavioral identification. This suggests that, compared to "visual graphics," behavioral and philosophical identifications are not as easily perceived by the participants. However, these two areas are crucial components of CIS. Therefore, it is necessary for designers and business owners to enhance the design of "behavioral identification" and "philosophical identification" in CIS, and consider how to present them using straightforward, concise "visual elements" to improve public understanding and recognition.

From Table 5, it can be seen that the most preferred case among participants is C1 (IBM), with 84 participants (52.17%) favoring it. Case C3 ranks second with 42 participants (26.09%) preferring it, and Case C2 is liked by 35 participants (21.74%).

Although Case C1 was introduced a long time ago, it remains very classic, with more than half of the participants favoring it. Moreover, C1 scored the highest among the three cases across all nine questions. This indicates that traditional, classic design concepts and styles still tend to be favored by more people. Case C2, being introduced more recently, may be more favored by participants in Taiwan due to local emotional factors. This suggests that local brands' CIS designs are more likely to be liked by local residents, and if they can better highlight local elements, they might become even more popular.

By examining the perceptual differences among participants of different aspects (gender, age, and highest educational attainment) and their scoring trends, the following inferences can be made:

1. Although there is only one perceptual difference between male and female participants, this can be further analyzed in subsequent research. It might also be due to the relatively conventional design of the cases, leading to consistent evaluations from both men and women.
2. The middle-aged and young groups gave relatively high evaluations, while the perceptual differences among participants with different levels of highest education were more complex. Generally, participants with higher education levels gave slightly higher evaluations than others, possibly because they evaluated the nine questions more rigorously or had deeper insights. This study initially anticipated that participants from the humanities and arts fields might make judgments significantly different from those of participants from other fields.
3. Surprisingly, the perceptual differences among participants from different backgrounds were not significant. Except for Case C2 (Startup Island TAIWAN) which was introduced relatively recently, the other two cases have more mature CIS, forming systems well-known to the public, resulting in less significant differences in their evaluations.
4. The average scores of participants for the three cases on the nine questions were all above 3 (Fig. 5), indicating that the three cases generally followed the basic principles of CIS design. However, a closer examination (Table A3) reveals that a certain proportion of participants disagreed or even strongly disagreed with some questions. Additionally, some participants chose "uncertain/no opinion," which warrants further investigation into the underlying reasons for these evaluations. These findings further confirm the complexity and diversity of CIS, which can serve as a basis for the continuous evaluation and updating of CIS.

5 Conclusions and Recommendations

The study found that the audience's experience with a company's CIS (Corporate Identity System) is not completely uniform. Some audience members may prefer certain companies' CIS, while others may be indifferent to it. Thus, a CIS does not necessarily produce the same experience for all audience members. The reason lies in the interaction process between the audience and the CIS, which depends on the individual's perceptual activities. The study draws the following conclusions about the differences in perception:

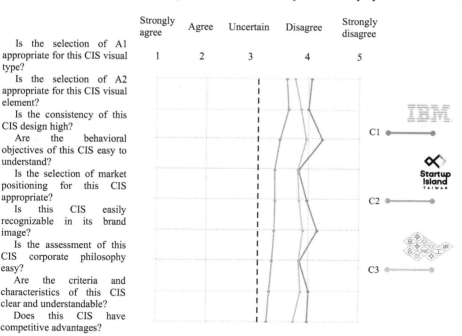

Fig. 5. Trends in Participants' Evaluations of Cases (Drawn by This Study).

1. Visual Elements Significantly Impact CIS.

    The questionnaire results clearly show that participants are more concerned with the visual effects of CIS. This is because "visual graphics" often serve as the first point of contact and sensation for observing CIS, while behavioral and philosophical identifications are less easily perceived. The emphasis on "visual graphics" has already become mainstream in modern CIS design, with a trend towards highlighting the Corporate Visual Identity System (CVIS). This includes choosing fonts, using bright colors, and employing 3D effects to capture attention, thereby creating a unique corporate image and enhancing brand value and competitiveness. The hidden content of CIS, namely "behavioral identification" and "philosophical identification," should be closely integrated with visual elements to ensure audience understanding and recognition.

2. Personal Factors Influence CIS Perception.

    The perception of CIS is influenced by personal factors such as gender, age, and highest educational attainment. Gender differences may be limited by case design, showing perceptual differences but not significantly. However, educational factors show more pronounced perceptual differences. This could be because individuals with higher education levels tend to think more independently and rigorously, leading to different evaluation results.

3. Design Style of CIS Influences Perception.

The perception experience of participants indicates that traditional and classic design concepts and styles are more likely to be favored. This may be because traditional or classic designs are closely related to the participants' growth experiences or living habits, thus more easily resonating with their perceptions.

4. In summary, each company has its unique CIS. Consumers facing various companies' CIS naturally exhibit perceptual differences due to factors such as gender, age, and field of study. How a company's CIS design creativity can accurately project to the audience should be positioned from the perspective of the audience's perceptual experience. Therefore, the perceptual differences of the audience regarding CIS are a practical method for reviewing and analyzing CIS creative design and application.

References

Balmer, J.M.: Identity based views of the corporation. Eur. J. Mark. **42**(9/10), 879–906 (2008)
Balmer, J.M.T., Wilson, A.: Corporate identity: there is more to it than meets the eye. Int. Stud. Manag. Organ. **28**(3), 12–31 (1998)
Harwood, J.: The Interface: IBM and the Transformation of Corporate Design, 1945–1976. University of Minnesota Press, Minneapolis, MN (2011)
Lee, C.W.: Corporate identity as strategic management communication: a working framework. Contemporary Thoughts on Corporate Branding and Corporate Identity Management, pp. 138–149 (2008)
Melewar, T.C., Jenkins, E.: Defining the corporate identity construct. Corp. Reput. Rev. **5**(1), 76–90 (2002)
Schacter, D.L., et al.: Psychology. Worth Publishers (2011)
Tinsley, H.E., Tinsley, D.J.: Uses of factor analysis in counseling psychology research. J. Couns. Psychol. **34**(4), 414–424 (1987)
Wu, M.L., Tu, J.T.: SPSS and Statistical Application Analysis. Wunan, Taipei (2011)
Li, X.F.: Research on corporate image and its communication issues. J. Natl. Taichung Univ. Sci. Technol. **4**, 323–342 (2003)
Lin, D.H., Chang, L.C.: Corporate Identity System. New Image Publishing, Taipei (1993)
Lin, P.S.: Advancing together, creating the future—a comparison of CIS development across the Taiwan Strait. Decor **4**, 6–8 (1994)
Lin, P.S.: The development and case analysis of CIS in Taiwan. Monthly Aesthetic Educ. **90**, 17–26 (1997)
Lin, P.S.: Corporate Identity System, 3rd edn. Art Wind Press, Taipei (2018)
Deng, C.L.: A study on the design strategy of corporate internal design activities. J. Des. **6**(2), 101–111 (2002)

Exploration of Mural Creation in Eastern Zhejiang Art Mural Villages

Bai-Hui Du(✉)

Graduate School of Creative Industry Design, National Taiwan University of Arts, New Taipei City 220307, Taiwan
baihuidu1@gmail.com

Abstract. This study focuses on the creation of fishermen's murals in Eastern Zhejiang's "Art mural villages" drawing on experiences from South Korea's "Mural villages" and Taiwan's "Painted mural villages". It explores combining community development with public art. Targeting common fishermen's murals, the research uses surveys to understand community residents' perceptions and preferences, offering mural creation suggestions to promote local economic development. Eastern Zhejiang fishermen's paintings, an intangible cultural heritage, showcase traditional maritime culture with bold colors and delicate brushwork. Depicting fishermen's daily lives, ocean scenes, deity worship, and myths, these paintings highlight the fishing culture's importance. The study conducts cognitive research through instinct, behavior, and reflection levels to identify audience preferences, aiding artists in planning murals more effectively. Methods include case studies, MDS and Multidimensional Scaling Analysis. Cognitive content was classified into nine categories based on instinctive, behavioral, and reflective levels to explore audience interpretation. Results indicate that themes such as p1(Crab God), p3(Emergency Repair), p4(Big Net Head), p5(Zhoushan Hairtail), and p9(Basket Pants) excel in visual appeal and emotional resonance. While fishing culture themes are popular, overly modern themes should be avoided. Recommendations for creators include emphasizing visual elements, incorporating marine culture, delicate styles, and integrating local history to enhance emotional resonance and cultural heritage, strengthening community cohesion.

Keywords: fishermen's paintings · cognitive communication theory · MDS · community development · art Mural village/Mural village/painted Mural village

1 Introduction

1.1 Research Background

The researcher hails from the eastern coastal region of Zhejiang, an area rich in maritime culture where fishing plays a vital role in residents' lives. Fishermen's paintings vividly reflect this culture, fishermen's paintings, cognitive communication theory, MDS, community development, art Mural village/Mural village/painted Mural village. Showcasing the maritime environment and deeply embedding local maritime traditions. These paintings play a crucial role in preserving and transmitting maritime culture. Through visual art, they enrich people's understanding of ocean life and serve as an effective means to protect the gradually disappearing fishing traditions. This allows people to glimpse the past scenes of fishing Mural villages and feel the profound heritage of maritime culture.

Traditionally, fishermen's paintings commonly use acrylic gouache as a medium on paper, canvas, or wood. The 1990s saw a flourishing period for fishermen's painting, attracting creators from diverse backgrounds and incorporating more varied forms such as sand painting and embroidery. The mediums also expanded to murals, clay, and paper art. In recent years influenced by the"rural revitalization policy", fishermen's paintings have gradually entered the tourism development trend. For instance, under government guidance and support various design competitions have been held such as the Zhoushan Xin Yi tourism product design competition in 2018 and the China Dong Fishermen's Painting Cultural and Creative Design Competition starting in 2022. Besides integrating fishermen's paintings as cultural and creative products into the tourism market, presenting these paintings as murals in Art mural villages or tourist residential areas in Eastern Zhejiang has become common. Examples include the Zhoushan Dandong Art Mural village, Shensi Tano Art Mural village, and Wenzhou Dong-tou Art Mural village. This form of presentation not only enhances people's understanding of ocean life but also plays a positive role in cultural dissemination and protection.

1.2 Research Motivation and Objectives

The phenomenon of fishermen's paintings as murals in" art "Mural villages" is prevalent in the Eastern Zhejiang region. In South Korea the "Mural village "project became widespread under the implementation of local autonomy in 1995(For example, in Korea, Ihwa-dong Mural village and Gamcheon Culture Mural village). Similarly, Taiwan introduced various local revitalization policies in the 1990s, such as "public art" "community comprehensive development" and "rural regeneration projects". These policies improved the living environment of existing communities and promoted the development of local culture. The "Painting Mural village" phenomenon rapidly developed under these policies, where creators painted murals on building exteriors, attracting social attention and visitors.

Taiwan recognized the importance of integrating external murals with local historical and cultural elements in its "painting Mural village" constructions over the years. This approach helped in developing local community economies and enhancing community cohesion. In summary this study starts from the existing fishermen's painting murals in "Art mural villages" in Eastern Zhejiang analyzing the community residents' perceptions of fishermen's painting styles. Starting from existing fishermen's painting works the study aims to identify the most popular artwork types and propose applying these popular styles to mural creations in Eastern Zhejiang's Art mural villages.

The objectives of this research are as follows: Explore the implementation experiences of community development and public art in Mural villages/introductions Mural villages understanding how community paintings integrate local historical and cultural elements Analyze the visual art attributes of fishermen's paintings, analyze the visual al art attributes of fishermen's paintings, through a questionnaire survey on fishermen's painting styles, understanding differences in responders' cognitive preferences, thereby promoting local community economic development and cultural prosperity.

2 Literature Review

2.1 Mural Village

Public Art and Painted Villages. "The term new types of public art" refers not only to sculptures displayed in public spaces but also to art forms that focus on public issues and actively involve community participation. These art forms can be tangible, such as murals, installations, photography, or video, or intangible, such as performance art or action art [19]. Among the common forms of public art found in communities, "mural art" is particularly prevalent [1]. The concept of the "Mural Village" originated in South Korea, with Seoul's Ihwa Mural Village (Ihwa Mural Village) being the birthplace of this idea. This village was originally a declining old community near downtown Seoul. In 2006, it was included in the "Art in the City 2006" policy under the "Naksan Project", which aimed to revitalize the area through cultural transformation to boost local tourism and promote both economic and community development. Artists painted murals on the exteriors of buildings in this community, making it a popular destination for tourists for a time. Influenced by the success of this "Mural Village" project and related media, Taiwan saw the development of similar community mural projects known as "Painted Villages" after 2010 [12]. However, the tourist boom in these "Painted Villages" was relatively short-lived, facing challenges like those encountered by Ihwa Mural Village in its later stages, namely the conflict between the influx of tourists and the daily lives of community residents.

Community Development in Mural Villages. The common form of public art in "Mural Villages" is known as community painting or murals. Community painting involves treating the community environment as a public space and incorporating public art into this space to facilitate community development [30], Community development also known as community development, was proposed as a development strategy by the Taiwan Cultural Affairs Council in the 1990s. It aims to establish local community culture and a sense of community identity by encouraging residents to actively participate in local public affairs, thereby fostering a unified community image [31].

Taiwan's early "Painted Villages" can be traced back to 2008, with Taichung's Rainbow Village being a notable example. After gaining popularity in 2010, various "Painted Villages" began to emerge. However, these community paintings often lacked thematic diversity and innovation, emphasizing the need to integrate local historical contexts and cultural characteristics [21]. Despite the initial tourist boom brought about by public art in some Taiwanese "Mural Villages", conflicts arose between tourists and residents. Taiwanese scholar Lee You-line [14] categorized public art plans for "Mural Villages" into three types: policy-led, artist-participatory, and 'resident-participatory.' For instance, in the "resident-participatory" model, residents voluntarily engage in public art projects and community maintenance to improve their living environment. Gamcheon Mural Village in South Korea is a prime example, emphasizing the involvement of various community groups (such as government agencies or artists or residents) [12, 15, 16], It has become a renowned tourist destination.

Therefore, this study attempts to draw on the implementation experiences of community residents' participation in public art. Focusing on the fisherman's mural art in "Art Villages" in eastern Zhejiang, it explores the mural styles that resonate best with the local audience.

2.2 Review of Fishermen's Paintings

Development History of Fishermen's Paintings. Fishermen's paintings along the coast of East Zhejiang belong to a form of artistic expression with a rich folk-art style, primarily focusing on the Zhoushan Archipelago region. There are various opinions on its specific historical origins, but according to the compilation by scholar Wang Yi-Qing [27], the academic community generally believes that it originated in the late 1970s and early 1980s. It went through an initial period of establishment in the early 1980s, a decline in the late 1990s, and a period of market development in the 21st century [7]. According to the research by scholar Luo Jiang-Feng [18], in the early stages, the creators of fishermen's paintings were mainly individuals with fishing experience or from fishing Mural village backgrounds. However, as the prosperity period began in the early 1990s, the backgrounds of the creators of fishermen's paintings became more diverse.

Themes for Fishermen Paintings. With the younger generation leaving their hometowns for urban development, the transmission of traditional cultural skills faces challenges and traditional fishing Mural village paintings suffer from talent drain. Insufficient creative resources in the region limit the development of fishing Mural village paintings making it difficult to attract new creators. In this context the creative forms and content of traditional fishing Mural village paintings have become rigid struggling to meet the demands of the modern art market. Limitations in creators' backgrounds and the homogenization of cultural products further weaken the market competitiveness of fishing Mural village paintings failing to effectively capture audience interest [27]. These issues not only affect the economic benefits of fishing Mural village paintings but also diminish their cultural influence. Therefore, in the process of innovating fishing Mural village paintings it is crucial to emphasize their cultural essence and find a balance between tradition and innovation [4]. This requires support from academic theories and government policies to guide innovation establish safeguard systems for subsequent creative teams rationalize interest distribution nurture new creators and maintain the maritime cultural essence of fishing Mural village paintings. It also aims to promote the tourism-oriented development trend of fishing Mural village paintings [17].

Despite differing opinions on the future of fishing mural village paintings, researchers mainly focus on the challenges faced by creators, including their backgrounds, which influence theme selection and presentation in their art. Therefore, this study selected 10 paintings by artists with and without fishermen experience, such as professional artists and teachers, to balance tradition and innovation. Fishermen experience was the criterion for selecting the artists, and the paintings were divided into two groups: 5 by artists with fishermen experience and 5 by those without. Later, we will analyze which factors appeal to audiences in East Zhejiang and drawing on Taiwan's "painted village" concepts of 'community construction' and 'public art' and their implementation strategies, we will propose recommendations for the composition of external murals in "Art Mural Villages. Please refer to Table 1 for detailed descriptions of the artwork.

Table 1. 10 Categories of Fishermen's Paintings.

Artwork	Author and Title	Artwork	Author and Title
Fishermen Experience Artists			
	(Crab Deity) Sun Yue-Guo		(One-year harvests one season, eat one fresh (world) He Yue-Ming
	(Emergency Repair) Cai Cheng-Shi		Bountiful Harvest Zhu Son-Xiang
	Zhoushan Ribbon Fish Jiang De Ye		
Artists without Fishermen Experience			
	Ocean Marshal Ma Shao-Hong		Around Happiness Li Zhi-Qin
	Fishing Village Zhou Man-Fei		Zhu Guo-An
	Approaching Typhoon Yao Mei-Fei		

2.3 Exploration of Painting Styles

Visual Arts and Painting Styles. Visual arts encompass various forms such as drawing, painting, photography, printmaking, lighting, and film [24]. Viewing artistic expression from a semiotic perspective reveals that once created, artworks undergo development beyond the creator's control [22]. This highlights that the value of artwork extends beyond its formal aesthetic presentation to encompass emotional and speculative dimensions. Artworks serve as responses to personal experiences and societal influences, fostering a dialogue between the audience and art. This interaction encourages interpretation and allows viewers to express their perspectives and emotions, thereby creating personalized artistic experiences. In the realm of visual arts, painting is categorized as an act that combines graphic elements, composition, and aesthetic methods to convey the artist's intended concepts and meanings [15]. The term style refers to the manner and distinctive characteristics of an object, influenced by factors such as era, region, lifestyle, cultural, background, and artistic trends [26]. As a result, paintings of the same type can exhibit variations due to these influences. Fisherfolk paintings, examined in previous literature, face challenges as regional cultures evolve, impacting their intrinsic meanings. Deconstructing these stylistic elements is essential to understanding audience preferences and adapting to the demands of the modern art market, while also preserving their cultural essence.

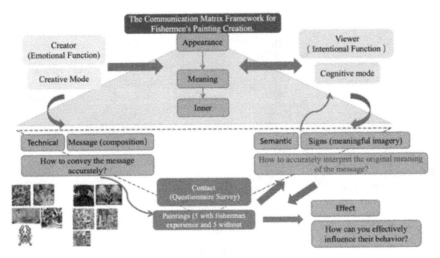

Fig. 1. The Communication Matrix Framework for Fishermen's Painting Creation.

The theories mentioned apply to both how artists create and how viewers interpret art. In painting cognitive attributes are constructed through formal principles like repetition, balance and contrast. Additionally, Jung Tsao identifies seven cognitive traits: image design, color planning, uniqueness, functionality, expressiveness, thematic elements and

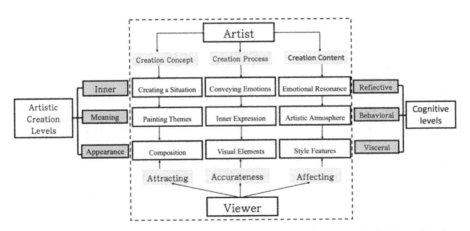

Fig. 2. The Communication Matrix Framework for Fishermen's Illustrative Painting Creation.

emotional meaning [11]. These principles and traits help artists effectively communicate with their audience. This study integrates these ideas, along with communication theories, to develop nine cognitive attributes for fishermen's paintings: visual elements, composition, painting themes, style features, atmosphere, emotional conveyance, situational creation, inner expression, and emotional resonance. Table 2 further explains the meanings of these attributes (Fig. 2).

Table 2. The Definitions of 9 Cognitive Attributes.

Creation and cognitive aspects	Attributes	Definition
Appearance	Composition	The overall visual structure is formed through the arrangement of elements by an artist, as well as the organization and combination of these elements
	Visual Elements	The fundamental visual elements used in visual arts are typically the visible components in artworks or design pieces, such as points, lines, surfaces, color, shapes, etc.
	Style Features	The distinctive artistic features expressed in an artwork, specifically manifested in aspects such as composition, utilization of visual elements, color selection, brushstroke techniques, etc., serve as points of individual creativity and artistic language formation
Meaning	Painting Themes	The specific content chosen by an artist in the creative process, such as landscapes, figures, abstract concepts, social issues, etc.
	Inner Expression	Refers to the abstract or implied meanings, emotions, thoughts, or symbols contained in an artwork
	Artistic Atmosphere	The overall sense and emotional atmosphere presented in an artwork are created by a combination of visual elements, color usage, brushstroke techniques, lighting effects, and other factors in the composition

(*continued*)

Table 2. (continued)

Creation and cognitive aspects	Attributes	Definition
Inner	Creating a Situation	The artist consciously establishes a specific scene or environment through skillful use of visual elements, subject matter, composition, and other techniques
	Conveying Emotions	Artists typically use various means, such as composition, color usage, brushstroke techniques, and subject selection, to express and convey the emotional content behind their works
	Emotional Resonance	Artworks enable viewers to deeply understand, feel, and connect with the emotions conveyed in the artwork, creating a sense of empathy to some extent

3 Research Methods

This study aims to explore the characteristics of suitable exterior murals for "Mural Villages" in the eastern coastal region of Zhejiang. The research methods include case analysis, questionnaire surveys, and the application of MDS regression matrices. The study analyzes nine cognitive elements of visual arts to understand how they are interpreted by the audience. Through this cognitive theory research, the study aims to establish the cognitive attributes of fishermen paintings and use MDS (Multidimensional Scaling Analysis) to understand the differences in respondents' preference factors. Based on these findings, the study will propose recommendations that align with audience recognition and preferences.

3.1 MDS Multidimensional Scaling Analysis

After the preliminary questionnaire survey, we received a total of 60 feedback questionnaires from residents in the eastern Zhejiang area of mainland China. We conducted multidimensional scaling (MDS)analysis using SPSS, obtaining Kruskal's Stress $= .05183$ and RSQ $= .99426$. The results of the analysis provided us with the average recognition data for nine types of visual elements, as shown in Table 4.

In Table 3, researchers focused on average attribute values. The highest-rated projest among fishermen painters was p5(Zhoushan Ribbon Fish), while for non-fishermen painters, it was p9(Pants with Dragon Patterns). Project p5 scored highest across all nine attributes, receiving positive feedback on instinctive, behavioral, and reflective levels, indicating strong visual appeal, user experience, and emotional response. "Style features" scored highest, highlighting p5's prominent ocean imagery expression. Project p9 excelled in "Composition", clearly conveying its artistic intent. Both projects were recognized primarily at the instinctive level, capturing audience attention with visual appeal that effectively showcases marine culture. Although they may prompt less reflection, they successfully convey core marine imagery, all ten works, appearance attributes were most valued, enhancing audience perception of marine culture through clear presentation of "Visual Elements" "Composition" and "Painting Themes".

Table 3. Average Values of 9 Categories of Style Attributes in Fishermen's Paintings.

	p1	p2	p3	p4	p5	p6	p7	p8	p9	p10
F1 (Composition)	4.03	3.87	3.95	4.27	4.45	3.60	3.65	4.02	4.25	3.95
F2 (Visual Elements)	4.33	4.35	4.25	4.33	4.50	3.63	3.62	4.07	4.22	3.97
F3 (Style Features)	4.00	3.73	4.10	4.05	4.40	3.78	3.77	3.95	4.08	3.90
F4 (Painting Themes)	4.07	3.93	3.97	4.18	4.37	3.60	3.50	3.83	4.02	3.95
F5 (Inner Expression)	4.08	4.00	3.88	4.18	4.37	3.60	3.52	3.93	4.02	3.90
F6 (Artistic Atmosphere)	4.13	3.85	3.98	4.03	4.37	3.90	3.68	3.93	4.03	3.95
F7 (Creating a Situation)	3.75	3.93	3.87	4.20	4.28	3.60	3.58	3.88	4.00	3.88
F8 (Conveying Emotions)	4.00	3.95	3.95	4.15	4.33	3.67	3.57	3.88	4.07	3.85
F9 (Emotional Resonance)	3.72	3.62	3.82	3.98	4.18	3.58	3.53	3.80	3.90	3.75

Using the data from Table 3, we performed a multidimensional scaling (MDS)analysis, resulting in the cognitive space diagram shown in Fig. 1. This diagram illustrates the style attributes and their three-dimensional cognitive space. As depicted in the diagram, the 10 fisher paintings are grouped into four clusters within the quadrants: Cluster 1: p8(Fishing Village), p10(Approaching Typhoon), p6(Ocean Marshal), p7(Around Happiness).Cluster 2: p4(Bountiful Harvest), p9(Pants with Dragon Patterns), p5(Zhoushan Ribbon Fish).Cluster 3: p3(Emergency Repair), p1(Crab Deity).Cluster 4: p2(One-year harvests one season, eat one fresh(world).

In Fig. 3, the cluster in the first quadrant consists of p8(Fishing Village), p10(Approaching Typhoon), p6(Ocean Marshal), and p7(Around Happiness). The second quadrant cluster is represented solely by p2(One-year harvests one season, eat one fresh(world). These two clusters show lower performance in the nine content features,

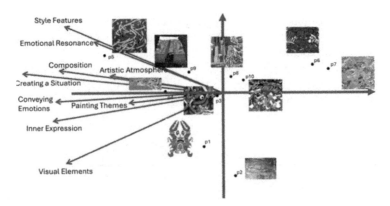

Fig. 3. Style Attributes and a Three-Dimensional Cognitive Space Map.

with p2(One-year harvests one season, eat one fresh(world)) being distinctly separated from the other clusters, located in the fourth quadrant, indicating significant differentiation. The second quadrant includes three items: p4(Bountiful Harvest), p9(Pants with Dragon Patterns), and p5(Zhoushan Ribbon Fish). This quadrant exhibits the richest content features, with p5(Zhoushan Ribbon Fish) being the best-performing item across the nine content features, particularly excelling in "Painting Themes" and "Emotional Resonance". However, it shows the lowest correlation in the "Visual Elements" feature among its nine content attributes. Similarly, p4(Bountiful Harvest) demonstrates balanced audience recognition across the nine content features without any single feature standing out prominently. The third quadrant includes p3(Emergency Repair) and p1(Crab Deity), with p1(Crab Deity) being notable for its "Visual elements" attribute. In summary, the projects p1(Crab Deity), p3(Emergency Repair), p4(Bountiful Harvest), p5(Zhoushan Ribbon Fish), and p9(Pants with Dragon Patterns) have overall garnered audience recognition across the nine attributes. In terms of visual presentation and audience comprehension, these artworks tend to focus on accessible themes related to marine life and oceanic lifestyles. The choice of subjects and their presentation methods allow the audience to more intuitively understand and feel the essence of the works, demonstrating a relatively unified character. In contrast, projects p6(Ocean Marshal), p7(Around Happiness), p8(Fishing Village), and p10(Approaching Typhoon) lean towards depicting diverse themes such as fishing village life and deity worship. These projects exhibit rich diversity and might evoke deeper resonance within specific groups, but overall, they do not perform as well as the more straightforward themes of marine life and oceanic lifestyles. Despite their less prominent performance across the nine attributes, the p6-p10 projects display greater flexibility in creativity and individuality, showcasing different cultural styles. The p2(One-year harvests one season, eat one fresh(world)) project is quite unique. It has lower recognition across the nine attributes, possibly due to its distinctive choice of subject and visual expression.

The subject of traditional fishing beliefs may be difficult for the audience to understand, affecting deeper resonance. Therefore, highly unique themes may limit comprehension, and a lack of visually engaging elements may diminish the work's impact. From the subsequent questionnaire analysis of the" preference "data, we compiled Table 4, which shows that the top five projects have high correlation in the "Visual Elements" content. This further confirms that when presenting clear and vivid marine imagery and related visual elements, audiences more easily progress from understanding to resonance. Combining the cluster analysis and preference data, p5(Zhoushan Ribbon Fish) achieved the highest recognition in both attribute correlation and preference, particularly excelling in "Painting Themes "and "Emotional "Resonance. This indicates that the artwork needs to convey meaning relevant to the thematic nature through highly correlated visual elements to evoke emotional resonance regarding marine culture. Project p1(Crab Deity) shows the strongest recognition in the "Visual Elements "attribute among its nine attributes. However, despite high recognition in "Visual Elements", it performs weakly in other attributes. Thus, balancing the "Visual elements" attribute is crucial. While a strong visual impact is eye-catching, it may lack depth in emotional and thematic expression. Projects p4(Bountiful Harvest) and p3(Emergency Repair) do not stand out in any single attribute but display balanced performance across all nine attributes. This stability helps attract and positively engage the audience.

Table 4. Average Preference Scores and Rankings.

	p5	p4	p9	p1	p3	p8	p10	p2	p6	p7
Ranking	1	2	2	3	4	5	6	7	8	9
Mean	4.35	4.00	4.00	3.97	3.02	3.83	3.75	3.70	3.55	3.48

3.2 Independent Samples T-Test

Based on the previous research results it was found that the five projects—p1(Crab Deity), p3(Emergency Repair), p4(Bountiful Harvest), p5(Zhoushan Ribbon Fish, and p9(Pants with Dragon Patterns)—were the most recognized by audiences in terms of multiple regression and preference ratings. Therefore, this section will discuss these five projects using independent samples t-tests to understand the preference differences among different audience groups based on professional background and age. The aim is to gain deeper insights into the instinctive, behavioral, and reflective levels involved in the cognitive process, to better understand how professional background and age differences affect the cognitive levels of fishermen's paintings. By analyzing questionnaire data, respondents were divided into two groups: 15 people with an art/design background and 45 people without an art/design background, as well as 43 people under the age of 40 and 17 people aged 40 and above.

Presence of Art/Design Background. The independent sample T-test analysis of the five professional backgrounds in the table below reveals a significant difference (p = .025*) in the "conveying friendship" attribute of p1(Crab God) between groups with and without a design background. "Conveying friendship" falls under the cognitive reflection level, indicating that individuals with an art and design background better understand the emotions and cultural characteristics conveyed by the p1(Crab God) project. This may be because such individuals are more sensitive to visual symbols or symbols. The p1(Crab God) is a highly visually appealing painting, with "visual elements" as its advantageous attribute. The visual elements are very clear and distinct, allowing those with an art and design background to interpret the composition and elements within the painting to gain a deeper understanding of its cultural connotations. However, for those without an art and design background, the emphasis on "visual elements" may hinder their interpretation at deeper levels. The audience tends to focus more on the overall intuitive impression of the work, neglecting its cultural significance, resulting in a less favorable perception of the project (Table 5).

Table 5. Presence of Art/Design Background in Style Perception Table.

Artwork	Evaluation criteria	t-test for Equality Means		
		t	df	Sig.
	Conveying Emotions (Inner)	-2.307*	58	.025

*p<.05

Age. Participants were divided into two age groups: under 40 and over 40. Table 6 shows a significant difference between these groups in their response to the p4 project (Bountiful Harvest) (p = .033*). The under-40 group demonstrated a better grasp of the story and situational context of p4(Bountiful Harvest). The project's unique overhead perspective may appeal particularly to younger audiences with diverse visual experiences, enhancing engagement and prompting deeper interpretation. p4(Bountiful Harvest) showed balanced performance across its nine attributes, without significant strengths or weaknesses, and ranked highly in preference, making it a well-rounded and appealing project.

Table 6. Differences in Style Perception Between Two Age Groups.

Artwork	Evaluation criteria	t-test for Equality Means		
		t	df	Sig.
	Creating a Situation (Inner)	2.180*	58	.033

*p<.05

Exploration of Mural Creation 427

4 Research Conclusions and Recommendations

This study explores the style perception of fishermen's mural paintings in the "Art village" of the Zhoushan region. It aims to determine the preference factors of the residents for these murals. The research conclusions will reveal relevant similarities and differences, providing recommendations for the creation of fishermen's mural paintings in the "Art village" of the Zhoushan region.

4.1 Research Conclusions and Recommendations

This study investigated the style perception of fishermen's mural paintings in the "Art village" of Zhoushan East, identifying preference factors among residents. By integrating the analysis of 9 style perception attributes through MDS Multidimensional Scaling Analysis and analyzing preference ratings, the study focused on four groups and their average preferences. The conclusion reveals that the items p1(Crab Deity), p3(Dock Repair), p4(Large Net Head), p5(Fisherman from Zhoushan), and p9(Basket Pants)are perceived as the best projects by the audience in terms of multidimensional regression and preference ratings. Consequently, the conclusion primarily analyzes these five items. Overall, the optimal aspects of these five items are predominantly found at the instinctual level, indicating that visual appeal is a critical factor determining the success of the artworks among local audiences.

p5(Fisherman from Zhoushan) received the highest preference rating. This project, along with p9(Basket Pants), excels in the categories of "stylistic features" and "emotional resonance", both known for their meticulous portrayal style. However, p5(Fisherman from Zhoushan) outperforms p9(Basket Pants) in these categories, likely because its cultural theme of fishing season culture resonates more with residents, enhancing emotional connection. On the other hand, p9(Basket Pants) depicts local deity worship, and it is noteworthy that p4(Large Net Head) ties with p9(Basket Pants) in preference rating, ranking second. p4(Large Net Head) similarly focuses on fishing season culture, employing a distinctive aerial perspective. Despite its average performance across various MDS attributes, it still received high preference ratings. p1(Crab Deity) stands out for its "visual elements", being the most visually appealing project. However, its preference rating ranks fourth, reflecting weaknesses in "stylistic features" and" emotional resonance." Independent sample T-tests for professional backgrounds revealed that an excessive focus on "visual elements" hindered residents from interpreting deeper content, as audiences tended to focus on the artwork's initial impression (such as a crab) while overlooking stylistic features and cultural connotations, thus slightly lacking in emotional resonance. p3(Dock Repair) ranks fifth in preference ratings. Despite its average performance across MDS attributes, p3(Dock Repair) depicted the unique "Green Eyebrow" ship repair theme exclusive to the Zhoushan East region. Its use of a unique frontal perspective like an aerial view adds novelty and visual interest to the artwork.

4.2 Research Limitations and Future Directions

This study explored the style perception of fishermen's mural paintings in the "Art village" of Zhoushan East, aiming to identify preference factors among residents. By employing a progression of instinctual, behavioral, and reflective cognitive levels, the study provided preliminary insights to assist creators in effectively planning mural artworks in the "Art Village". For residents, the audience is first drawn to the immediate visual impact of the painting, content at the instinctual level ("composition" "visual elements" "stylistic features") emerged as crucial in determining visual appeal. In designing murals, it is essential to emphasize visual elements such as color, composition, and perspective angles, integrating local maritime cultural elements to evoke initial intuitive responses to marine imagery. Incorporating meticulous painting styles helps convey profound cultural information, allowing the audience to perceive distinctive maritime cultural traits. However, it's crucial to balance the use of these elements to avoid over-reliance on any single element, which could potentially impact subsequent cognitive levels. At the behavioral cognitive level ("painting themes""expressing content" "atmosphere"), themes related to fishing season culture were well-received by residents. Therefore, focusing on themes related to fishing season culture and traditional life subjects is advisable. Avoiding overly modern or culturally disconnected themes maintains close ties with local cultural backgrounds. Lastly, at the reflective cognitive level ("scenario creation" "emotional conveyance" "emotional resonance" "emotional resonance") emerged as highly valued among Zhoushan residents. While artworks can evoke emotional responses, there's a need to delve deeper into and integrate local history and cultural values more effectively. Therefore, when selecting themes, creators need to consider their relevance to local culture. Although traditional subjects are still well-received by audiences, the intricate details often present in traditional fishermen's paintings can sometimes disrupt viewer perception. As a result, the visual elements in the artwork should be clear, easily understandable, and straightforward. This approach helps residents feel the power of cultural heritage in murals, enhancing local cohesion, this study was limited to the "Art mural villages" in Zhoushan East, focusing on artworks from creators with two different backgrounds. Future research could expand the local sample size and explore more factors, integrating cultural elements more deeply into the creative process for richer stylistic outcomes. Furthermore, it's suggested that creators gather ongoing community feedback to refine their creative strategies, ensuring mural artworks better reflect and serve local cultures.

References

1. Chen, C.P.: Exploring urban beautification issues through mural creation. J. Urban Stud. **11**(1), 143–165 (2021)
2. Chang, C.J.: Exploring the benefits of community painting for local tourism content: a case study of Meilin and Erlun communities in ZheDong township Pingtung County. J. Leisure Tourism Sports Health **9**(2), 48–65 (2019)
3. Tides, C.: Cherishing Memories. Xu, Z.-F., (ed.) The Spiritual Journey of Fishermen Painter. http://www.360doc.com/content/17/1120/18/42824938_705625024.shtml, Accessed 3 Nov 2023
4. Chen, Y.I.: The Characteristics and developmental journey of Zhoushan fishermen paintings. China Arts **4**(5), 119–123 (2014)
5. Chen, S.J., Yen, H.Y., Lee, S., Lin, L.C.: Applying design thinking in curating model a case study of the exhibition of turning poetry into painting. J. Des. **21**(4), 1–24 (2016)

6. Fishermen paintings: inspiration from the sea, https://www.sohu.com/a/411864874_999 53422, Accessed 26 Oct 2023
7. He, L.F., Wang, Y.: Prospects and Ideas for the industrialization of Zhoushan fishermen paintings. J. Zhejiang Ocean Univ. **28**(1), 55–59 (2011)
8. Hsu, F.H.: A study on the signification of visual art communication (master's thesis). National Taiwan University of Arts (2008)
9. Huang, M. H.: Zhoushan Fishermen Paintings Born from Island Soil, Zhejiang Art News 11 (2019)
10. Jin, T.: A discourse on Zhoushan maritime art. J. Zhejiang Ocean Univ. **16**(1), 38–43 (1999)
11. Jung, T.: Study on the visual style of digital illustration design using vector skills. J. Bus. Des. **14**, 253–272 (2010). https://doi.org/10.29514/TJCD.201011.0015(2010)
12. Liu, C.L.: A comparison of participatory mural villages between Taiwan and South Korea: a case study of Gamache Culture Village in South Korea and His-tree Community in Taiwan (master's thesis). Department of Korean Language and Literature, National Cheng chi University, Taipei City (2019)
13. Liao, K.M.: First Volume of Contemporary Chinese Folk Painting, Science Press, China (1994)
14. Lee, Y.: A study on revitalizing local culture through the utilization of public art projects: Focusing on mural villages (master's thesis). Department of Korean Language and Literature, National Cheng chi University, Taipei City (2015)
15. Li, X.: Revised Mandarin Chinese Dictionary. The Commercial Press, Ltd, Taiwan (1994)
16. Lin, C., Chen, S., Lin, R.: Efficacy of virtual reality in painting art exhibitions appreciation. Appl. Sci. **10**(9), 3012 (2020)
17. Luo, J.F.: Research on the inheritance and development of Zhoushan fishermen paintings. J. Zhejiang Normal Univ. **34**(1), 79–84 (2009)
18. Luo, J.F.: Maritime Folk Art: Zhoushan Fishermen Paintings. China Arts, pp. 107–111 (2001)
19. Liao, P.Y.: A study on the images of agricultural community mural paintings: a case study of Kengkou community in Luzhu District, Taoyuan City (master's thesis). National Tsing Hua University, Hsinchu City (2018)
20. Norman, D.A.: The Design of Everyday Things. Basic Books, USA (2013)
21. Overwhelming and featureless murals, New Taipei needs to review. China Times News. https://tw.news.yhoo.com/%E5%BD%A9%E7%B9%AA%E6%B0%BE%E6%BF%AB%E6%B2%92%E7%89%B9%E8%89%B2-%E6%96%B0%E5%8C%97%E8%A6%81%E6%AA%A2%E8%A8%8E-201000520.html last accessed 2024/3/2
22. Principles of beauty. https://pedia.cloud.edu.tw/Entry/Detail/?title=%E7%BE%8E%E7%9A%84%E5%8E%9F%E7%90%86. Accessed 9 Oct 2023
23. Shih, H.R.: Examining the essential differences between performing arts and visual arts from the perspective of the "creativity of the present moment "in art. J. Aesthetics Art Manage. **2**, 74–80 (2006)
24. Shen, W.Q.: Portrait of the Sea God. Zhejiang People's Fine Arts Publishing House, China (2019)
25. Sheng, W.-Q.: Fishermen Paintings Chronicle. |Tianya New Publication. https://www.sohu.com/a/411864874_99953422. Accessed 17 Oct 2023
26. Shih, H.J.: Examining the inherent differences between performing arts and visual arts from the perspective of the 'moment of creation' in art. J. Aesthetics Arts Manage. **2**, 74–80 (2006)
27. Torossian, A.: A Guide to Aesthetics. Oxford University Press, London (1937)
28. Wang, Y.Q.: Review of the current research situation and problem analysis of fishermen's painting. Packaging Des. **6**(2), 174–175 (2022)
29. Wang, Y.F., Fu, J.Q.: Research on inheritance and development of island folk culture. J. Baotou Vocat. Tech. College **20**(3), 88–92 (2019)

30. What Are the Visual Arts? https://www.thoughtco.com/what-are-the-visual-arts-182706. Accessed 27 Oct 2023
31. Zhang, C.C.: Discussion on the effect of community painting on local content tourism-take Milan and Erlang communities in Jhutian township, Pingtung county as examples. J. Leisure Tourism Sport Health **9**(2), 48–65 (2019)
32. Zhu, Y.S.: The beauty and sorrow of community overall construction. Journal of Social Sciences, Manhua University (2002)

Author Index

A
Abdullah, Munaisyah I-3
Ahmad, Azran I-93
Ahmad, Nik Azlina Nik I-3
Ahmad, Zaiha I-53
Anh, Khuat Duc II-3
Anh, Vu Kim II-3
Ariawan, I Made Bambang I-104
Aziz, Azhar Abd I-28
Azmi, Nur Nafishah I-53

C
Cao, Xin II-171
Chang, Chang-Wei I-348, I-362
Chang, Chia-Ling I-276
Chang, Chung-Yi II-277
Chang, Hung Yu II-220
Chang, Teng-Wen I-191, I-232
Chang, Ting-Cheng II-92
Chang, Yao-Xun I-143
Chang, Yu-Tien II-129
Chao, Yu II-159
Chen, Cheih Ying II-80
Chen, Chun-Wei I-263
Chen, Han II-103
Chen, Hao-Yang I-304
Chen, Huang-Yin I-191
Chen, Jen-Feng I-374
Chen, Tien-Li II-266
Chen, Yen-Ting II-39
Chen, Young-Long I-155, I-166
Cheng, Chi-Cheng II-183
Cheng, Tain Junn II-306
Cheng, Tain-Junn II-277, II-294
Cheng, Yong-Yi II-103
Chiu, Min-Chi I-177
Choi, Jeewon II-254

Chu, Yun-Song I-374
Chua, Jing Lun II-115
Chuang, Yu II-266
Chung, Chuan-Cheng I-166
Ciou, Jing-Fong I-155

D
Daud, Nur'aina I-28
Du, Bai-Hui I-415

F
Fan, Yuk-Wa II-254
Fann, Shih-Cheng II-14

H
Hamidi, Saidatul Rahah I-78
Ho, Ling-Yu II-206
Hou, Chun-Yueh II-266
Hsiang, Ping Chia II-80
Hsiang, Tsu-Min I-289
Hsieh, Wei-Her II-67
Hsu, Tse-Wei I-247
Hsu, Yi-Fu I-348, I-362
Hu, Pey-Yune I-304
Huang, Cailin II-194
Huang, Hsin-Wei II-28
Huang, Tze-Fei I-389
Huang, Wen-Tsong II-144
Huang, Yinghsiu I-216
Hung, Hei-Man I-143
Hung, Phan Duy II-3
Hussein, Surya Sumarni I-78, I-93

I
Ibrahim, Nordiana I-53
Ibrahim, Zuraeda I-53
Isa, Indra Griha Tofik I-129

Ishak, Khairul Khalil   I-116
Ismail, Afiza   I-104
Isna, Ade   I-15

J
Jiang, Jhih-Ling   I-403

K
Kadir, Shamsiah Abd   I-104
Kato, Toshikazu   II-243
Khoa, Luong Nguyen Viet   II-3
Kojima, Taiyo Sunny   II-243
Kuo, Jo-Yu   II-103
Kuo, Yi Chun   II-306

L
Lee, Chein-Hui   I-177
Lee, Jeffrey   II-294
Lee, Meng Chieh   II-306
Lee, Meng-Chieh Jeffrey   II-277
Lee, Wei-Lun   II-294
Lee, Yu-Hsu   II-28, II-39
Liao, Shao-Han   I-323
Liao, Wei-Ming   I-289
Lien, Shih-Sheng   I-155
Lin, Chen-Syuan   I-232
Lin, Chih-Han   I-155
Lin, Chih-Kuan   II-183
Lin, Chih-Long   I-362
Lin, Po-Hsien   I-337, I-374
Lin, Rungtai   I-403
Lin, Yi-Cheng   II-67
Lin, Yi-Hang   I-337
Lin, Yi-Tung   II-294
Lo, Chia Hui Nico   I-204
Lo, YiLing   II-51
Lokman, Anitawati Mohd   I-3, I-15, I-28, I-53, I-104
Lu, Shih-Yun   II-67
Lung, Syue-Ting   I-216
Luo, Ding-Xiang   II-206

M
Ma, Ki-Yan   I-143
Meng-Chieh   II-294
Minh, Ngo Truong   II-3

Minh, Nguyen Duy   II-3
Muraki, Satoshi   II-254
Mustafa, Nur Batrisyia Damia   I-78

N
Nakata, Toru   II-243
Naufal, Raditya   I-15

O
Ogasawara, Etsuko   II-243
Ono, Shugo   II-243

P
Pan, Chang-Yu   II-232
Pei, Cunfeng   II-171
Permana, Rafi Ramdan   I-41, I-67
Pramesti, Rachmadita Dwi   I-41
Prastiwinarti, Wiwi   I-15, I-41, I-67
Putri, Lytta Yennia   I-41

Q
Qin, Li-Hong   I-166

R
Razali, Noor Afiza Mat   I-116
Rosli, Nur Hanis Solehah Mohd   I-93

S
Sakamoto, Takashi   II-243
Sakaori, Fumitake   II-243
Saputri, Evelyn   I-177
Sari, Novi Purnama   I-15, I-41, I-67
Saw, Jo Anne   I-28
Shahibi, Mohd Sazili   I-53
Shen, Kai-Shuan   II-183
Shih, Ying-Shueh   I-348
Shuhidan, Shuhaida Mohamed   I-78
Song, Ke   II-194
Suhaimi, Ahmad Iqbal Hakim   I-3
Sun, Shih Chiao   II-306
Sung, Wen-Tsai   I-129
Suzuki, Atsunari   II-243

T
Takahashi, Naoki   II-243
Tan, Jenn-Ann   II-277

Author Index

Tasi, Su-Ting I-323
Teng, Chien-Kuo II-115, II-159
Tsai, I-Chia II-129
Tsai, Tzu-Wei I-323

W
Wan, Hok Kun II-194
Wang, Yu-Han II-254
Wu, Lien-Shang I-304

Y
Yamin, Iqbal I-15
Yennia, Lytta I-15, I-67
You, Hsiao-Chen II-92, II-206

Z
Zaabar, Liyana Safra I-116
Zabani, Fatin Nadhirah I-28
Zhu, Rui II-266

Printed in the USA
CPSIA information can be obtained
at www.ICGtesting.com
CBHW051509241124
17930CB00007B/59